W9-CGQ-815

ENVIRONMENT 92/93

Eleventh Edition

Editor

John L. Allen
University of Connecticut

John L. Allen is professor of geography at the University of
Connecticut. He received his bachelor's degree in 1963 and
his M.A. in 1964 from the University of Wyoming, and his
Ph.D. in 1969 from Clark University. His special area of
interest is the impact of contemporary human societies on
environmental systems.

A Library of Information from the Public Press

Cover illustration by Mike Eagle

The Dushkin Publishing Group, Inc.
Sluice Dock, Guilford, Connecticut 06437

The Annual Editions Series

Annual Editions is a series of over 55 volumes designed to provide the reader with convenient, low-cost access to a wide range of current, carefully selected articles from some of the most important magazines, newspapers, and journals published today. Annual Editions are updated on an annual basis through a continuous monitoring of over 300 periodical sources. All Annual Editions have a number of features designed to make them particularly useful, including topic guides, annotated tables of contents, unit overviews, and indexes. For the teacher using Annual Editions in the classroom, an Instructor's Resource Guide with test questions is available for each volume.

Printed on Recycled Paper

VOLUMES AVAILABLE

Africa
Aging
American Government
American History, Pre-Civil War
American History, Post-Civil War
Anthropology
Biology
Business and Management
Business Ethics
Canadian Politics
China
Commonwealth of Independent States and Central/Eastern Europe (Soviet Union)
Comparative Politics
Computers in Education
Computers in Business
Computers in Society
Criminal Justice
Drugs, Society, and Behavior
Early Childhood Education
Economics
Educating Exceptional Children
Education
Educational Psychology
Environment
Geography
Global Issues
Health
Human Development
Human Resources
Human Sexuality

International Business
Japan
Latin America
Life Management
Macroeconomics
Management
Marketing
Marriage and Family
Microeconomics
Middle East and the Islamic World
Money and Banking
Nutrition
Personal Growth and Behavior
Physical Anthropology
Psychology
Public Administration
Race and Ethnic Relations
Social Problems
Sociology
State and Local Government
Third World
Urban Society
Violence and Terrorism
Western Civilization, Pre-Reformation
Western Civilization, Post-Reformation
Western Europe
World History, Pre-Modern
World History, Modern
World Politics

Library of Congress Cataloging in Publication Data
Main entry under title: Annual Editions: Environment. 1992/93.
1. Environment—Periodicals. 2. Ecology——Periodicals.
I. Allen, John L., *comp.* II. Title: Environment.
ISBN 1–56134–088–X   301.31′05   79–644216

Eleventh Edition

Manufactured by The Banta Company, Harrisonburg, Virginia 22801

Editors/ Advisory Board

EDITOR

John L. Allen
University of Connecticut

ADVISORY BOARD

STAFF

To the Reader

In publishing ANNUAL EDITIONS we recognize the enormous role played by the magazines, newspapers, and journals of the *public press* in providing current, first-rate educational information in a broad spectrum of interest areas. Within the articles, the best scientists, practitioners, researchers, and commentators draw issues into new perspective as accepted theories and viewpoints are called into account by new events, recent discoveries change old facts, and fresh debate breaks out over important controversies.

Many of the articles resulting from this enormous editorial effort are appropriate for students, researchers, and professionals seeking accurate, current material to help bridge the gap between principles and theories and the real world. These articles, however, become more useful for study when those of lasting value are carefully *collected, organized, indexed,* and *reproduced* in a *low-cost format,* which provides easy and permanent access when the material is needed. That is the role played by *Annual Editions.*

Under the direction of each volume's *Editor,* who is an expert in the subject area, and with the guidance of an *Advisory Board,* we seek each year to provide in each ANNUAL EDITION a current, well-balanced, carefully selected collection of the best of the public press for your study and enjoyment. We think you'll find this volume useful, and we hope you'll take a moment to let us know what you think.

During the last two decades, and particularly during the late 1980s and beginning of the 1990s, the environmental predicament foreseen by environmental scientists has begun to emerge in a number of guises such as population/food imbalances, problems of energy resource scarcity, acid rain, toxic and hazardous wastes, water shortages, massive soil erosion, global atmospheric pollution, forest dieback and tropical deforestation, and the highest rates of plant and animal extinction the world has known. The last half of the 1980s and the opening of the 1990s have been characterized by drought and famine in Africa, a major environmental chemical accident in Bhopal, India, the burning and cutting of thousands of square miles of tropical rain forest, a near-meltdown of a nuclear power generator in Chernobyl in the Soviet Union, unprecedented heat waves, drought, and wildfires in the United States, several serious oil spills, and an energy-related military crisis in the oil-rich Persian Gulf that produced unprecedented environmental disruptions. Moreover, the last few years have brought scientific validation of the concern that the life-protecting ozone layer is being destroyed, and that the long-term global climate changes scientists have warned about may have already begun. These and other problems have surfaced in spite of the increased environmental awareness and legislation that characterized the decade of the 1970s. They have resulted, in part, from a misguided environmental "counterrevolution" that characterized much of the 1980s and favored the short-term, expedient approach to problem-solving over longer-term economic and ecological good sense. The drive to produce enough food to support a growing population, for example, has resulted in the use of increasingly fragile and marginal resources in Africa, which has produced the desert expansion that brings famine to that troubled continent. Similar social and economic problems have contributed to massive deforestation in Latin America and Southeast Asia. The economic problems caused by resource scarcity have caused the relaxation of environmental quality standards that have become viewed as too costly. The decrease in standards has been particularly apparent in Third World countries, striving to become economically developed, and has contributed to accidents such as that at Bhopal. In addition, concerns over energy availability have prompted increasing reliance on technological quick fixes—a Faustian bargain that creates conditions under which a terrifying Chernobyl accident can occur. There are signs, however, that a new environmental consciousness is awakening. The dissolution of the Iron Curtain and the environmental horror stories that have emerged from Eastern Europe have given new incentives to international cooperation. Several major publications have claimed the 1990s "The Decade of the Environment," and there is growing public clamor that something must be done about environmental quality before it is too late.

The articles contained in *Annual Editions: Environment 92/93* have been selected for the light they shed on these and other problems and issues. The selection process was aimed at including material that will be readily assimilated by the general reader. Additionally, every effort has been made to choose articles that do not engage in futile vilification of the species Homo sapiens as a fouler of its own nest. Handwringing or haranguing for birth control, new visions, or repentance is of little purpose. Rather, what is needed is an understanding of the nature of the environmental problems that beset us as a species and how, with wisdom and knowledge and the proper perspective, they can be solved—or at least mitigated. Accordingly, the selections in this book have been chosen more for their intellectual content than for their emotional tone. They have been arranged into an order—the global environment, population and food, energy, pollution, resources, and the biosphere—that lends itself to a progressive understanding of the causes and effects of human modifications of the Earth's environmental systems. It is hoped that this collection of articles will not be used simply as passive "readings" but as a source for discussion and debate, reorientation of thought, and active participation in learning more about environmental issues. We will not be protected against the ecological consequences of human actions by remaining ignorant of them.

Readers can have input into the next edition by completing and returning the article rating form in the back of the book.

John L. Allen
Editor

Unit 1

The Global Environment: An Emerging World View

Six selections provide information on the current state of the Earth and the changes we will face.

To the Reader   iv
Topic Guide   2
Overview   4

1. **Historical Perspectives on Sustainable Development,** 6
   Clive Ponting, *Environment,* November 1990.

   The history of human activities since the origin of agriculture has included major **environmental disruptions** as far back as the ancient civilizations of Mesopotamia and the Maya. Like many of today's societies, ancient peoples found it difficult to achieve **sustainable development** of the **population** and **resource** bases without significant alteration of **ecological systems.**

2. **Managing As If the Earth Mattered,** James E. Post, 12
   Business Horizons, July-August 1991.

   Managers of both public and private sector enterprises cannot ignore **environmental problems** in their decision-making processes. Learning how international trade, competitiveness, and **global resources** are interrelated is a challenge to managers whose businesses and agencies will be affected by both global and local **environmental issues.**

3. **Debating Gaia,** Stephen H. Schneider, *Environment,* May 18
   1990.

   The **Gaia hypothesis,** the notion that the world's **atmosphere** is an integral part of the world's **biological systems,** is no longer just an argument of those on the mystical fringe of scientific disciplines. The argument that the Earth's physical and chemical environment and life itself are closely linked has become the subject of critical scientific debate.

4. **War & the Environment,** Malcolm W. Browne, Michio 24
   Kaku, James M. Fallows, and Eric A. Fischer, *Audubon,*
   September/October 1991.

   After the end of the **Persian Gulf** conflict between Iraq and the allied forces, **environmental devastation** lingered in the forms of burning oil wells, oil slicks on the Gulf's waters, and massive **ecological destruction.** This article presents four varying viewpoints on the ecological consequences of war. However, even the holders of divergent opinions agree that dealing with the **environmental impact of war** must become an integral part of any war planning.

5. **The Environment of Tomorrow,** Martin W. Holdgate, 30
   Environment, July/August 1991.

   Throughout history, humans have adapted to **environmental changes.** In the world of today and tomorrow, when potential changes can occur with startling rapidity, the kind of adaptation that took place in the past may no longer be possible. If humanity is to have a **sustainable future** on this diverse planet, it can only be through a process of **international cooperation.**

6. **Global Change Ecology,** John J. Magnuson and Jennifer 39
   A. Drury, *The World & I,* April 1991.

   Predicting the **environmental consequences** of human activities requires long-term research and larger areas of study than what was used in past analyses. **Global change ecology** requires putting the places we know in our restricted human scale of **environmental perception** into regional and global contexts. This will bring global ecology into focus so that fact, not fiction, can guide the development of future global **environmental policies.**

The concepts in bold italics are developed in the article. For further expansion please refer to the Topic Guide, the Index, and the Glossary.

Unit 2

The World's Population: People and Hunger

Five selections examine the problems the world will have in feeding the ever-increasing population.

Overview   46

7. **World Population Continues to Rise,** Nafis Sadik, *The*   48
   Futurist, March/April 1991.

   The world's population is expected to increase by another billion in the 1990s. Choices made during this decade about investment in economic growth in *Third World countries* will determine *population growth* for much of the next century. These choices will determine whether *world population* will triple or merely double before it finally stops growing.

8. **Sheer Numbers: Can Environmentalists Grasp the Net-**   54
   tle of Population? Garrett Hardin, *E Magazine,* November/ December 1990.

   During the two decades between Earth Day 1970 and Earth Day 1990, the issue of *population growth* somehow was dropped from the environmental agenda. During this same period, however, *world population* expanded by 47 percent. Part of the problem seems to be that population growth is a chronic problem rather than a critical one, and people tend to pay less attention to chronic problems.

9. **Population Politics,** Werner Fornos, *Technology Review,*   59
   February/March 1991.

   No challenge facing the world's decisionmakers is more urgent than the need to stabilize *world population.* Both the knowledge and the technology exist to reach the objective of *zero population growth* (ZPG), but the will to do so seems to be absent. In the face of declining per capita *food production,* this absence is troubling but explainable in ideological terms.

10. **Feeding Six Billion,** Lester R. Brown, *World • Watch,*   68
    September/October 1989.

    The *global environment* has now reached the point at which *population growth* is exceeding the ability of farmers to produce food. The central question for humanity is "how do we achieve a balance between food and people?" The only rational solution, according to the author, is for every nation in the world to develop a strategy of *population control* that would limit families everywhere to two children.

11. **Forecast: Famine?** Françoise Monier, *Ceres,* September/   77
    October 1990.

    The hypothesis of *global climate change* became a reality several years ago, and we now know that our children will live on a planet transformed by the consequences of *the greenhouse effect.* Global warming will produce enormous changes in *food production* as the productive capacity of semi-arid regions (which produce much of the world's grain) drops dramatically.

The concepts in bold italics are developed in the article. For further expansion please refer to the Topic Guide, the Index, and the Glossary.

Unit 3

Energy: Present and Future Problems

Five articles consider the problems of meeting present and future energy needs. Alternative energy sources are also examined.

Overview   82

12. **Energy for the Next Century,** Will Nixon, *E Magazine,*   84
    May/June 1991.

    All environmental indicators point toward the necessity to move to a *postpetroleum society*—one in which *alternative energy* sources such as *solar power*, *wind power*, and *geothermal energy* are used to a higher degree than energy from oil. Yet the U.S. *national energy strategy* (NES) reveals a country still locked into the oil habit. If the flood of carbon emissions into the atmosphere that causes *global warming* is to stop, this habit must be broken.

13. **Balance Sought: Energy, Environment, Economy,** William   90
    H. Miller, *Industry Week,* April 1, 1991.

    The *national energy strategy* of the United States will become an issue in the 1992 presidential campaign, as will diametrically opposed prescriptions for the U.S. *energy policy.* Rather than relying exclusively on the development of *renewable energy* on the one hand, or *nuclear energy* on the other, to ease U.S. reliance on foreign oil, energy needs should be linked with both economic and environmental considerations.

14. **The Negawatt Revolution,** Amory B. Lovins, *Across the*   95
    Board, September 1990.

    A leading expert on *electrical energy* argues that, using existing technology, the United States can save three-fourths of all the electricity used today. This can be achieved through improvements in *energy efficiency* rather than reduction in *energy consumption,* and it could stimulate a trillion-dollar-a-year global market in efficient energy devices.

15. **Here Comes the Sun,** Christopher Flavin and Nicholas   101
    Lenssen, *World • Watch,* September/October 1991.

    The *energy technology* exists today to produce most of the energy needs of the world from *solar power*, *wind power,* and *geothermal energy.* Tapping into these *alternative energy* resources, however, will require a vigorous public commitment to push *renewable energy* into the mainstream. The key to overcoming political barriers to that commitment is the ability to demonstrate the advantages of alternative energy over *fossil fuels.*

16. **Energy Crops for Biofuels,** Janet H. Cushman, Lynn L.   109
    Wright, and Kate Shaw, *The World & I,* August 1991.

    The development of *biofuels* (crops grown for energy) may become as important as the development of more productive food crops. Research is now underway that aims to make cultivated grasses and trees an important source of fuel for transportation and energy generation. The development of *biotechnology* as related to *energy crops* could have an enormous impact on both economic and environmental systems.

The concepts in bold italics are developed in the article. For further expansion please refer to the Topic Guide, the Index, and the Glossary.

Unit 4

Pollution: The Hazards of Growth

Six selections weigh the environmental impacts of the disposal and control of toxic waste, agricultural pesticides, unwanted radioactive side products, and the environmental damage to Eastern Europe and Russia.

Overview   **116**

17. It's Enough to Make You Sick, Susan Q. Stranahan, *National Wildlife,* February/March 1990.   **118**

After two decades of research, much more is known about the relationship between *pollution* and *public health.* Unfortunately, the research has brought to light environmental dilemmas no celebrants of *Earth Day* 1970 even dreamed of. Each step taken along the path of increasing knowledge of the *technological hazards* of our society brings to light more problems—but not necessarily more solutions.

18. The Greening of Industry, Ken Geiser, *Technology Review,* August/September 1991.   **122**

The chief environmental concern of many industries is the safe disposal of *toxic wastes.* As this concern increases with growing accumulation of waste materials, it becomes clear that the most viable approach is the development of *sustainable industry* that uses safer materials and cleaner technologies. This transition to clean production offers an opportunity to unite *public health* and industrial productivity.

19. From Ash to Cash: The International Trade in Toxic Waste, Ron Chepesiuk, *E Magazine,* July/August 1991.   **130**

Tough *environmental laws* in their own countries encourage many developed nations to ship their wastes abroad to *Third World countries* for storage, disposal, or recycling. The poorer nations at the receiving end of these *toxic wastes* are becoming increasingly sensitive about their role as the receptacles of the leftovers of industrial civilization, even when the *international trade* in toxic waste means more money for the poor.

20. Will the Circle Be Unbroken? David Weir and Constance Matthiessen, *Mother Jones,* June 1989.   **135**

Nearly a decade of research and investigative reporting has demonstrated the threat to *developing nations* of *agricultural chemicals* banned for use in the United States but sold abroad where they pose a health threat to farm workers and, ultimately, to consumers in the United States who eat imported foods. In spite of new *pesticide* safety programs, Central American farm workers continue to be exposed to deadly chemicals.

21. Eastern Europe: Restoring a Damaged Environment, Richard A. Liroff, *EPA Journal,* July/August 1990.   **141**

The lifting of the Iron Curtain has revealed truly appalling *environmental impacts* in an Eastern Europe savaged by *economic development* policies that were totally indifferent to the environment. In addition to supporting the political and economic transition of Eastern Europe, the United States should offer a balanced, integrated program of *environmental assistance* that would foster a full restoration of a healthy environment in the region.

22. Environmental Devastation in the Soviet Union, James Ridgeway, *Multinational Monitor,* September 1990.   **148**

A survey of the environmental situation in Russia and other republics of the former Soviet Union reveals a region in desperate straits. Over 16 percent of the country and 25 percent of the population is at *environmental risk* from *air pollution, water pollution,* and other environmental ills. There was no such thing as an *environmental impact* statement in the former Soviet system, and no one can be certain who is responsible for most of the polluting.

The concepts in bold italics are developed in the article. For further expansion please refer to the Topic Guide, the Index, and the Glossary.

Unit 5

Resources: Land, Water, and Air

Seven selections discuss the environmental problems affecting our land, water, and air resources.

Overview   152

23. **23rd Environmental Quality Index: The Year of the Deal,** *National Wildlife,* February/March 1991.   154
As more information about **environmental hazards** surfaces, Americans are showing increasing concern. Although being an "environmentalist" seems to be both good business and good politics, official U.S. government **environmental policy** is mostly an attempt to perform a precarious balancing act between **environmental protection** and the environmental indifferences that characterized the 1980s.

A. LAND

24. **A New Lay of the Land,** Holly B. Brough, *World • Watch,* January/February 1991.   163
The uneven distribution of good farmland has pushed poor peasants into the **marginal resources** of deserts and forests, and this has caused increased **environmental deterioration.** In **Third World countries,** in particular, strategies of **land reform** or finding a way to parcel good land fairly to all farmers might spell a break for threatened ecosystems.

25. **U.S. Farmers Cut Soil Erosion by One-Third,** Peter Weber, *World • Watch,* July/August 1990.   170
As an important piece of **environmental policy** legislation, the Farm Act of 1985 has succeeded in reducing **soil erosion** on American farms from 1.6 billion to 1 billion tons per year. The philosophy behind this far-reaching legislation is that farmers receiving federal subsidies should in return conserve the nation's **soil resources.**

B. WATER

26. **Holding Back the Sea,** Jodi L. Jacobson, *The Futurist,* September/October 1990.   172
The threat of **global warming** could mean a greater chance of **rising sea levels** that would represent an **environmental threat** of unprecedented proportions. Coastal communities the world over could face two fundamental choices: retreat from shoreline locations or hold back the sea using expensive technologies.

27. **Water, Water, Everywhere, How Many Drops to Drink?**   180
William H. MacLeish, *World Monitor,* December 1990.
On a global scale, the shortage of oil gets most of the publicity, but water crises such as **water pollution** and **water shortage** will cause greater shocks to human and economic systems unless new **environmental technology** and **conservation measures** are adopted in time. Global monitoring of water problems is far from complete, but a sampling of reports from the media and from the United Nations suggests that the dimensions of the problem are enormous.

The concepts in bold italics are developed in the article. For further expansion please refer to the Topic Guide, the Index, and the Glossary.

C. **AIR**

28. Global Climate Change: Fact and Fiction, S. Fred Singer,   **184**
The World & I, July 1991.
Although **greenhouse gases** have been increasing in the atmosphere, the climate record does not show the **global warming** predicted by the climate models. Even if modest global warming does occur, the available evidence suggests that the net impact of **climate change,** particularly for **agricultural production,** might well be beneficial.

29. Gazing Into Our Greenhouse Future, Jon R. Luoma,   **190**
Audubon, March 1991.
Scientists from the National Center for Atmospheric Research agree that there are many uncertainties about the extent of changes in the **global climate,** but that even small changes could have dramatic **environmental impact.** Even if the "worst case predictions" are wrong, reducing emissions of **greenhouse gases** would improve other environmental problems such as **acid precipitation** and air pollution.

Unit 6

Biosphere: Endangered Species

Six articles examine the problems in the world's biosphere. Not only are plants and animals endangered, but so are many human groups who are disastrously affected by deforestation and primitive agricultural policies.

Overview   **196**

30. The Origin and Function of Biodiversity, Otto T. Solbrig,   **198**
Environment, June 1991.
There is an enormous variety of plants and animals on Earth. This **biodiversity** is the ultimate source of human sustenance. Yet humans are endangering the immense richness of species, and a reduction in the **genetic variety** of crops and wild species could seriously affect human welfare. Increased public awareness of humanity's depletion of biodiversity is necessary to stimulate national and international efforts to learn more about the role of diversity in the **ecosystem** function.

A. **PLANTS**

31. Conserving the Tropical Cornucopia, Nigel J. H. Smith,   **207**
J. T. Williams, and Donald L. Plucknett, *Environment,* July/August 1991.
The plight of **tropical forests** has moved to the forefront of public attention around the world. People everywhere are increasingly alarmed at the widespread **forest destruction** that has wiped out species, accelerated **soil erosion,** and threatened to disrupt the **global climate.** Largely ignored, however, has been the shrinking population of tropical plants that are important for both subsistence and commercial activities.

The concepts in bold italics are developed in the article. For further expansion please refer to the Topic Guide, the Index, and the Glossary.

32. **Timber's Last Stand,** John C. Ryan, *World • Watch,* July/   **212**
August 1990.

Because of the increasing rate of ***deforestation,*** the timber industry is rapidly running out of natural forests to exploit. The adoption of a ***new forestry,*** one that seeks to use and maintain the complexity of forest ***ecosystems*** rather than eliminate it, could help to save the destructive timber industry from itself. If societies can learn that jobs and profits based on ***ecological destruction*** cannot continue, there is still time to make the transition to ***sustainable ecosystems.***

33. **Forests Under Siege,** Jeffrey L. Chapman, *USA Today*   **219**
Magazine (Society for the Advancement of Education),
March 1991.

While Americans clamor for an end to tropical ***deforestation,*** the native forests of the United States are being cut at an alarming rate. The ***environmental impact*** of lumbering in the old-growth forests of the Pacific Northwest and Alaska is no less significant than that of deforestation in the tropics. U.S. forests are being cleared not for the country's lumber needs but for the timber's cash value on the international market.

B.   *ANIMALS*

34. **Wildlife as a Crop,** Dick Pitman, *Ceres,* September/Octo-   **222**
ber 1990.

Few conservationists or planners in developed countries can bring themselves to regard ***wildlife*** as an economic resource rather than solely an aesthetic one. In Africa, however, ***wildlife management*** strategies are being developed that will make wildlife an agricultural option to complement crop production and livestock raising. The successful integration of wildlife management and agricultural policy could bring benefits to both humans and wildlife.

35. **How the West Was Eaten,** George Wuerthner, *Wilderness,*   **226**
Spring 1991.

The list of ***endangered species*** in the American West continues to grow, at least partly as the result of the demand for land by the western livestock industry. The potential ***extinction of species*** is only one cost of livestock grazing on public land. The industry has come under fire from a broad array of critics, ranging from conservative economists to liberal ecologists.

Environmental Information Retrieval   **232**
Glossary   **237**
Index   **241**
Article Review Form   **244**
Article Rating Form   **245**

The concepts in bold italics are developed in the article. For further expansion please refer to the Topic Guide, the Index, and the Glossary.

Topic Guide

This topic guide suggests how the selections in this book relate to topics of traditional concern to students and professionals involved with environmental studies. It is useful for locating articles that relate to each other for reading and research. The guide is arranged alphabetically according to topic. Articles may, of course, treat topics that do not appear in the topic guide. In turn, entries in the topic guide do not necessarily constitute a comprehensive listing of all the contents of each selection.

TOPIC AREA	TREATED IN:	TOPIC AREA	TREATED IN:
Acid Precipitation	29. Gazing Into Our Greenhouse Future	Environmental Changes	5. Environment of Tomorrow
Agricultural Chemicals	20. Will the Circle Be Unbroken?	Environmental Consequences	6. Global Change Ecology
Agricultural Production	28. Global Climate Change: Fact and Fiction	Environmental Deterioration	24. A New Lay of the Land
Air Pollution	22. Environmental Devastation in the Soviet Union 29. Gazing Into Our Greenhouse Future	Environmental Devastation	4. War & the Environment
Alternative Energy	12. Energy for the Next Century 15. Here Comes the Sun	Environmental Hazards	23. 23rd Environmental Quality Index: Year of the Deal
Atmosphere	3. Debating Gaia	Environmental Impact of War	4. War & the Environment
Biodiversity	30. Origin and Function of Biodiversity	Environmental Impacts	21. Eastern Europe: Restoring a Damaged Environment 22. Environmental Devastation in the Soviet Union 29. Gazing Into Our Greenhouse Future 33. Forests Under Siege
Biofuels	16. Energy Crops for Biofuels		
Biological Systems	3. Debating Gaia		
Biotechnology	16. Energy Crops for Biofuels		
Climate Change	28. Global Climate Change: Fact and Fiction	Environmental Issues	2. Managing as if the Earth Mattered
Conservation Measures	27. Water, Water, Everywhere, How Many Drops to Drink?	Environmental Laws	19. From Ash to Cash
Deforestation	32. Timber's Last Stand 33. Forests Under Siege	Environmental Perception	6. Global Change Ecology
Earth Day	17. It's Enough to Make You Sick	Environmental Policy	6. Global Change Ecology 23. 23rd Environmental Quality Index: Year of the Deal 25. U.S. Farmers Cut Soil Erosion by One-Third
Ecological Disruption	4. War & the Environment 32. Timber's Last Stand		
Ecological Systems	1. Historical Perspectives on Sustainable Development	Environmental Problems	1. Historical Perspectives on Sustainable Development 2. Managing as if the Earth Mattered
Economic Development	21. Eastern Europe: Restoring a Damaged Environment		
Ecosystem	30. Origin and Function of Biodiversity 32. Timber's Last Stand	Environmental Protection	23. 23rd Environmental Quality Index: Year of the Deal
Electrical Energy	14. Negawatt Revolution	Environmental Risk	22. Environmental Devastation in the Soviet Union
Endangered Species	35. How the West Was Eaten	Environmental Technology	27. Water, Water, Everywhere, How Many Drops to Drink?
Energy Consumption	14. Negawatt Revolution	Environmental Threat	26. Holding Back the Sea
Energy Crops	16. Energy Crops for Biofuels	Extinction of Species	35. How the West Was Eaten
Energy Efficiency	14. Negawatt Revolution	Food Production	9. Population Politics 11. Forecast: Famine?
Energy Policy	13. Balance Sought: Energy, Environment, Economy		
Energy Technology	15. Here Comes the Sun	Forest Destruction	31. Conserving the Tropical Cornucopia
Environmental Assistance	21. Eastern Europe: Restoring a Damaged Environment	Fossil Fuels	15. Here Comes the Sun

TOPIC AREA	TREATED IN:	TOPIC AREA	TREATED IN:
Gaia Hypothesis	3. Debating Gaia	Public Health	17. It's Enough to Make You Sick 18. Greening of Industry
Genetic Variety	30. Origin and Function of Biodiversity	Renewable Energy	13. Balance Sought: Energy, Environment, Economy
Geothermal Energy	12. Energy for the Next Century 15. Here Comes the Sun		15. Here Comes the Sun
Global Change Ecology	6. Global Change Ecology	Resource	1. Historical Perspectives on Sustainable Development
Global Climate	29. Gazing Into Our Greenhouse Future 31. Conserving the Tropical Cornucopia	Rising Sea Levels	26. Holding Back the Sea
		Soil Erosion	25. U.S. Farmers Cut Soil Erosion by One-Third
Global Climate Change	11. Forecast: Famine?		31. Conserving the Tropical Cornucopia
Global Environment	10. Feeding Six Billion	Solar Power	12. Energy for the Next Century 15. Here Comes the Sun
Global Resources	2. Managing as if the Earth Mattered	Sustainable Development	1. Historical Perspectives on Sustainable Development
Global Warming	12. Energy for the Next Century 26. Holding Back the Sea 28. Global Climate Change: Fact and Fiction 29. Gazing Into Our Greenhouse Future	Sustainable Ecosystems	32. Timber's Last Stand
		Sustainable Future	5. Environment of Tomorrow
Greenhouse Effect	11. Forecast: Famine?	Sustainable Industry	18. Greening of Industry
Greenhouse Gases	28. Global Climate Change: Fact and Fiction 29. Gazing Into Our Greenhouse Future	Technological Hazards	17. It's Enough to Make You Sick
International Cooperation	5. Environment of Tomorrow	Third World Countries	7. World Population Continues to Rise 19. From Ash to Cash 20. Will the Circle Be Unbroken? 24. New Lay of the Land
International Trade	19. From Ash to Cash		
Land Reform	24. New Lay of the Land	Toxic Wastes	18. Greening of Industry 19. From Ash to Cash
Marginal Resources	24. New Lay of the Land	Tropical Forests	31. Conserving the Tropical Cornucopia
National Energy Strategy	12. Energy for the Next Century 13. Balance Sought: Energy, Environment, Economy	Water Pollution	22. Environmental Devastation in the Soviet Union 27. Water, Water, Everywhere, How Many Drops to Drink?
New Forestry	32. Timber's Last Stand	Water Shortages	27. Water, Water, Everywhere, How Many Drops to Drink?
Nuclear Energy	13. Balance Sought: Energy, Environment, Economy	Wildlife	34. Wildlife as a Crop
Persian Gulf Crisis	4. War & the Environment	Wildlife Management	34. Wildlife as a Crop
Pesticide	20. Will the Circle Be Unbroken?		
Pollution	17. It's Enough to Make You Sick	Wind Power	12. Energy for the Next Century 15. Here Comes the Sun
Population	1. Historical Perspectives on Sustainable Development	World Population	7. World Population Continues to Rise 8. Sheer Numbers 9. Population Politics
Population Control	10. Feeding Six Billion		
Population Growth	7. World Population Continues to Rise 8. Sheer Numbers 10. Feeding Six Billion	Zero Population Growth	9. Population Politics
Postpetroleum Society	12. Energy for the Next Century		

The Global Environment: An Emerging World View

The celebration of Earth Day 1990, the twentieth anniversary of the original Earth Day, came at a time when public apprehension over the environmental future of the planet reached levels unprecedented even during the activist days of the late 1960s and early 1970s. No longer were those concerned about the environment viewed as "eco-freaks" and "tree-huggers" as many serious scientists joined the rising clamor for environmental protection, as did the more traditional environmentally-conscious public interest groups. There are a number of reasons for this increased environmental awareness. Some of these reasons arise from environmental events such as drought, heat wave, fire, and famine. But more arise simply from the increase in information and ideas about the global nature of environmental processes. For example, the raising of the Iron Curtain that separated East and West since the end of the World War II and the disintegration of the Soviet Union have brought visions of the end of the cold war, a reawakening of the democratic spirit in Eastern Europe, and a hope for a more integrated global economy promising both peace and prosperity for the world's peoples. But that same raising of the barrier to the flow of people, goods, and services, has also allowed information and ideas to pass more freely between East and West. Much of what has been learned through this increased information flow, particularly by Western observers, has been an environmentally-ravaged Eastern Europe and Russia—a chilling forecast of what other industrialized nations will become in the near future unless strict international environmental measures are put in place. As distressing as the pictures and descriptions of forest destruction in eastern Germany and Czechoslovakia or the devastation of the Aral Sea have been, they have had a positive value. For perhaps the first time ever, countries are beginning to recognize that environmental problems have no boundaries and that international cooperation is the only way to solve them. The subtitle of this first unit, "An Emerging World View," is an optimistic assessment of the future: a future in which less money will be spent on defense and more on environmental protection and cleanup. In a continuation of the cautiously optimistic tone of their last edition, the authors of the Worldwatch Institute's *State of the World 1991* (a publication that has assumed a near-official status as the annual assessment of the global environment) describe a New World Order in which political influence will be based more upon leadership in environmental and economic issues than upon military might. Perhaps it is far too early to make optimistic predictions that the decade of the 1990s will, indeed, be "The Decade of the Environment," and the world's nations—developed and underdeveloped—will begin to recognize that the Earth's environment is a single unit. Nevertheless, there is growing international recognition that we are all, as environmental activists have said for decades, inhabitants of "Spaceship Earth" and that, as such, we will survive or succumb together.

The articles selected for this unit have been chosen to illustrate this increasingly global perspective on environmental problems and the degree to which environmental problems and their solutions must be linked to political, economic, and social problems and solutions. In the lead piece of the section, Clive Ponting, author of a new book on "green history," places the current environmental crisis in an appropriate historical perspective. In "Historical Perspectives on Sustainable Development," Ponting points out that major environmental disruptions have been a feature of human history since the development of agriculture and the transition from spatially mobile hunting-gathering societies to sedentary agricultural ones. The major difference between the environmental impacts of the Maya, Mesopotamians, and Meditarraneans of past millennia and our modern industrial society is that now, with the creation of a global economy, humanity must contend with damage to global mechanisms that make life on Earth possible. This global perspective is shared by James E. Post's "Managing as If the Earth Mattered" in which the author, a professor of management and public policy, suggests that for managers of commercial enterprises to continue to ignore environmental problems at a global scale is not only environmentally short-sighted but economically self-defeating. As with Ponting, Post's primary concern is for something that past managers (or societies) have not been able to create: sustainable economic development. In a sharp departure from the historical and management orientation of the first two articles in unit 1, the third article moves toward the natural sciences by describing the nature of one of the twentieth century's most intriguing scientific debates. In "Debating Gaia," Stephen H. Schneider, head of the Interdisciplinary Climate Systems Section of the National Center for Atmospheric Research, tells us that the Gaia hypothesis—the idea that the Earth's lower atmosphere is an integral part of life itself—has moved out of the shadows of fringe science to become the center of critical discussion. This discussion may lead to new approaches to long overdue research on the interaction between the organic and inorganic components of the Earth's environment. The fourth article in the unit also deals with relationships, not between the different components of the natural environment but between human groups. In "War & the Environment," four different authorities on human conflict and its attendant environmental impact take some intriguing and not-so-predictable stands on where we have been and where we are going in terms of the environmental impact of war. Obviously prompted by the consequences of the war in the Persian Gulf region, this compelling article concludes that any future war planning must include environmental issues. The fifth article in the unit ties together many of the threads in the previous articles, taking an historical approach and, at the same time, focusing on the future. In "The Environment of Tomorrow," Martin W. Holdgate, Director General of the World Conservation Union, notes that people have always adapted to environmental change and that this adaptation has often been imperceptible except in retrospect. Such unconscious adaptation is no longer possible, given the rapidity of human-induced environmental change. If our species is to have a sustainable future on a diverse planet with many environmental inequalities, Holdgate asserts, it can only be through a process of international cooperation that transcends anything we see today. Finally, a biological scientist and a free-lance writer conclude this unit on the global perspective by discussing "Global

Change Ecology." John J. Magnuson and Jennifer A. Drury point out that much of past research on environmental change has been locally-based and temporally-restricted. The newer global focus of environmental science requires that research itself incorporate larger geographical areas (up to the entire Earth) and longer time frames. A proper understanding of even local environmental change in a short-run time context can be properly achieved only by the longer and broader view they recommend.

Looking Ahead: Challenge Questions

What lessons on the development of sustainable societies can be learned by examining the relationships between past human societies and the environments they occupied and modified?

How can the mangers of business enterprises reconcile their business-management goals of higher profits and greater productivity with the public interest goals of preserving nature's ecological balance and a clean, healthy environment?

How do proponents of the "Gaian hypothesis" view the relationship between the organic and inorganic components of the Earth's environmental systems, and how could proving the hypothesis aid in protecting the planet against the negative consequences of human actions?

What are some of the most obvious recent examples of the impact of war on environmental systems? How can the environmental effects of war be compared with those of other human activities such as population growth, expansion of agricultural lands, or industrialization?

How can the process of international cooperation increase significantly the chances for the development of sustainable human society? Are there differences between the goals of developed and developing countries that make the achievement of international cooperation on environmental issues difficult, if not impossible?

What is meant by the term "global change ecology," and how does it relate to both the environmental consequences of human activity and the development of future global environmental policies?

HISTORICAL PERSPECTIVES

On Sustainable Development

Clive Ponting

Clive Ponting is an honorary research fellow in the Department of Political Theory and Government at the University College of Swansea in Wales. His book, A Green History of the World, *will be published by Sinclair Stevenson, Ltd., London. This article is based on a paper he presented at an August 1990 meeting of the British Association in Swansea.*

Most discussions of environmental issues today contain very little historical perspective. It is generally assumed that environmental problems have only affected contemporary societies. In some cases, this view may be justified. For example, the use of highly toxic pesticides and other chemicals and the resulting pollution problems are essentially a phenomenon of the last 40 years. Most people recognize that acid rain and the enhanced greenhouse effect are the result of industrial processes and fossil-fuel burning during the past century or so. But is even this longer time scale really adequate for considering environmental problems?

How can someone put into perspective the current deforestation of the Amazon basin without considering what has happened in Europe, China, and North America? Originally, 95 percent of western and central Europe was covered in forest, but that amount has now fallen to about 20 percent. Ten thousand years

ago, China was 70 percent forest; it is now about 5 percent. In the 100 years after the 1790s, about three-quarters of the forests in the United States were cleared. How can anybody understand the current problems of soil erosion, desertification, and the salinization and waterlogging of irrigated land without studying the historical examples of all these events? Is it possible to understand the present situation in the Third World—its poverty and dependence on cash crops and commodity exports—without understanding how Europe remade not just the political but also the economic and social relationships in the world after 1500? How can one understand the current energy crisis without considering the long-term changes in energy availability over the last 10,000 years? A thorough discussion of these issues is beyond the scope of this article; instead, it examines three societies affected by environmental problems and considers how they reacted.[1]

Origins of Ecosystem Disruption

To give an accurate chronological account of human history in 30 minutes, one would have to spend 29 minutes and 51 seconds on gathering and hunting groups, a little more than 8 seconds describing settled agricultural societies, and a fraction of a second considering the problems of the modern industrial world. These proportions illustrate how relatively recently and quickly the major

changes in human society have come about. For about 2 million years, humans lived in small bands of perhaps 25 people, moving around a territory according to the seasonal availability of food and congregating in larger groups for ceremonial and other social activities when food supplies permitted. Without doubt, it was the most long-lasting and well-adapted way of life in human history. It also was one of great ecological stability, involving minimal environmental alteration and damage. And it was adaptable enough to enable humans to settle almost every area of the globe and find enough food to survive.

The most fundamental alteration in human history occurred only 10,000 years ago: The development of agriculture, which occurred in at least three separate locations—southwest Asia, China, and Mesoamerica—led directly to the first settled societies. These societies were characterized by the expropriation of surplus food grown by farmers to feed and support a growing class of nonproducers—priests, rulers, bureaucrats, and soldiers.

Agriculture involves a massive disruption of natural ecosystems, which are cleared to provide fields for growing crops and grazing domesticated animals. Nutrient recycling is disrupted and extra inputs in one form or another are required to sustain the system. Settled societies also increase environmental pressure in the occupied areas

From *Environment*, Vol. 32, No. 9, November 1990, pp. 4-9, 31-33. Reprinted with permission of the Helen Dwight Reid Educational Foundation. Published by Heldref Publications, 1319 Eighteenth St., N.W., Washington, DC 20036-1802.

through deforestation to provide space for fields, construction materials for houses, and firewood for cooking and heating.

The most significant factor in widening the scale of human destruction of natural ecosystems has been the growing population, which forces settlement into new areas and increases the demand for resources. Seven thousand years ago when the first settled societies emerged, the total population of the world was about 5 million, equivalent to a large city today. It is now more than 5 billion, or 1,000 times greater. Indeed, the rate of increase in the world's population is now about 95 million per year, equivalent to the global population only 2,500 years ago.

Apart from this generalized pressure on the environment, the spread of agriculture and the rise of settled societies placed immense strain on those ecosystems that were particularly sensitive to disruption. Such was the situation in Mesopotamia, where the first civilizations and the most extensive modifications to the natural environment occurred; in the Mediterranean basin; and in the lowland jungles of Mesoamerica, where the Mayan civilization briefly flourished.

Mesopotamia

The valley of the twin rivers, the Tigris and the Euphrates, was not the most hospitable environment for an emerging society of the third and second millennia B.C.E. (B.C.), especially in the south. The rivers flowed highest in the spring, following the melting of the winter snows near their sources, and lowest between August and October, when the newly planted crops needed the most water. In northern Mesopotamia, the problem was eased by late autumn and winter rains, but farther south these rains were very sparse and often nonexistent. Thus, in the southern state of Sumer, water storage and irrigation were essential to agriculture.

Because of the local climate and geology, however, these practices incurred costs as well as benefits. The advantages would have outweighed the disadvantages at first, but several major problems would slowly have become

apparent. In summer, temperatures were high, often up to 40° C, which increased evaporation from the soil surface and, consequently, the amount of salt in the soil. Water retention in the deeper layers of the soil and, hence, the likelihood of waterlogging increased because of two factors: very low soil permeability and the slow rate of drainage caused by the very flat land. The poor drainage was exacerbated by the amount of silt carried down the rivers, which probably was caused by deforestation in the highlands. The river piled the silt up at the river mouth to a depth of about 5 feet every millennium, which extended the delta of the two rivers by about 15 miles a millennium.

As the land became more waterlogged and the water table rose, more salt was brought to the surface, where the high evaporation rates left a thick layer of salt. Modern agricultural knowledge suggests that the only way to avoid the worst of these problems would have been to leave land fallow and unwatered for long periods to allow the water table to fall. But internal pressures within Sumerian society made that impossible and brought about disaster. The limited amount of land that could be irrigated, along with a rising population and increasing competition among the city-states, increased the pressure to intensify the agricultural system. The overwhelming need for more food made it impossible to leave land fallow for long periods. Short-term demands outweighed any considerations of the need for long-term stability and the maintenance of a sustainable agricultural system.

About 3000 B.C.E., Sumerian society became the first literate society in the world. The detailed administrative records kept in the temples of the city-states provide a record of the changes in the agricultural system and the development of major problems. In the Early Dynastic period, which lasted a little more than 600 years until 2370 B.C.E., the major city-states (Kish, Uruk, Ur, and Lagash) were militaristic, hierarchical societies that used the food surplus produced by irrigation to feed both their bureaucracies, which ran the states, and their armies, which continually competed for domination of the

area. These states depended on the large-scale production of wheat and barley, which was slowly being undermined by the environmental degradation wrought by irrigation.

About 3500 B.C.E., roughly equal amounts of wheat and barley were grown in southern Mesopotamia. But wheat can tolerate a salt level of only 0.5 percent in the soil, whereas barley can grow in soil with a salt level of 1 percent. The increasing salinization of the region's soil can be deduced from the declining amount of wheat cultivated and its replacement by the more salt-tolerant barley. By 2500 B.C.E., wheat constituted only 15 percent of the cereal crop. By 2300, it was no longer grown in the Sumerian city-state of Agade. By 2100, Ur also had abandoned wheat production, and overall wheat represented just 2 percent of the crop in the Sumerian region. By 2000, Isin and Larsa no longer grew wheat, and, by 1700 B.C.E., salt levels in the soil throughout the whole of southern Mesopotamia were so high that no wheat was grown at all.

Even more significant than the replacement of wheat by barley was the declining yield from crops throughout the region. In the Early Dynastic period, agricultural areas that went out of production because of salinization were replaced by newly cultivated fields. Growing population and the rising demand for a greater food surplus to maintain armies, as competition among states increased, reinforced the need for new land. But the amount of new land that could be cultivated was limited, even with the more extensive and complex irrigation works that were becoming common at this time. Until about 2400 B.C.E., crop yields remained high. In some areas, they were at least as high as those in medieval Europe and possibly even higher. Then, as the limit of arable land was reached and salinization took an increasing toll, the food surplus began to fall rapidly. Crop yields per cultivated hectare fell 42 percent between 2400 and 2100 B.C.E. and were reduced by 65 percent by 1700 B.C.E. By 2000 B.C.E., there were reports of "earth turned white," a clear reference to the drastic impact of salinization.

7

1. GLOBAL ENVIRONMENT

The consequences for a society so dependent on a food surplus were predictable. The size of the bureaucracy and the army that could be fed and maintained fell rapidly, making the state very vulnerable to external conquest. What is remarkable is the way that the political history of Sumer and its city-states so closely follows the steady decline of the agricultural base. The independent city-states survived until 2370 B.C.E., when the first external conqueror of the region, Sargon, established the Akkadian empire. That conquest was concurrent with the first serious decline in crop yields following widespread salinization. During the next 600 years, the region saw the Akkadian empire conquered by the Guti nomads from the Zagros mountains; a brief revival of prosperity under the Third Dynasty of Ur between 2113 and 2000 B.C.E.; the dynasty's collapse under pressure from the Elamites to the west and Amorites to the east; and, about 1800 B.C.E., conquest by the Babylonian kingdom of northern Mesopotamia.

From the end of the once flourishing and powerful city-states in 2370 B.C.E. to the Babylonian conquest, crop yields continued to fall, making it very difficult to sustain a viable state. By 1800 B.C.E., when yields were only about one-third of those obtained during the Early Dynastic period, the agricultural base of Sumer had effectively collapsed. The focus of Mesopotamian society shifted permanently to the north, where a succession of imperial states controlled the region, and Sumer declined into insignificance as an underpopulated, impoverished backwater of the kingdom.

The Mediterranean

The consequences of steady and continual tree cutting can be seen most clearly in the Mediterranean region. Many tourists now regard the landscape of olive trees, vines, low bushes, and strongly scented herbs, such as the maquis of southern France, as one of the main attractions of the region. The Mediteranean flora is, however, the result of massive environmental degrada-

tion brought about not by the creation of an artificial system such as irrigation but by the relentless pressure of long-term human settlement and growing population. (For more on the region's environment, see Michel Batisse's article, "Probing the Future of the Mediterranean Basin," in the June 1990 *Environment*.)

The natural vegetation of the Mediterranean basin was a mixed evergreen and deciduous forest of oak, beech, pine, and cedar. The forest was cleared bit by bit to provide land for agriculture, fuel for cooking and heating, and construction materials for houses and ships. Because sheep, cattle, and goats grazed on the young trees and shrubs that sprang up after clearing, the forests did not regenerate. Gradually, flocks of animals cleared the land of edible plants and reduced the vegetation to a low, largely inedible scrub. Removal of the tree cover, especially on steep slopes, led to severe soil erosion that ruined agricultural land, which already was short of manure because the farmers practiced transhumance, moving the flocks of animals to different areas for summer and winter. In addition, the large amount of silt carried down the rivers blocked water courses and caused large deltas and marshes to form at river mouths.

Long-term environmental decline can be traced in every area of the Mediterranean and the Near East. Overall, it is now estimated that no more than 10 percent of the original forests that once stretched from Morocco to Afghanistan even as late as 2000 B.C.E. still exist. One of the first areas to suffer deforestation was the hills of Lebanon and Syria. The natural climax forests there were particularly rich in cedars, and the cedars of Lebanon became famous throughout the ancient Near East for their height and straightness. They were prized by the states and empires of Mesopotamia as building materials, and control of the area or trade with its rulers was a high priority for all neighboring states. Later, the cedars became a mainstay of the trade of the Phoenicians and were sold over a wide area. Thus, the renowned cedars of Lebanon gradually were reduced to a pathetic

remnant—there are now just four small groves left in the region—maintained as a symbol of former glory.

In Greece, the first signs of widespread environmental destruction began to appear about 650 B.C.E., as the population grew and settlements expanded. The root of the problem was overgrazing on the 80 percent of the land that was unsuitable for cultivation. Although the Greeks were well aware of techniques for preserving soil, such as the use of manure to maintain the fertility and structure of the soil and terracing to limit erosion on hillsides, the pressure from a continually rising population proved too great. The hills of Attica were stripped bare of trees within a couple of generations, and, by 590 B.C.E., the great reformer of the constitution, Solon, was arguing in Athens that cultivation on steep slopes should be banned because of the amount of soil being lost. A few decades later, the tyrant of Athens, Peisistratus, introduced a bounty for farmers to plant olives, the only tree that would grow on the badly eroded land because it had roots strong enough to penetrate the underlying limestone rock.

The same problems occurred in Italy a few centuries later, as the population rose and Rome grew from a small city into the center of an empire encompassing the Mediterranean and most of the Near East. About 300 B.C.E., Italy and Sicily were still well forested, but the increasing demand for land and timber resulted in rapid deforestation. The inevitable consequence was severe soil erosion and, as the earth was carried down the rivers, the gradual silting up of ports in the estuaries. The port of Paestum in southern Italy silted up completely and the town declined, while Ravenna lost its access to the sea. Ostia, the port of Rome, survived only by the construction of new docks. Elsewhere, large marshes developed around river mouths as the water deposited soil eroded from the deforested hills. The Pontine marshes were created about 200 B.C.E. in an area that had supported 16 Volscian towns just 400 years earlier. The creation of marshes increased the breeding grounds for mosquitoes and, therefore, the incidence of malaria, which became common in Rome in the second century B.C.E.

The growth of the Roman empire increased the pressure on the environment in other areas of the Mediterranean as the demand for food increased. Many of the empire's provinces were turned into granaries to feed the population of Italy, particularly after 58 B.C.E., when the citizens of Rome started to receive free grain for political reasons. North Africa, for example, contains many impressive Roman remains, such as the great city of Leptis Magna, of what were once some of the most flourishing and highly productive provinces of the empire. The area had continued to flourish even after the final destruction of Carthage in 146 B.C.E., but the growing Roman demand for grain pushed cultivation further into the hills and onto vulnerable soils that were easily eroded when deforested.

There is no single date that marks the decline of the North African provinces. It was a long, drawn-out process of increasing environmental strain and deterioration, as soils eroded and the desert slowly encroached from the south. The process was intensified after the fall of Rome, when tribes such as the Berbers moved into the cultivated areas with their large flocks of grazing animals, which completed the work of removing the remaining vegetative cover. Today, the Roman ruins lie surrounded by vast deserts, a memorial to widespread environmental degradation brought about by human actions.

Similar pressures occurred in Asia Minor, where the interiors of the old Roman provinces of Caria and Phrygia were completely deforested by the first century C.E. (A.D.). A few decades later, the emperor Hadrian had to restrict all access to the remaining forests of Syria because of the amount of deforestation. Some regions in Asia Minor were less seriously affected and continued to prosper as food exporters to the main imperial cities; towns such as Antioch and Baalbek flourished until the early Byzantine period. But both towns are now ruins. Antioch is under 28 feet of waterborne silt from hillsides ravaged by deforestation, and some limestone hills in the area have lost up to 6 feet of soil.

The Maya

Mayan society was remarkable in that it developed in the lowland tropical jungle of what is now parts of Mexico, Guatemala, Belize, and Honduras, an area in which obtaining enough food for a large population would pose major problems. The earliest settlements date from about 2500 B.C.E. The population rose steadily, and settlements grew in size and complexity so that, by about 450 B.C.E., there were separate ceremonial areas and buildings within the settlements. Two hundred years later at Tikal in Guatemala, a complex hierarchical society (easily identified by the large differences among burial arrangements) emerged, and steep pyramids more than 100 feet tall with temples on their summits were built out of the local limestone in the north acropolis. During the next two or three centuries, several major settlements developed throughout the area with a remarkably uniform culture displayed in the architectural styles and common script.

The considerable intellectual achievements of the Maya were particularly reflected in their studies of astronomy—they made detailed and accurate calculations of the positions and phases of the sun, moon, and some planets—and in their highly complex and extremely accurate calendar, which was based on a 52-year cycle that counted from a fixed date in the past equivalent to 3114 B.C.E. (although the significance of this date remains unknown). All the Mayan sites have a large number of stone steles inscribed with a series of dates, which can be translated, and texts, which remain largely undeciphered. However, the main phases of Mayan history are clear. By the first few centuries C.E., a large number of elaborate ceremonial centers had developed throughout the region. For a couple of centuries after 400 C.E., the Mayan culture was strongly influenced by the city of Teotihuácan in central Mexico, but when that city declined after 600, the Maya entered their most spectacular period. Huge pyramids, often aligned toward significant astronomical points, were built at all the population centers, and large numbers of steles were erected. Then, within a few decades after 800, the whole society began to disintegrate. No steles were erected, the ceremonial centers were abandoned, population levels fell abruptly, and these magnificent cities soon were covered by the encroaching jungle.

Until the 1960s, historians believed that the Maya were virtually unique in the world in that they were peaceful and governed not by secular rulers and a military elite but by a religious caste obsessed with the intricacies of their calendar and astronomical observations. Because only the dates on the steles could be understood, it was assumed that these recorded various events associated with astronomical and calendrical cyles. How the Maya obtained their food and supported the priestly elite in a lowland jungle environment remained a puzzle.

Studies of the 20th-century Maya suggested that the only viable strategy would have been to use swidden agriculture, which would have involved clearing a patch of jungle with stone axes during the dry season between December and March and then setting fire to the area just before the start of the rainy season, when maize and beans would have been planted with a digging stick to be harvested in the autumn. The cultivated patches would have been abandoned after a couple of years as soil nutrients were depleted and as weeds reinvaded and made clearing too difficult. (Clearing jungle is far less laborious than clearing grass and scrub.) This agricultural system is widely used in tropical areas and is highly stable over the long term, but it can only support a small population because of the need to have a large amount of land for each farmer. The cleared patches cannot be reused for 20 years or more until the jungle has regrown. Therefore, it was assumed that the Maya lived in small, shifting settlements scattered throughout the jungle and came together only at the ceremonial centers (where the small priestly caste lived permanently) for part of the year.

1. GLOBAL ENVIRONMENT

In the last 30 years, these assumptions about Mayan society have been abandoned, replaced by a radically different picture of the Maya that helps explain why the society collapsed so abruptly. The new picture stems from a new understanding of the texts engraved on the steles. It is now clear that these are not religious texts; rather, they are monuments to the different secular rulers of the cities, with the dates of their birth, accession, and death together with the major events of their reign. All the rulers at Tikal between 376 and about 800 C.E., when the site was abandoned, have now been identified, as have the rulers of Palenque from 603 to 799 and those of many other cities. The signs on the steles of the different cities have been deciphered, and, although the texts still cannot be read in full, the conquest of one city by another and, hence, the existence of warfare can be deduced.

The picture of a peaceful, religious society has been replaced by a view of Mayan society dominated, in the same way as other early societies, by a secular elite supported by armies that were engaged in fairly continuous warfare among the different cities. Recent archaeological work also has made the nature of these cities much clearer. They were not merely ceremonial centers occupied by a small elite but true cities with large permanent populations. At the cities' centers were huge ceremonial areas with magnificent temples and palaces built around a plaza. Beyond were complexes of thatched huts on platforms grouped around courtyards where most of the people lived in extended family units. They provided the labor force that constructed the public buildings and those for the elite. Recent archaeological work in the outer areas of Tikal suggests that, at its height, the population was at least 30,000 and possibly as high as 50,000 (on the same order as the great cities of Mesopotamia). Other cities, though not quite so large, would have followed the pattern of dense urban settlement, and it seems likely that the total population in the Mayan lowland jungle at its peak might have been near 5 million, whereas today that area supports only a few tens of thousands.

This new knowledge about the nature of Mayan society has been complemented by new information about the way the Mayans obtained their food. Obviously, swidden agriculture could not support such a large population. Not enough land was available between the cities, which in some cases were no more than about 10 miles apart, to make this system feasible. Hunting and fishing could have provided little more than useful supplements. Although the *ramon*, or breadnut tree, whose nuts can be ground to make flour, grows in profusion in the Mayan area, studies of the current day Maya suggest that the nuts would be used only as a food of last resort.

Recent archaeological work has discovered that a much more intensive agricultural system was used by the ancient Maya. On the hillsides, they cleared jungle and made fields using extensive terracing to try to contain the inevitable soil erosion. Equally important was the construction of raised fields in swampy areas. The raised fields followed the same principle as the *chinampas* of central Mexico except that they were not built out into lakes. Grids of drainage ditches were dug in the swamps and the material from the ditches was used to form raised fields. Traces of the huge areas once covered with these fields have been found in the jungle from Guatemala across to Belize. Mayan farmers grew such crops as maize and beans for food together with others such as cotton and cacao.

This intensive cultivation system was the foundation for all the achievements of the Maya. However, when too much was demanded, the system could not withstand the strain. The crucial period came after the waning of Teotihuácan influence about 600 C.E. Warfare increased among the Mayan cities, and the elite demanded construction of more and larger ceremonial buildings, requiring huge amounts of labor. Population continued to rise steadily, and a higher proportion lived in the cities where they were available to fill the armies and work on construction projects.

Cultivation became even more intense. The ecological basis to support such a massive infrastructure was simply not there. The soils in tropical for-

ests are easily eroded once the tree cover is removed. Mayan settlements clustered, not surprisingly, around the areas of fertile soil, but three-quarters of the fertile soil in the area occupied by the ancient Maya is today classified as highly susceptible to erosion. Around Tikal, for example, 75 percent of the soil is considered very fertile, but nearly 60 percent of it would be highly vulnerable to erosion if it were cleared of trees. Clearing the forest, therefore, ran the risk of bringing about soil deterioration and declining crops yields. The Maya also lacked domesticated animals, which would have provided manure to maintain soil structure and fertility. The forest was cleared not just to provide land for agriculture, but also to provide timber for fuel, construction, and the making of lime plaster to coat ceremonial buildings with stucco. Population pressure pushed fields into ever more marginal areas. Across the region, the vulnerable soils increasingly were exposed to wind and rain and subsequently eroded.

The exact sequence of events that brought about the fall of the Maya has not yet been pieced together, but it seems clear that increasing environmental degradation played a major role. In particular, it prevented sufficiently high levels of food production from being sustained. Not only would soil erosion caused by deforestation have reduced crop yields, but the associated higher levels of silt in the rivers would have seriously damaged the extensive raised fields in marshy areas by altering the water level and by making the ditches much more difficult to keep clear.

The first signs of declining food production are evident in the period before 800 C.E. Skeletons from burials of the period show higher infant and maternal mortality and increasing levels of nutritional deficiency. The reduction of the food surplus on which the ruling elite, the priestly class, and the army depended would have had major social consequences. Attempts probably were made to increase the amount of food taken from the peasant cultivators and may have led to internal revolt. Conflicts among the cities over the declining resources would have intensified, leading to more warfare. The fall in food supplies and the increasing competition for what was available would have brought about very high death rates and a catastrophic fall in population, making it impossible to sustain the elaborate superstructure the Maya had built on their limited environmental base.

Within a few decades, the cities were abandoned, construction ceased, and no more steles were erected. Only a small number of peasants continued to live in the area. The deserted fields and cities, buried under dense jungle, were not found again until the 19th century, when the temples, palaces, and steles of what seemed to be mysterious, lost cities in the jungle were uncovered by two U.S. explorers, John Stevens and Frederick Catherwood, between 1839 and 1840.

Today's Questions

What happened to the Maya is one example of how many of the earliest settled societies overreached themselves. By using the natural resources readily available, by finding ways of exploiting these more fully, and, in some cases, by creating artificial environments, the Maya were able to build a complex society capable of great cultural and intellectual achievements. For a considerable period, they appeared to be highly successful, but the demands of an increasingly complex society began to overstretch the ability of the agricultural base to support the large superstructure that had been erected. In the end, the unwanted and unexpected side effects of what at first appeared to be solutions to environmental difficulties became overwhelming problems. In short, the Maya ended up destroying the environment on which they relied for food and, hence, for the whole complex social structure that had developed.

At the core of any such structure is the need for a continuing food surplus to support all the nonproducers. In Mesopotamia, intensive, irrigated agriculture and the consequent salinization and waterlogging of the fields destroyed the basis for Sumerian society. In the Indus valley and the Mayan lowland jungle of Mesoamerica, large-scale deforestation leading to soil erosion brought about a similar collapse. In some ancient lands, the results were less dramatic than the downfall of a once-powerful empire or a unique culture. For example, northern China and the Mediterranean suffered from long-term environmental decline caused by deforestation and soil erosion stemming from continual human occupation. The demands of burgeoning populations gradually reduced the resources available and left large areas relatively impoverished.

The histories of Mesopotamia, the Mediterranean, and the Maya suggest some questions about current human societies and their development. Are contemporary societies any better than ancient ones at controlling the drive toward ever greater use of resources and heavier pressure on the environment? Is humanity too confident about its ability to avoid ecological disaster?

Viewed over the long term, human history has been a story of rising numbers and greater pressures on the environment. In ancient times, those pressures occurred on a local scale, as in Mesopotamia and other areas. Now, with the creation of a global economy to exploit resources, humanity for the first time must contend with damage to the global mechanisms that make life on Earth possible—the ozone layer and the concentration of carbon dixoide in the atmosphere. Given the 2-million-year history of humans on Earth, it is still an open question whether the 10,000-year-old development of agriculture and settled societies and the more recent dependence on nonrenewable fossil fuels constitute an ecologically sustainable strategy.

NOTE

1. The data and ideas presented here are extracted from the author's forthcoming book: C. Ponting, *A Green History of the World* (London: Sinclair Stevenson, Ltd., 1991). See also, T. P. Culbert, ed., *The Classic Maya Collapse* (Albuquerque, N. Mex.: University of New Mexico Press, 1973); J. D. Hughes, *Ecology in Ancient Civilizations* (Albuquerque, N. Mex.: University of New Mexico Press, 1973); J. Rzoska, *Euphrates and Tigris: Mesopotamian Ecology and Destiny* (The Hague: W. Junk, 1980); and R. McC. Adams and I. Jacobsen, "Salt and Silt in Mesopotamian History," *Science* 128 (1958):1251-8.

Managing as if the Earth Mattered

James E. Post

James E. Post is a professor of management and public policy in the School of Management, Boston University.

Alverino Gomez lives and works in Mexico City. He is the assistant plant manager for an American multinational corporation manufacturing industrial machines. The company's Mexico City facility is a major production center for its North American operations. The company has another Mexican facility near the U.S. border and other operations in the United States, Europe, and Asia. Mr. Gomez has a promising career in this company and hopes to continue his career development in Mexico, where his family is located. On this day, Mr. Gomez has received a telephone call from a government official from SEDUE, the Mexican environmental protection agency. The official ordered the plant to shut down because air quality had reached emergency levels. Mexico City is home to the world's largest urban population and lives with one of the world's greatest urban air pollution burdens. This is not the first emergency declared by environmental officials, but others have been widely ignored by industry. Business leaders often talk about the severe impact of a closing on production schedules and customer deliveries. Labor leaders are also unhappy with such emergencies because workers are sent home and receive no compensation for the hours not worked. Because the plant manager is away this day, Mr. Gomez must make the decision whether or not to close the plant.

Several thousand miles away, Peter Crumholz faces a different type of dilemma. He is an assistant product manager for a large consumer products firm and is responsible for maintaining the

product's market share. The product is trademarked, dominates the consumer segment of the market, and has high buyer loyalty. Competition is based on price, image, and product quality. As his boss likes to say, "We have a mission and there is no room for error." A new study of the product's packaging has pointed out that customers are attracted to the presentation of the product but that the package contains substantial quantities of cardboard, polystyrene, and insulation made with ozone-depleting chlorofluorocarbons (CFCs). The study group recommends a major redesign to eliminate the CFC materials but believes it is unwise to reduce packaging size because of the marketing effects, despite the ability to reduce solid waste by 15 percent. Peter is to be the "point person" in developing a course of action for the company. He is scheduled to make a presentation, with recommendations, to the senior product manager and his counterparts in two days.[1]

Alverino Gomez and Peter Crumholz are both facing the increasingly common dilemma of reconciling routine business activity with emerging environmental concerns that force hard choices. For Mr. Gomez, there is the choice of closing the facility—thereby forcing economic injury on his work force and disrupting manufacturing processes and customers' plans—to meet an unevenly enforced emergency order that will not solve the air quality problems in any event. For Mr. Crumholz, there is the choice of endorsing packaging changes that are either too little to meet environmental needs or too much to guarantee the product will not be hurt in the marketplace.

Readers no doubt have ideas about clever ways to reconcile the conflicting dimensions of these issues. Considerable imagination can be—and actually was—brought to both of these cases. The dilemmas faced by these managers are not unique, and they raise unfamiliar problems of

Reprinted from *Business Horizons*, Vol. 34, No. 4, July/August 1991, pp. 32-38. Copyright © 1991 by the Foundation for the School of Business at Indiana University. Used with permission.

corporate responsibility. This article examines a variation on the central theme of this issue: What are a corporation's environmental responsibilities in a world where environmental problems are growing in number, severity, and complexity?

Business schools have largely ignored both this question and natural resource and environmental issues in their curricula (Post 1990). To the extent attention has been given to the dilemmas, it has been in the study of business-government and business-society relations. It is not sufficient that only such courses address these issues. The prevalence and seriousness of environmental problems and crises is becoming increasingly evident to managers everywhere. I believe environmental issues will be to business in the 1990s what quality issues were to business in the 1980s: a force of such power as to literally transform the way managers manage their businesses and think about the relationship of the firm to its internal and external stakeholders. This article develops the argument from two perspectives: (1) an analysis of why environmental problems are a basic new reality for managers and corporations in the 1990s; and (2) an assessment of what types of decisions managers and firms will be forced to make in reconciling environmental and economic considerations.

ENVIRONMENTAL CHALLENGES[2]

The 1990s are being called the "decade of the environment." Public interest in preserving nature's ecological balance and a clean, healthy environment is high and growing. This means that business must strive to reconcile these goals with other, equally demanding goals.

In the desert near Tucson, Arizona, there exists what appears to be a giant greenhouse. When it is lighted at night, it shines against the dark, star-filled sky like an enormous beacon. Inside, eight human beings are living and working in a series of climatological zones called "biomes" that include a tropical rain forest, an ocean (complete with tides), a savannah, and a desert. Living in each zone are appropriate animals and plants, balanced in the best self-sustaining ways human beings can imagine. This environment will be sealed for two years, during which time the people, animals, plants, and other living organisms will try to survive. It is called "Biosphere 2," and it is a $30 million experiment to learn about ecological interdependencies. The scientists who developed Biosphere 2 say it will inform us about what must happen if humans are ever to build space colonies on distant planets. Also important are the lessons it may teach about our own relationship with Biosphere 1—the planet Earth (Allen 1991).

Thousands of miles away, crews of workers toil on the smoky desert and oily shores of the Persian Gulf. They are trying to cap the wells and clean the remains of one of the world's worst oil disasters, the intentional dumping of oil by Iraqi military officials during the 1990-91 war. Environmental experts helplessly watched the ecological carnage, aware it was taking place but unable to stop it. Many experts believe it will take years, even decades, for the desert and the Gulf to be ecologically restored. Some believe it will never happen.

People living near Prince William Sound, Alaska, the Rhine River that flows through Switzerland and Germany, and the Russian city of Kiev, a few miles north of Chernobyl, understand the importance of the biosphere and the terrible consequences of human damage to the environment. The Exxon Valdez oil spill, the explosion of a Sandoz AG factory in Basel, Switzerland, and the nuclear reactor fire at Chernobyl are tragic milestones in recent environmental history. Today the world is more aware than ever of the limited ability of earth to accept such ecological shocks. There is a new urgency to halting environmental damage, cleaning up the effects of past practices, and creating a new relationship between economic activity and environmental concerns.

Tragic events symbolize the major collision between industrial technology and nature's ecological systems. Agricultural chemicals, nuclear power, and the many other advances made possible by modern science and technology bring enormous benefits to humankind. But the human price and the pressures on the earth's ecological systems are sometimes unacceptably high. Finding a balance between industrial benefits and life-sustaining ecological systems is a major challenge facing business managers, government policy makers, and society in general. The work of the "biospherians" in Biosphere 2 is intended to advance our scientific understanding of humans and the environment. The work of business and government leaders, and society in general, is critical to meeting what has been called the challenge of "managing planet earth" (*Scientific American* 1989).

Ecology, the study of how living things interact with one another and their environment, is certain to become part of the modern managerial consciousness. In a general sense, business cannot be conducted today without an understanding and appreciation of the stakeholders and interdependencies that exist between the corporation and others. Specifically, managers in business, government, and nonprofit organizations are realizing the impact, power, and transforming potential of ecological issues.

The ecological challenge requires managers to formulate strategies, for the present and the future, that (1) make the most efficient use of scarce resources; (2) reduce wastes that pollute the environment; and (3) keep industrial production and other human activities within the limits set by nature's ecological systems. In the 1990s, this challenge is more formidable and more critical than at any time in human history. As global natural resources are depleted, the survival potential of the planet itself is at stake. Never before has Earth itself become a stakeholder of such significance to corporations and managerial decision making.

The Global Commons

Throughout history, communities of people have created "commons." A commons is shared land on which, for example, a herder can graze his or her animals. The limited carrying capacity, or ability to sustain population on a given quantity of land that comprises the commons, is exceeded as each herder adds more animals to the land. If short-term decisions dominate each herder's thinking, the commons will be destroyed because each herder gets near-term advantage from grazing the maximum number of animals. In the long run, of course, the herder also loses because the commons is destroyed. The only solution in the near term is restraint, either voluntarily or through some form of coercion such as a law, that would limit the maximum number of animals. As the author of "Tragedy of the Commons" writes, "freedom in a commons brings ruin to all" (Hardin 1968).

The managerial dilemmas facing Mr. Gomez and Mr. Crumholz described at the beginning of this article are problems of the commons. We live on a global commons. As scientific evidence has demonstrated that we are testing—and surpassing in some instances—the earth's carrying capacity for pollution, our behavior is as threatening to the planet as that of the herders to the commons. Present dangers exist at two levels: (1) local environmental damage, such as toxic waste dumping, that leaves areas of the earth unable to support living organisms; and (2) global systems of climate, atmospheric protection, and food resources that are breaking down as the result of cumulative pollution. Depletion of the ozone layer (recently revealed to be occurring at a rate even faster than that previously understood), destruction of the rain forests, and desertification of land from topsoil loss are but a few of the ominous global environmental transformations now underway. These systems cannot be damaged or destroyed without affecting everyone. The deliberate destruction of wells and spilling of oil into the Persian Gulf damaged not just Iraq's enemies, but everyone in the region. Such events reinforce the message that preservation of the global commons is a new imperative for institutions, their managers, and all citizens.

Sustainable Development

The World Commission on Environment and Development, including leaders from many industrialized and developing nations, has described the need for balance between economic and environmental considerations as sustainable development—"development that meets the needs of the present without compromising the ability of future generations to meet their own needs" (World Commission . . . 1987). There are two concepts within this idea that bear directly on business and society:

• First, the concept of "needs," in particular the essential needs of the world's population, rich and poor, to survive;

• Second, the concept of "limitation" imposed by the state of technology or social organization on the environment's ability to meet present and future needs.

Reconciling human needs, which are met through economic activity, with limitations imposed by ecological systems is the practical challenge that now confronts all managers. Protection of the global commons through responsible environmental management is a vital step toward sustainable development. But for companies and governments, and all their managers, the diagnosis is easier than the solution.

MANAGING AS IF THE EARTH MATTERED

The twin ideas driving the environmental movement are preservation and conservation of natural resources on the one hand and control of pollution on the other. Conservationist thinking dates to the early 1900s in the U.S., when conservation leaders such as John Muir led campaigns to save natural resources, and political leaders such as President Theodore Roosevelt took actions to establish nature preserves. In the 1990s, this strain of thought is once again ascendant as biodiversity, preservation of wilderness, and animal habitat protection are sought by environmental advocates.

Concern about pollution has taken new forms in the 1990s. Scientific understanding of risks to human health and natural resources is more refined. The scientific measurement of exposures surpasses that of the past. Whereas "parts per million" was once the standard for quantifying toxins in air and water, it is now possible to state those exposures in parts per billion or parts per quadrillion. Whether such minute exposures are meaningful to humans is a question that provokes debate among scientists and policymakers. But as long as public fear of toxins is high, political pressure can be effectively exerted on business and government to reduce such perceived risks. Some have questioned the ethics of advocates—such as those who succeeded in banning Alar, a pesticide used on apples—as capitalizing on media hype and public fear. Whatever the merits of that criticism, it underscores the problem of creating reasoned dialogue about scientific information with a public that is highly emotional about toxic risks.

Figure
Effects of Environmental Issues on Business Functions

Area	Example
Human Resources	Workplace risk exposures
Marketing	"Green" products
Finance	Liabilities; investment criteria; full environmental cost accounting
Operations/ Manufacturing	Waste reduction; energy use; process design
Product Development	Environmental life cycle; packaging
Research and Development	Use of animals; product specifications
Transportation	Vehicle mileage; alternative fuels; hazardous contents

A complex set of scientific, social, and political values has also made environmentalism a powerful political philosophy today. "Green politics" describes a view of the world that uses environmental impact as a litmus test for all types of industrial, social, and technological decisions. In such European nations as Holland and Germany, green political parties have emerged, held office, and shaped the political debate. Hedrick Smith, author of *The New Russians* (1990), notes that if a green political party were permitted in the Soviet Union, it would almost surely be a significant political power given public concern in the aftermath of the Chernobyl disaster, the death of the Aral Sea, and innumerable toxic sites that have come to light since the fall of the Iron Curtain.

Global Issues and Business Response

The preservation/conservation theme is especially powerful in our emerging understanding of global resource issues. Three global issues have profound consequences for business and society in the near and intermediate term: ozone depletion, global warming, and biodiversity. Each is a global commons-type issue; each presents itself to managers and firms with a degree of scientific uncertainty (Buchholz, Marcus, and Post 1992).

Ozone depletion. Since the 1970s, scientific concern for the depletion of stratospheric ozone has grown. The rapidly escalating estimates of skin cancers and agricultural damage caused by increasing amounts of ultraviolet rays from the

sun have prompted nearly unprecedented international action. The development of the Montreal Protocol as a political framework to reduce and eliminate chlorofluorocarbons (CFCs) that deplete stratospheric ozone has been signed by more than 100 nations. The industrial producers of CFCs—primarily located in the U.S., Europe, and Japan—have begun the phaseout of CFC production, and industrial users, including manufacturers of packaging, air conditioning, and computers, are moving to safer substitutes.

Global warming. The increasing temperature of the earth's atmosphere is subject to more uncertainty and debate than ozone depletion. Nevertheless, a broad scientific concern about global warming has moved business and government to consider the consequences if theories prove correct. Foremost among the effects is climate change, with direct effects on agriculture, food supplies, and human starvation. The release of carbon dioxide is a primary contributor to global warming, along with the release of other greenhouse gases used by industry. In the 1990s, efforts are underway to develop an international agreement like the Montreal Protocol to limit greenhouse gas emissions and slow the pace of warming. If not done, experts fear that polar ice caps will melt, raising sea levels and flooding coastal plains such as Bangladesh. The human and habitat costs of such climate change could be economically and socially devastating to many of the world's nations.

Biodiversity. Genetic diversity is essential for healthy species of plants, animals, and human beings. As in the Biosphere 2 example discussed above, habitats must be carefully balanced if they are to survive. As areas of rich natural resources, such as the tropical rain forests of Brazil, Malaysia, Costa Rica, and Mexico, are destroyed, many species of plant and animal life are endangered or eliminated. This has serious environmental balancing effects and high human costs as well. The pharmaceutical industry, for example, each year develops new medicines based on newly discovered plants from rain forest areas. As they are destroyed, so too is the stock of potential new medicines.

Local Pollution and Business

Pollution is felt most immediately and acutely by local communities. Fouling of the air and water, for example, usually has its most direct effects on those citizens and communities living downriver or downwind of the source. As the example of Mexico City suggests, managers still face dilemmas when responding to these issues. In nations where governments have established environmental standards and engaged in vigorous and vigilant enforcement, progress in reducing pollution has been made. In nations where laws are weak or nonexistent and enforcement is lax, serious local air and water pollution problems plague communities. For those who make decisions, whether in business or government, the tradeoffs between cost and environmental standards remain difficult to make without visible public policy that encourages action favoring environmental protection.

Recent studies (Environmental Protection Agency 1990) indicate that United States' citizens and institutions spent more than $100 billion on pollution abatement in 1990. This is projected to double to $200 billion by 1995. In the U.S. and some European nations, approximately 2 percent of gross national product is now directed toward environmental protection. These data demonstrate that managers are making environmental investments every day. Government has moved to permit more incentive-oriented initiatives, such as "bubble concepts," pollution charges, and tradable emission rights. The EPA's policy decision to emphasize voluntary pollution prevention by industry is a landmark in U.S. environmental policy. Such approaches can only succeed, however, if companies make creative use of the flexibility to harmonize business decisions on plant location, product and process design, and operational standards with environmental goals. If voluntary action does not lead to improved environmental performance, more "command and control" regulation is likely to occur given the high societal importance placed on pollution control. The interplay of corporate policy and public policy thus directly shapes managerial behavior.

THE GREENING OF MANAGEMENT

Environmental concerns touch all aspects of a business' operations. As illustrated in the **Figure**, modern environmental problems affect the management of a company's operations, marketing, human resources, and other activities. Even areas such as finance and accounting are directly affected. For example, federal rules now require a company to account for its toxic materials with an elaborate system of reports and an internal audit of environmental compliance. Financial officers recognize the increasing power of institutional investors, using environmental criteria such as the Valdez Principles (CERES 1989), to select companies for investment or damage those that fail to meet the criteria. The rise of environmental mutual funds and the attention of limited partnerships and venture capitalists to environmental businesses are also significant to the finance function. Some courts have ruled that the purchaser of property assumes full environmental liability, affecting the economics of some mergers and acquisitions and adding significantly to due diligence expectations.

Strategic and operational decisions are both affected. Management decisions about where to locate facilities, what product lines to develop, and environmental, health, and safety standards in all the company's facilities are major decisions. How a firm fuels its fleet of cars and trucks (gasoline, alternative fuels, electricity), designs energy efficiency into facilities, organizes employee transportation services, minimizes toxins in manufacturing, and communicates about all of these to communities and government officials affects its environmental profile.

Corporate environmentalism, a term that is sometimes used to describe responsible management responses to these issues, is neither a fad nor peripheral to the "real business" of decision making. Rather, it represents the commitment to environmental responsibility and a translation of that commitment into action. Research into the criteria that guide such efforts and the actions that distinguish an effective corporate environmental program is still relatively anecdotal. Yet a few studies have shown that corporate culture, reward and evaluation systems, and location and staffing of the environmental, health, and safety functions have some degree of influence on the effectiveness of environmental responses. Research programs now in progress should shed further light on these factors. They are important, for a commitment to environmental responsibility without a capability to translate rhetoric into action seems likely to lead to disappointing results.

More than 30 years ago, Edward Mason (1960) posed the central issues of managerial responsibility as two questions: To whom, and for what, is the modern corporation accountable? These questions still remain at the heart of today's discussion of what is expected of managers and the institutions they direct. Each attempt to extend corporate responsibility to meet the expectations of new stakeholders forces a reassessment of corporate accountability theory and practice. The emergence of environmental issues as the most prominent points on the modern political, economic, and social agenda in the U.S. and abroad poses a new type of challenge to theory and practice.

For some academics, environmental problems are unlike other social issues. The analysis of stakeholders that is so easily organized for most problems is radically different when earth itself is a stakeholder and the "stake" is nothing less than planetary survival. The stakeholder map one can draw for analyzing the water resources issue in Southern California is complicated, but it pales by comparison to that needed to assess global warming, climate change, or ozone depletion. In this context alone, environmental issues force thinking at levels of abstraction not normally used in managerial decision making.

It is not only to introduce value considerations that abstraction is required. It is vital to even define the nature of the problems we now face. What does global resource depletion mean in human terms? How can a manager like Alverino Gomez make a wise decision to shut down a plant that emits little but employs many? Is his company really accountable for Mexico City's air pollution problems? How can such a responsibility be met if it exists at all? Do conventional evaluation criteria suffice when global survival is at issue?

For managers and their companies, the problems are also complicated. Introducing good environmental management requires technical staff, commitment, and a receptive organizational culture. Meeting these needs in an era of global competitiveness exacerbates the challenge. Scarce resources become even less plentiful, and capital and human training needs are even greater. The greening of management is ultimately a matter of infusing environmental concerns, understanding, and commitment into each person's thought process. But it is also a matter of supporting those who wish to act through values that recognize the importance of environmental action, through systems that reward responsible behavior and deter unthinking action.

To manage as if the earth matters is not the same challenge as managing planet earth. But it requires no less a commitment to environmental preservation and no less an understanding of how international trade, competitiveness, and global resources are connected. Managers, like Alverino Gomez and Peter Crumholz, whose lives are not devoted to global and local environmental problems will still be affected by these issues. That is the nature of our world as this century closes. One hundred years ago, business was on the verge of defining a scientific way to manage enterprises. Today, we stand at the edge of another transformation, in which the planet imposes the boundaries within which efficiency and abundance are understood. It is a time of both promise and consequence. The promise is that we will find an environmentally sustainable path into the twenty-first century. The consequence of not doing so is, as Churchill said of defeat, unthinkable.

Notes

1. Both incidents actually occurred. The first example was described to the author shortly after the events occurred. The names of both managers have been changed but other facts are presented as described by the participants.

2. This discussion is an adaptation of Frederick, Post, and Davis 1992, Chapter 19.

References

John Allen, *Biosphere 2: The Human Experiment* (New York: Penguin Books, 1991).

R. Buchholz, A. Marcus, and J. Post, *Managing Environmental Issues: A Casebook* (Englewood Cliffs, N.J: Prentice-Hall, 1992).

CERES (Coalition for Environmentally Responsible Economies), "The Valdez Principles," 1989.

Corporate Conservation Council, "Environmental Education: A Statement for Business Management," National Wildlife Federation/Corporate Conservation Council, 1991.

Environmental Protection Agency, *Environmental Investments: The Cost of a Clean Environment* (Washington, D.C.: U.S. Environmental Protection Agency, 1990).

William C. Frederick, James E. Post, and Keith Davis, *Business and Society: Corporate Strategy, Public Policy, and Ethics*, 7th ed. (New York: McGraw-Hill, 1992).

Garrett Hardin, "Tragedy of the Commons," *Science*, December 13, 1968, pp. 1243-1248.

W.M. Hoffman, R. Frederick, and E.S. Petry, Jr., eds., *The Corporation, Ethics, and the Environment* (Westport, Conn.: Quorum Books, 1990).

Edward Mason, ed., *The Corporation in Modern Society* (Cambridge, Mass.: Harvard University Press, 1960).

"Noah's Ark—The Sequel," *Time*, September 24, 1990, p. 72.

Robert Paehlke, *Environmentalism and the Future of Progressive Politics* (New Haven, Conn.: Yale University Press, 1989).

James E. Post, "The Greening of Management," *Issues in Science and Technology*, Summer 1990, pp. 68-72.

Scientific American, Special Issue: "Managing Planet Earth," September 1989.

Hedrick Smith, *The New Russians* (New York: Random House, 1990).

Peter Stillman, "The Tragedy of the Commons: A Re-Analysis," *Alternatives: Perspectives on Society and Environment*, Winter 1975, p. 12.

World Commission on Environment and Development, *Our Common Future* (New York: Oxford University Press, 1987).

DEBATING GAIA

Stephen H. Schneider

Stephen H. Schneider, a climatologist, is head of the Interdisciplinary Climate Systems Section at the National Center for Atmospheric Research in Boulder, Colorado. He is editor of the interdisciplinary journal Climatic Change *and author of the book* Global Warming: Are We Entering the Greenhouse Century?

The Gaia hypothesis, as defined by its prime advocates, states that Earth's lower atmosphere is an integral, regulated, and necessary part of life itself and that, for hundreds of millions of years, life has controlled the temperature, chemical composition, oxidizing ability, and acidity of the Earth's atmosphere.[1] As the 1990s dawn, the Gaia hypothesis is finally out of the shadows of fringe science. The notion that life wields active control over the planet's physical and chemical environment has become the subject of critical scientific debate some 20 years after British scientist and inventor James E. Lovelock and U.S. microbiologist Lynn Margulis first described the Gaia hypothesis.

One point in the debate—whether a biotic control mechanism somehow maintains the Earth's climate at a temperature suitable for life—has attracted the interest of those who discount the forecasts of a depleted ozone layer and an enhanced greenhouse effect. Indeed, the Gaia hypothesis won some of its earliest support from polluting industries who interpreted Gaia to mean

that nature could counter the effects of pollution and keep the planet inhabitable. (The other major supporters were environmental spiritualists looking for oneness in nature—an idea for which Gaia is a marvelous symbol.) But, until recently, the Gaia hypothesis was virtually ignored by most of the scientific community.

Lovelock was aware in the 1970s that Earth scientists would be skeptical of a Gaian perspective. The predominant view in the natural sciences was that life on Earth is primarily passive, responding to nonliving forces like volcanic eruptions, severe storms, droughts, and even drifting continents. In return, life can modify the local environment and, to some extent, the chemical environment—through the exchange of gases in photosynthesis, for example.

But the Gaia hypothesis goes further and holds that the biota can effectively and directly manipulate the environment *for its own purposes,* or that life optimizes its environment to suit itself. This is the most radical idea to grow out of the Gaia hypothesis and the one whose criticisms are most difficult to answer. Nevertheless, by promoting the profound realization that climate and life mutually influence each other, the Gaia hypothesis provides an important counterpoint to the predominant view that environment dominates life.

The Gaia hypothesis evolved from Lovelock's work as a consultant for the U.S. National Aeronautics and Space Administration (NASA) in the 1960s.

NASA scientists were preparing to launch the Viking spacecraft for a mission to examine the possibility of life on Mars. Lovelock, working with philosopher Dian Hitchcock, argued that there was a simpler way of detecting whether life existed on Mars. Telescopic observations from Earth had already revealed that the Martian atmosphere is predominantly composed of carbon dioxide (CO_2), with relatively little oxygen, methane, or other reactive gases that, on Earth, are the product of photosynthesis and other biological processes. Thus, Lovelock deduced that the probability of life existing on Mars was extremely small. He postulated that, on a lifeless planet, one might expect such gases to be rare because, without constant replenishment by plants and bacteria, these gases would react with other gases and with minerals on the planet surface and disappear, except in minute amounts. Lovelock argued that the inorganic and organic processes of a planet are not independent and that the absence of these gases in the Martian atmosphere indicated the absence of life (assuming that life on Mars would biochemically resemble life on Earth). Nevertheless, the Viking mission went on, analyzed Martian soils, and found no evidence of life.

Lovelock's work at NASA eventually led him to his important association with microbiologist Lynn Margulis, who, at that time, also worked at NASA. They postulated the Earth—its biota and environment—to be a self-

From *Environment*, Vol. 32, No. 4, May 1990, pp. 5-9, 29-30, 32. Reprinted with permission of the Helen Dwight Reid Educational Foundation. Published by Heldref Publications, 1319 Eighteenth St., N.W., Washington, DC 20036-1802.

regulating system able to maintain both the climate and chemical composition of the planet in a state favorable to life. The idea that life shapes the physical environment is an old one, articulated at least since Victorian times. For example, in 1877, T. H. Huxley wrote:

Since the atmosphere is constantly receiving vast volumes of carbonic acid from various sources, it might not unnaturally be assumed that this gas would unduly accumulate, and at length vitiate the entire bulk of the atmosphere. Such accumulation is, however, prevented by the action of living plants.[2]

The innovative and controversial part of the Gaia hypothesis is that life somehow maintains control mechanisms for its own good—that is, that life achieves a sort of homeostasis through negative feedback, or cybernetic control. The name Gaia, after the classical Greek word for Mother Earth, was the suggestion of Lovelock's neighbor, the Nobel Prize-winning novelist William Golding. Said Lovelock, "It is a more convenient term than biological cybernetic system with homeostatic tendencies."[3]

Lovelock and Margulis suggested that the Earth as a whole be viewed as a physiological system—the study of which Lovelock has recently called the new science of geophysiology—wherein complex but not yet well understood mechanisms maintain a stable environment beneficial for life on the planet. Just as a person's body maintains its temperature or the thermostat in a home turns on a furnace or an air conditioner to maintain a set temperature range, the Earth may have its own internal feedback control system. The Gaia hypothesis simply states that such mechanisms of physical and chemical control are embedded in the totality of life on Earth.

Climate Control

Consider the Gaian argument for planetary-scale control of the climate. It is widely believed that the sun has been heating up since its formation many billions of years ago. This belief is based on known principles of nuclear physics that indicate that the hydrogen

in the sun fuses into helium. This process very likely requires the sun to emit more radiative heat energy over time. Calculations suggest that, some 4 billion years ago when primitive life first appeared on Earth, the sun was perhaps 25 percent less luminous than it is today. Modern climatic theory suggests that, given such low solar energy, the Earth should have been a frozen ball. Yet sedimentary rocks that could have been formed only by water flowing on the planet's surface have been dated as having been formed as long ago as 3.8 billion years. Fossil evidence of bacteria also has been dated as more than 3 billion years old. Therefore, at least some part of Earth supported both life and liquid water when the sun was perhaps 25 percent less luminous than at present.

One plausible explanation of this "faint early sun paradox" was offered in 1971 by astronomer Carl Sagan of Cornell University and George H. Mullen of Mansfield University.[4] They suggested that the gases methane and ammonia, which are efficient absorbers of infrared radiation, could have been present in the Earth's atmosphere in concentrations sufficient to trap radiative heat and, through the greenhouse effect, prevent the loss of energy from Earth to space. In other words, these gases could have created a super greenhouse effect that kept the Earth's temperatures equable while the sun was still relatively faint. However, methane and ammonia are removed so quickly from the atmosphere that they probably could not have reached the levels that Sagan and Mullen suggested. Scientists have since proposed that CO_2 would have been the best candidate to create a super greenhouse effect, but the basic idea is still attractive.[5] Indeed, Lovelock and Margulis have argued that both the emission and removal of greenhouse gases, such as ammonia, CO_2, and water vapor, by various organisms are part of the Gaian planetary temperature control mechanism.

But why has the planet not subsequently overheated, since the sun presumably has increased its luminosity by some 25 percent over the past 4 billion years? Lovelock and Margulis find a

Gaian explanation in the tiny ocean phytoplankton that incorporate CO_2 from the atmosphere into their calcium carbonate shells. As the Earth warmed up, the plankton should have taken up CO_2 more efficiently. When the plankton died and sank to the ocean bottom, their carbonate shells would have become sediment, thus removing CO_2 from the system. Moreover, increased rain from warmer conditions would have created more run-off from the land. Run-off provides nutrients to feed plankton and removes CO_2 from the air through the weathering process. Lovelock and Margulis suggest that the net loss of CO_2 would have been enough to compensate for the warming sun. Thus, they argue, Gaia is actively maintaining a fairly constant climate temperature as the sun heats up.

Several possible problems with this scenario have emerged, however. First, phytoplankton, although biologically primitive relative to organisms like trees, are still much more sophisticated than the simple bacteria that were the only living things for the first two-thirds of the past 4 billion years. Indeed, eucaryotes such as phytoplankton that contain distinct nuclei evolved only about a billion years ago. Thus, phytoplankton could not have been the primary CO_2 sink during the 2 to 3 billion years in which simpler life forms were dominant and the sun continued to heat up. Perhaps certain photosynthesizing bacteria were involved in removing CO_2, but the mechanisms and any quantitative assessment of the magnitude of the removal are yet to be shown.

A more serious criticism, however, is the assertion by inorganic geochemists that temperature control through CO_2 removal could be accomplished inorganically, without any biological mechanism. In 1981, University of Michigan geochemists James C. G. Walker, Paul B. Hays, and James Kasting developed an elaborate model for feedback control of climate temperature but with a different mechanism for CO_2 removal. Instead of plankton removing CO_2 by depositing calcium carbonate, the geochemists postulated an inorganic competitor for this process involving the weathering of silicate minerals on land. In this model it is assumed that, as the sun heats up, the climate and oceans become warmer. More water would evaporate from the warmer oceans and eventually rain back down on Earth. CO_2 in the air dissolves in water droplets and forms weak carbonic acid, which reacts with calcium silicate in rocks to form carbon-containing sediments. The geochemists hypothesized that, as the sun heated up, this weathering could have effectively removed enough CO_2 from the air to maintain climatic stability.

The Lovelock and Margulis scenario acknowledges the role of weathering.

As Lovelock has commented, "The Gaian variant of Walker's model assumes that the biota are actively engaged in the process of weathering and the rate of this process is directly related to the biomass of the planet. If conditions are too cold the rate of weathering declines, and as a conse-

THE MANY FACES OF GAIA

The Gaia hypothesis received perhaps its most serious critique to date at the March 1988 American Geophysical Union Chapman Conference in San Diego, California. Presenting the critique was James Kirchner, a physicist, philosopher, and—at that time—graduate student of the Energy and Resources Group at the University of California at Berkeley. The essence of his criticism was that there are many Gaia hypotheses rather than a single one. Kirchner described five hypotheses, each supported by a quote from Lovelock, Margulis, or one of their close colleagues:[1]

• Influential Gaia simply asserts that the biota has a substantial influence over certain aspects of the abiotic world, such as the temperature and composition of the atmosphere. According to Lynn Margulis and science writer Dorian Sagan, "The Gaia hypothesis . . . states that the temperature and composition of the Earth's atmosphere are actively regulated by the sum of life on the planet."

• Coevolutionary Gaia asserts that the biota influences its abiotic environment and that the environment in turn influences the evolution of biota by Darwinian processes. In the words of Andrew Watson and James Lovelock, "The biota have effected profound changes on the environment of the surface of the Earth. At the same time, that environment has imposed constraints on the biota, so that life and the environment may be considered as two parts of a coupled system."

• Homeostatic Gaia asserts that the biota influences the abiotic world and that it does so in a way that is stabilizing. In the language of systems analysis, the major linkages between the biota and the abiotic world are negative feedback loops. Lovelock and Margulis have called the Gaia hypothesis "the notion of the biosphere as an active adaptive control system able to maintain the Earth in homeostasis."

• Teleological Gaia holds that the atmosphere is kept in homeostasis, not just by the biosphere, but by and for (in some sense) the biosphere. According to Lovelock and Margulis, "The Earth's atmosphere is more than merely anomalous; it appears to be a contrivance specifically constituted for a set of purposes."

• Optimizing Gaia holds that the biota manipulates its physical environment for the purpose of creating biologically favorable, or even optimal, conditions for itself. "We argue that it is unlikely that chance alone accounts for the fact that temperature, pH and the presence of compounds of nutrient elements have been, for immense periods, just those optimal for surface life. Rather we present the 'Gaia hypothesis,' the idea that energy is expended by the biota to actively maintain these optima," wrote Lovelock and Margulis.

Kirchner differentiated among the relatively weak-acting hypotheses, such as Influential or Coevolutionary Gaia, which seem to state only that the biota and the physical environment have something to do with one another, and those that imply a stronger connection. "If we all talk about 'the Gaia hypothesis' without specifying which Gaia hypothesis, we can create a lot of confusion," said Kirchner. This confusion can appear in different guises. One of the most serious lies in claiming that evidence for one of the weaker versions of the hypothesis somehow proves the much stronger versions of the hypothesis as well. Said Kirchner:

You may believe, as I do, that the biota affect the physical environment. You may also think, as I do, that the physical environment shapes biotic evolution. You are in good company, because scientists have thought these things for over a hundred years. So if you ask me whether I believe in the Gaia hypothesis, and you mean that Gaia hypothesis, I would say that I do. But if you then say that I must believe that the biota are part of a global cybernetic control sys-tem, the purpose of which is to create biologically optimal conditions . . . well, that's another matter entirely.[2]

Kirchner argued that the weak forms of Gaia are not new and that the strong forms are either incorrect or untestable. Ultimately, he classified the strong forms of Gaia as a metaphor, not a testable hypothesis. In concluding, Kirchner commented that "the common perception is that Gaia means 'the Earth is alive' or the biota are trying to make themselves a nice home here. Given that the public doesn't understand the risks of treating poetic statements as scientific propositions, the common perception is that a bunch of scientists are busy trying to figure out whether the earth *really is* 'alive' and I don't think that perception helps any of us."[3]

Two years later, Kirchner's critique remains a stunning challenge to the Gaia hypothesis. Lovelock recently responded to Kirchner's "spirited attempt to demolish all notions of Gaia. Like some figure of the Inquisition, he publicly burned several imaginary Gaias, and his pyrotechnic demolition of the strong Gaia stole the show. But when the sparks faded, the real system Gaia was still there hidden only by the smoke. The flux of papers inspired by Gaia, and now appearing in the journals, are the real proof of the value of the conference. It has not stopped peer review from censoring any mention of Gaia by name."[4] Of course, Kirchner never intended to demolish the Gaia hypothesis but, rather, to sharpen the debate.[5] In that quest, he clearly succeeded.

1. James W. Kirchner, "The GAIA Hypotheses: Are They Testable? Are They Useful?" in S. H. Schneider and P. J. Boston, eds., *Science of Gaia* (Cambridge, Mass.: MIT Press, forthcoming).

2. Ibid.

3. Ibid.

4. James E. Lovelock, "Hands Up for the Gaia Hypothesis," *Nature* 344 (1990):100-02.

5. James W. Kirchner, conversations with the author.

quence of the constant input of CO_2 by degassing from the Earth's interior the CO_2 partial pressure rises."[6]

Kasting, now at Pennsylvania State University, and NASA planetary scientists Owen Toon and James Pollack have argued that, even without life, inorganic oceanic chemical reactions would produce carbonate sediments, but this feedback control system would result in global temperatures about 10° C warmer than would result if life were the significant CO_2-removing agent.[7] More recently, Tyler Volk at New York University and David Schwartzman at Howard University argued that soil biota facilitates the rate of weathering reaction to such an extent that a lifeless Earth would be as much as 45° C warmer than it is today.[8] This hypothesis, however, has been challenged by Harvard University geochemist Heinrich Holland.[9] Clearly, the scientific debate over Gaia is in full bloom.

Unfortunately, there is as yet no sufficient evidence, empirical or theoretical, to resolve the question of whether organic or inorganic CO_2 removal would have dominated in this potential mechanism of climate control. One of the most tenuous aspects of the argument for Gaian climate control has to do with the Earth's temperature at the time life got a toehold on the planet some 3.5 to 4 billion years ago. That the Earth's temperature has been reasonably constant since its inception, as Gaian supporters have stated, is not well supported by the paleoclimatic evidence. Direct temperature measurements, upon which quantitative knowledge of Earth's climate history is based, exist for only the past few hundred years. Nevertheless, physical evidence of the many life forms that have existed provide a proxy thermometer from which past temperature conditions can be estimated. For example, fossiliferous sediments may indicate the distribution of warm- and cold-loving species. Also, the ratio of various isotopes of oxygen in the fossilized shells of clams and plankton may indicate something about temperature conditions at the time the animals secreted their shells. Thus, about a half a billion years' worth of evidence is available

from which to draw crude but reasonable conclusions about the mean temperature of Earth.

It is quite possible that the planet was very cold by present Earth standards and that life existed only in limited domains in warm, tropical regions or in areas heated by upwelling lava or other flows from the interior. On the other hand, it is also plausible that the average global temperature was exceedingly hot and that the fossiliferous sedimentary rocks dated to 3.8 billion years ago were deposited in what were temperate or polar regions at the time. Paleoclimatologists have not yet resolved to better than plus or minus 25° C the Earth's mean temperature 3 to 4 billion years ago. For more recent geologic history, the evidence is more abundant, and for the past billion years, the record of life is sufficient to suggest that Earth's mean temperature has probably not been more than 10° to 15° C warmer or about 5° C cooler than it is today. This range at least suggests the existence of partial climate control, whether through organic processes, inorganic processes, or, more likely, both.

Plankton and Cloud Albedo

Another Gaian mechanism of potential climate control was debated in March 1988 at the American Geophysical Union Chapman Conference in San Diego, California, the first major scientific meeting on the Gaia hypothesis.[10] Certain marine phytoplankton produce dimethylsulfide (perhaps as part of a process to help maintain their internal osmotic pressure), which is subsequently emitted as a waste product and—the relevant hypothesis—may somehow influence the climate. This seemingly fantastic scenario was proposed when Lovelock visited atmospheric chemist Robert Charlson at the University of Washington. Also involved were oceanic chemist Meinrat Andreae of Florida State University and climate theorist Stephen Warren of the University of Washington. Their hypothesis evolved from a synthesis of Lovelock's interdisciplinary ideas, Charlson's knowledge of the effect of

atmospheric particles on cloud formation, Andreae's understanding of sulfur-cycle chemistry, and Warren's knowledge of climate modeling. They proposed that, after the dimethylsulfide (DMS) released into the ocean by phytoplankton is outgassed into the air, it is chemically converted to sulfur dioxide and then to sulfuric acid particles. These particles are incorporated into low-altitude clouds, which are made up of relatively few large water drops because of the lack of dust particles (which serve as cloud-condensation nuclei) over oceans. The extra sulfur particles substantially increase the number of water droplets, which, in turn, scatter more sunlight. Thus, the albedo, or sunlight reflectivity, of clouds increases, and this process, other things being constant, would cool the climate. (Indeed, when James Coakley, then working at the National Center for Atmospheric Research in Boulder, Colorado, saw bright streaks in clouds over the oceans in satellite pictures, he attributed them to the extra droplets condensed around smoke particles from ships crisscrossing the ocean.)[11]

Unfortunately, although the Charlson group could show that relatively small changes in the concentration of sulfuric acid in the atmosphere could cause substantial changes in cloud albedo—enough to vary temperature by a few degrees Celsius perhaps—the group could not show that this process was necessarily negative feedback. That is, they could not demonstrate how a cooling of the climate, for example, would reduce the plankton's DMS production and thus decrease cloud albedo, thereby opposing the cooling. In fact, recent studies (as yet unpublished) have found more sulfate in Antarctic ice cores at the height of glacial cooling, which suggests a positive feedback —namely, that more DMS is produced in cold times, causing more cloud seeding and greater cloud albedo. The increased reflection of solar radiation would have produced colder temperatures. However, as Warren has pointed out, it is also possible that the increased sulfate in Antarctic ice is not a result of globally increased DMS emissions, but simply that of a local, cold-water species of phytoplankton that expanded its numbers near the Antarctic continent, thereby inducing the enhanced cold-era Antarctic sulfur deposition.[12]

In short, while the DMS climate change hypothesis has not yet been demonstrated to include positive or negative feedback, without Lovelock or the Gaia hypothesis, it probably would not have been investigated for quite a while. The same seems true for the hypothesis of Schwartzman and Volk mentioned earlier about the effect of soil biota on Earth's temperature, which was also presented at the 1988 conference. The San Diego meeting signaled that, after 20 years, the Gaia hypothesis has moved from nonscientific forums to where it fundamentally belongs—in the turbulent flow of mainstream science. The most serious critique of the Gaia hypothesis to date was presented at the conference in a session on the philosophy of science and is summarized in the box on page 8.

The Question of Optimization

The realization that climate and life mutually influence each other is profound and provides an important counterpoint to the parochial view of the world as a physical environment dominating life. Nonetheless, to say that climate and life "grew up together," or coevolved, is not the same as to say that life somehow optimizes its own environment to suit itself.[13] It is the latter idea, the most radical proposed by Lovelock and Margulis, that is most open to criticism.

The early physical environment largely determined the ecological niches in which early life forms had to live. Life altered the physical environmental constraints on itself by changing the composition of the atmosphere. To be sure, this modification changed the competitive balance of species and forced evolutionary change—indeed, coevolutionary change—between the organic and inorganic parts of the environment. Change, yes; but to say "optimization" is problematic. If one simply asserts "optimum" in terms of the current biota, then life is at its optimum by definition. But what about the losers?

Extinct life forms probably would not view the current environment as having been optimized.

One of the principal confusions with the whole idea of life's self-regulation, or optimization, is what is "life"? Is life's self-regulation a matter of maintaining for the longest period of time the stability (that is, the survival) of extant species? Does optimization entail the maintenance of maximum biomass or the maintenance of maximum diversity of species? All of these goals seem to be legitimate definitions of optimization of life, yet to optimize each one is probably inconsistent—the three goals are not necessarily compatible.

Consider a specific example. Orbital variations of the Earth are thought to drive the ice age cycles, but changes in CO_2 and methane concentrations may amplify the cycles substantially. Near the end of the Pleistocene some 15,000 years ago, when the last ice age receded rapidly, the CO_2 content of the atmosphere was about one-third less than it was just before the Industrial Revolution. This difference suggests that a weakened greenhouse effect made the ice age colder than it otherwise would have been. A principal explanation for the decrease in CO_2 during the height of the last ice age has to do with a change in the biochemistry of the oceans, perhaps from a planktonic response to altered nutrient availability. In other words, ocean life probably helped alter the chemical composition of the atmosphere, the climate, and its own environment. (Another possible contributing factor to the decrease in atmospheric CO_2 is the storage of carbon in bogs.)

But it is hard to imagine how making an ice age even colder could increase the total biomass, let alone be any general statement of homeostasis, or self-regulation of life. If the atmospheric CO_2 content was diminished by life during an ice age, a monkey wrench is thrown into any hypothesis that environmental conditions were somehow being altered by life for self-optimization; the preponderance of evidence suggests that, relative to today, terrestrial biomass was at least 10 to 20 percent less at the last glacial maximum, some 15,000 to 20,000 years ago. Be-

cause the vast bulk of biomass is on land, even if oceanic organisms thrived during the ice age by sequestering more carbon for themselves, their proliferation most likely replaced only a small portion of the land biomass lost to the cold that was enhanced by a decrease in CO_2 for which the oceanic organisms were at least partially responsible. But, that the plankton take care of themselves without regard for the wider consequences for life on land sounds very Darwinian rather than Gaian. Indeed, Lovelock recently reformulated the question of optimization:

In the early days when it was a bit poetic one thought of life as optimizing conditions on Earth for its survival. Now that I understand the theory behind Gaia very much more than I did then, I recognize that this is not so, that it's nothing as highly contrived or as complicated as that. There is no foresight or planning involved in the part of life in regulating the planet. It's just a kind of automatic process.[14]

Gaia and Global Change

Lovelock, Margulis, and their collaborators have illuminated the very important role of feedback mechanisms between the organic and inorganic components of planet Earth. In particular, they have challenged scientists to view the Earth from a new perspective, one that Lovelock calls geophysiology. Gaian supporters believe that such a whole-system view should be as legitimate a scientific pursuit as the more traditional reductionist views. Yet the level of aggregation at which complex systems can be most profitably viewed cannot be determined without detailed empirical investigation. Both "bottom-up" and "top-down" approaches are valuable, especially for a science that is developing. Unless one has strong ideological views about how the physical and biological worlds are organized, it seems that the most practical way to study complex systems is to try to analyze them at those levels that provide the greatest explanation of system behavior for the least expenditure of effort.[15] In this context, the Gaian scientists' call for a geophysiological approach to the interaction among organic and inorganic components of Earth is both welcome and overdue. But this does not mean that scientists should accept the proposition that such feedbacks are always negative —that is, self-regulatory for life, once life has been defined.

Feedback processes are not just interactions that tend to stabilize a system; they also can be interactions that tend to destabilize—like those that appear to have operated between biomass and CO_2 during the last ice age. Life and the environment have coevolved, but their interactions have not always been optimal for all forms of life or even for the overall biomass. Such interactions simply lead to mutual changes— some beneficial and some detrimental for some forms of life at some times. The importance of whole-system studies alone is sufficient to justify looking beyond the narrow disciplines of biology, climatology, geophysics, chemistry, and so forth, in what has come to be called the "global change" movement. Clearly, more scientists should insist that the organic and inorganic parts of the planet be viewed as coupled systems that can be studied at various levels of aggregation.

At a 1975 meeting organized by anthropologist Margaret Mead on threats to the atmosphere, Lovelock applied Gaian ideas in a controversial statement: "Our capacity to pollute on a planetary scale seems rather trivial by comparison and the system does seem to be robust and capable of withstanding major perturbations."[16] Subsequently, a large debate broke out in which the author and energy analyst John Holdren from the University of California at Berkeley countered that, although no human intervention, probably not even nuclear war, could be powerful enough to threaten all of life on Earth, when dealing with human beings, of which a billion or so already are suffering from severe malnutrition, even slight disturbances in the environmental systems that produce food and recycle wastes can be catastrophic to some humans and many other species. "Though we probably are not threatening the survivability of the entire biosphere, changes of 5 or 10 percent in the carrying capacity of the earth for human beings must be viewed as having enormous social and political consequences on a global scale."[17] From the human point of view, prevention of such consequences is essential.

Lovelock responded that, indeed, pollution could be an enormous problem, but that, if scientists approached environmental problems from the perspective that nature has feedback on the system, they might very well propose different solutions than would otherwise be offered. Nevertheless, it is speculation at best and environmental brinkmanship at worst to believe that Gaia, through self-regulation, will somehow protect the planet from the negative consequences of all human intervention. Neither Jim Lovelock nor Lynn Margulis believes in this mystical protection.

Today, the unknown extent of negative and positive feedback poses a dilemma that applies to many environmental questions, including the greenhouse effect, ozone depletion, acid rain, and toxic waste disposal. It is extremely unlikely that scientific assessment can answer all of the questions relevant to enlightened policymaking before human-induced experiments unfold on "laboratory Earth,"[18] with humanity and every other living thing as passengers along for the ride.

NOTES

1. Lynn Margulis and James E. Lovelock, "Is Mars a Spaceship, Too?" *Natural History*, June/July 1976, 86–90.

2. T. H. Huxley, *Physiography* (London: MacMillan, 1877), 83.

3. William W. Kellogg and Margaret Mead, eds., *The Atmosphere: Endangered and Endangering*, Fogerty International Center Proceedings No. 39 (Washington, D.C.: U.S. Government Printing Office, 1976).

4. Carl Sagan and George H. Mullen, "Earth and Mars: Evolution of Atmospheres and Surface Temperatures," *Science* 177 (1972):52–56.

5. T. Owen, R. D. Cess, and V. Ramanathan, "Enhanced CO_2 Greenhouse to Compensate for Reduced Solar Luminosity on Early Earth," *Nature* 277 (1979): 640.

6. James E. Lovelock, "Geophysiology: A New Look at Earth Science," in R. E. Dickinson, ed., *The Geophysiology of Amazonia: Part I, Vegetation and Climate Interactions* (New York: John Wiley & Sons, 1987), 11.

7. James F. Kasting, Owen B. Toon, and James B. Pollack, "How Climate Evolved on the Terrestrial Planets," *Scientific American*, February 1988, 90–97.

8. David W. Schwartzman and Tyler Volk, "Biotic Enhancement of Weathering and the Habitability of Earth," *Nature* 340 (1989):459.

9. Heinrich D. Holland, "The Mechanisms That Control the CO_2 and the O_2 Content of the Atmosphere," in S. H. Schneider and P. J. Boston, eds., *Science of Gaia* (Cambridge, Mass.: MIT Press, forthcoming).

10. Robert J. Charlson, James E. Lovelock, Meinrat O. Andreae, and Stephen G. Warren, "Oceanic Phytoplankton, Atmospheric Sulphur, Cloud Albedo and Climate," *Nature* 326 (1987):655–61.

11. James A. Coakley, Jr., Robert L. Bernstein, and Philip A. Durkee, "Effect of Ship-Stack Effluents on Cloud Reflectivity," *Science* 237 (1987):1020–22.

12. Stephen G. Warren, University of Washington, letter to the author, January 1989.

13. Stephen H. Schneider and Randi Londer, *The Coevolution of Climate & Life* (San Francisco: Sierra Club Books, 1984).

14. James E. Lovelock, as quoted in "Goddess of the Earth," *NOVA* (Boston, Mass.: WGBH television program transcript, 1986).

15. Kenneth C. Land and Stephen H. Schneider, *Forecasting in the Social and Natural Sciences* (Dordrecht, the Netherlands: D. Reidel, 1987).

16. Kellogg and Mead, note 3 above.

17. Ibid.

18. Stephen H. Schneider, *Global Warming: Are We Entering the Greenhouse Century?* (San Francisco: Sierra Club Books, 1989).

W A R
& THE ENVIRONMENT

LONG AFTER THE SHOOTING has stopped, haunting images of the environmental devastation of the Persian Gulf war linger—funnels of black smoke billowing from oil-field infernos; oil slicks spreading for miles, fouling the gulf's once-teeming waters; the corpses of birds so drenched with oil they look like lumps of blackened sand on tar-ball beaches. Indeed, with the barrage of media coverage and speculation on the long-term effects of Saddam Hussein's tactics, the American public has been forced to confront—really for the first time—war's ecological toll.

But this is not a new phenomenon. Throughout history, the environment has been one of war's worst casualties. In the Second Century B.C., the invading Romans spread salt on the fields of Carthage to destroy crops and poison the soil. In medieval Europe, soldiers tossed the corpses of diseased animals over the walls of besieged towns to contaminate the water supplies. Scorched-earth tactics have been used repeatedly in modern times, from Sherman's ravaging march through Georgia, to the United States' relentless defoliation of the Vietnamese countryside, to the Soviet Union's destruction of the Afghan rebels' fields.

In the aftermath of the Persian Gulf

Four not-so-predictable views of where we have been and where we are heading.

war, *Audubon* asked two award-winning journalists and two prominent scientists to write essays on the environmental impact of war. Our contributors are Malcolm W. Browne, a Pulitzer Prize-winning war correspondent who now is a senior science writer for *The New York Times*; James M. Fallows, Washington editor of *The Atlantic Monthly,* who has written extensively on the U.S. military and on Asia; zoologist Eric A. Fischer, a senior vice-president of the National Audubon Society, and Michio Kaku, professor of nuclear physics at the City University of New York.

We allowed each writer to chart his own course, and they provided us with four very different perspectives. We think their essays are both disturbing and surprising. They detail the ecological costs of war—from the slaughter of European bison to near-extinction in World War II to the crushing of the desert's protective crust by armored vehicles in the Gulf War. They also stress the need for nations to agree to control such damage in future wars. But these writers also remind us that humankind often wreaks far more havoc on the environment in pursuing progress than in waging war. Indeed, much more of the world's rain forest has been destroyed by bulldozers than by bombs.

By Malcolm W. Browne, Michio Kaku, James M. Fallows, and Eric A. Fischer. From *Audubon,* September/October 1991, pp. 88-96, 98-99. Reprinted by permission.

WAR & THE ENVIRONMENT

MALCOLM W. BROWNE

MALCOLM W. BROWNE, now a senior science writer for The New York Times, *won the Pulitzer Prize in 1964 for his coverage of Vietnam for the Associated Press. He has reported on wars on three continents and was the* Times' *last bureau chief in Saigon when the city fell to the North Vietnamese in 1975.*

THE BEST LOBSTERS in Asia used to be caught in shallow water off the South Vietnamese town of Nha Trang, but along with so much else, they were done in by the war. The Nha Trang lobsters fell victim to the enduring principle that war is an activity in which normally rational people behave as if there were no tomorrow.

Before 1965 fishermen who worked the South China Sea set just enough traps to catch a few lobsters for local restaurants catering to the carriage trade—provincial officials, French planters, foreign diplomats, a few tourists. Ordinarily, the fishermen were too busy with their major catches to consider lobsters more than a luxury sideline.

But when legions of Americans began arriving in 1965, they delivered an economic jolt from which Vietnam never really recovered. In Nha Trang, appreciative GIs quickly discovered the renowned lobsters, and in a matter of weeks, the price of these delectable crustaceans escalated several thousand percent. Suddenly, lobsters meant big money, and to cash in, the fishermen abandoned their nets and traps, harvesting the lobster beds by exploding grenades and scooping up the casualties.

For six months the GIs gorged themselves. But then the lobsters were irretrievably gone, killed or driven from their breeding areas. Even fish avoided the Nha Trang area, and catches fell drastically. A decade later, when I boarded one of the last helicopters to leave Saigon, the Nha Trang lobsters were no more.

The loss of a lobster population, trivial in comparison with the grander trag-

edies of war, typifies a category of military insult to habitat I have seen in a half-dozen conflicts—an insult even more devastating than shells or bombs. It has nothing directly to do with combat. It is, rather, a result of the collective action of great masses of men and women for whom war is a dispensation to ignore normal restraints.

For many people, indeed, war seems to be a license to gorge and lay waste. People change under war's malign influence. I once knew a young Californian who at home had been an activist for environmental causes but who, in Vietnam, flew for Operation Ranch Hand, spraying defoliant over hundreds of square miles. When I asked him about the paradox he replied with a shrug, "It's a question of priorities."

As a matter of fact wartime havoc can actually create new habitats, as the bombing of North Vietnam demonstrated. In Vietnam and in other former Asian and European battlegrounds, farmers may curse the buried munitions that sometimes blow up in their fields. But I have seen the pockmarks left on North Vietnam's Red River delta by the rain of 500-pound bombs, and few farmers today would wish to fill them in. The old craters now are ponds, bountifully breeding crayfish and other delicacies.

War can benefit habitats in many unexpected ways. One of the oddest cases I discovered was the submerged German battle cruiser *Dresden*, sunk by its crew off Más a Tierra, a Chilean island in the Juan Fernández group, after the Battle of the Falklands in 1914. The turrets of that hulk now provide a breeding habitat for some of the most highly prized lobsters the Pacific Ocean produces.

On balance, of course, war depletes habitats and species. Armies traditionally supply themselves by foraging or buying up what they want—from fresh vegetables to young women. And the effects on wildlife can be staggering.

In 1961 a trickle of American troops began arriving in Saigon. Among the many businesses that sought their trade was the Vietnam Handicraft Center, a store on fashionable Tu Do Street owned by representatives of the ruling Ngo family. Besides the usual lacquer ware and ao dai dresses, the store sold ivory bric-a-brac carved from the tusks of the nation's shrinking elephant herds, as well as handbags and

women's accessories made of the shells of endangered tortoises.

Worst of all, the store offered tiger skin rugs at $300 each. These skins attracted so many American buyers that the shop's supply soon ran out, as Vietnam's tigers were virtually wiped out by hunters supplying the souvenir trade. Among the tiger slayers were some American shikaris [big game hunters], including a U.S. Embassy press officer, whose "bag," he proudly told me, included a half dozen skins.

In war few bother to tote up the animal casualties. But, especially during World War II, news film has captured haunting vignettes—the eating of dogs by starving residents of Leningrad; the forlorn carcasses of animals at the bombed-out Berlin zoo (and, more recently, at the war-ravaged Kuwait City zoo); vistas of dead birds and mammals along oil-drenched beaches at Tunis, Truk Lagoon, Normandy, and countless other battlegrounds.

One of the scourges of 20th Century war—the oil spill—exceeded all precedent in the recent Persian Gulf conflict. The hemorrhaging of Kuwait's oil-fields continues as this is written, and history will record the destruction of seabirds, mammals, and fish in the Persian Gulf and the Indian Ocean as one of the blacker episodes of environmental rape.

War feeds on life. One species—the European bison—was slaughtered nearly to extinction to supply the mess kitchens of German and Soviet troops in eastern Poland at successive stages of World War II. By the spring of 1945 only a score of bison were left—sufficient, as it turned out, to begin a breeding and conservation program that has finally removed the European bison from the danger list.

Sometimes cast as weapons, animals become military targets. Hannibal's elephants were the Punic equivalent of modern tanks and were treated as such by their Roman enemies. Elephants also figured in the Vietnam War. For a time the communist Pathet Lao and Viet Cong guerrillas used them to move supplies along the Ho Chi Minh Trail in southern Laos, and when the U.S. Air Force learned of this, the elephants were duly attacked and slaughtered.

But there are things worse for wildlife than war.

I have a feeling that if some intelligent panda could evaluate the real menace posed by the human race to other species, he might wish for more war, rather than less. From the non-human standpoint war at least has the merit of slightly slowing the human population explosion, the most devastating biological event in the planet's history.

I have witnessed guerrilla war in Indochina, South America, and Bangladesh and conventional war in the Middle East, North Africa, and Korea. For two years I was a soldier myself and, for a score more, a war correspondent. From what I have seen I cannot imagine even a nuclear war as destructive to wildlife as the effects of unchecked human reproduction, which is now doubling the numbers of *Homo sapiens* in less than 30 years. I have seen its pernicious effects on rainforest habitat in Brazil, on grazing land in Kenya, on habitat and species in China and the South Pacific, even on microscopic shrimp at the base of the food chain in Antarctic waters.

War does indeed place the creatures of the Earth in mortal peril. But it seems to me that the deadliest war is not between men. It is the war mankind is mindlessly waging against other living things in the name of what Nazi Germany called *Lebensraum* ["living space"]. It is a war in which victory will impoverish all of us beyond belief.

MICHIO KAKU

MICHIO KAKU is professor of nuclear physics at the City University of New York and writes and lectures extensively on the environment. He has published six books and numerous scientific articles and is on the advisory board of National SANE/Freeze, the Washington, D.C.-based anti-war organization.

IN THE AFTERMATH of World War II, after the raucous tickertape parades down Broadway and the spontaneous outbursts of joy and relief had subsided, a disturbing series of articles appeared, first in *The New Yorker* magazine and eventual-

ly as John Hersey's landmark book, *Hiroshima.*

At once moving and deeply unsettling the book chronicled to an unsuspecting world the birth of a powerful new technology. From then on the victory celebrations were tempered with the frightening knowledge that a new weapon with almost Biblical implications had been unleashed on the planet.

Perhaps no book as powerful as Hersey's will emerge in the aftermath of the Gulf War. But after the fanfare surrounding Desert Storm fades, perhaps a more sober, thoughtful analysis of the Gulf War will take hold in the public's consciousness, and with it, the realization that victory in war has meant a devastating defeat for the environment. For by almost any yardstick, the Gulf War was the greatest environmental setback in years.

It is not an exaggeration to compare the environmental and human destruction unleashed in Iraq and Kuwait to the atomic bombing of Hiroshima.

In Hiroshima the atomic bomb released the explosive equivalent of 13,000 tons of TNT. By comparison the cumulative bombing of Iraq and Kuwait by U.S. forces totaled 88,500 tons of explosives, or roughly seven times the destructive force of the Hiroshima bomb. Although most Americans remember the Gulf War as the first "clean" Nintendo war in history, the Pentagon now admits only 7 percent of the bombs dropped in the Gulf were "smart" bombs; fully 93 percent were gravity bombs, or "dumb" bombs, with no midcourse guidance systems whatsoever. Overall, 70 percent of the bombs missed their targets. The "clean" war was a myth. The astonishing video tapes aired during the war were a victory of editing and public relations over truth.

In Hiroshima over 100,000 Japanese perished because of the radiation and fire of the atomic bomb. Yet perhaps as many as 150,000 to 200,000 Iraqis lost their lives in the Gulf War, according to the estimates of the Sunday *London Times* and the Chaldean Catholic Church of Iraq. And a Harvard medical team reported in May that tens of thousands of children might later die because of the United Nations embargo and the collapse of basic sanitation and the medical infrastructure.

In Hiroshima the radioactive fallout

was limited because the bomb was detonated high in the atmosphere. But in Kuwait, there is a massive plume fed by the burning of 1,250 oil wells, nearly 500 of which still blaze. The plume's effects have spread more than 1,500 miles, darkening parts of Kuwait and Iran during the day, sending temperatures plunging 20 degrees in some areas, and creating acid rain as far away as Bulgaria, India, and Pakistan.

The fire may be spewing out as much as 50,000 tons of sulfur dioxide and 100,000 tons of sooty smoke per day. The pollution created by the burning of about five million barrels of oil per day is difficult to grasp. By the estimates of some scientists it is ten times the oil-related pollution generated by the entire United States.

Our planet has now become a laboratory for atmospheric scientists: Physicists are already analyzing the distribution and size of the particles of soot. Some scientists even subscribe to the "nuclear winter scenario." (Fortunately, preliminary studies show that most of the soot has not yet flooded the upper atmosphere, where it may trigger serious planetary disturbances.)

In Hiroshima radioactive levels began to drop dramatically after the first few weeks. However, in the Gulf, the fires may rage anywhere from two to four years. Capping them proceeds at an agonizing snail's pace: Only one oil well can be capped per crew, per week. Furthermore, the pollution caused by the massive slick in the Persian Gulf, which is several times the size of the Exxon Valdez oil slick, may last decades. Because the Persian Gulf has an average depth of only a few hundred feet and is partially sealed off at the Strait of Hormuz, it may take as long as a century before all the oil is completely washed out. But within a few years, many of the roughly 450 species of animal life in the Gulf may be wiped out, leaving a dead sea.

Equally ominous is the precedent set by the U.S. Air Force's destruction of two Iraqi nuclear reactors and chemical plants. In future wars, nuclear power plants will be fair game to the military. Ironically, the United States has most to lose by this: In particular, the new rules may open the country up to an increased threat of attacks against its 111 commercial nuclear plants. The nuclear ash contained within a single commercial U.S. nuclear plant (over 10 billion curies per plant) is several hundred times larger than the fallout released by a large hydrogen bomb.

The Hiroshima bombing had at least one positive legacy: Its horrific damage created a deep psychological inhibition against the wanton use of nuclear weapons.

It is to be hoped the Gulf War will likewise serve as a grim warning to military planners who view the devastation of the environment and civilians as mere "collateral damage." If environmentalists can successfully educate the public about the massive ecological impact of the Gulf War, then planners will be forced to consider the environment as a factor in the calculus of war.

Encouragingly, there are small, hopeful signs of a new era emerging. Throughout recorded history, nations have engaged in predatory wars in the quest to secure power, resources, and wealth. But we may now be seeing the dawning of a New Environmental Order, where the overarching necessity of controlling worldwide pollution and environmental desecration will force unwilling nations to cooperate on an unprecedented basis.

Pollution knows no boundaries. Greenhouse gases, ozone, acid rain, and the destruction of the rainforest have forced nations to reexamine priorities, such as energy use, consumption, industrialization, and social development. These are all pressing issues pushing nations to engage in cooperation, not confrontation.

For example, vivid satellite pictures of the large atmospheric hole above the South Pole have galvanized previously reluctant nations to restrict domestic use of chlorofluorocarbons. Most developed nations have agreed to stop CFC use by the year 2000, and most developing nations are set to stop soon afterward.

However, the transition to the New Environmental Order is still very fragile and must be nurtured carefully. Like John Hersey after World War II, environmentalists must now alert the public to the profound threat the world's vast military arsenal poses to the environment.

As with Hiroshima, this may be the real legacy of the Gulf War—rather than getting ready for the next war, perhaps we will become determined to prevent it.

WAR & THE ENVIRONMENT

JAMES M. FALLOWS

JAMES M. FALLOWS, Washington editor of The Atlantic Monthly, *won the American Book Award in 1982 for* National Defense, *which focused on reforming the U.S. military. He recently has lived in Asia and is working on a book about the economic rise of East Asia. Since 1987 he has been a commentator for National Public Radio's* Morning Edition.

FEW AMERICANS can be comfortable visiting Hiroshima, but many feel they should. Viewing the hulks of ruined buildings, riding the city streetcar to the stop named "A-Bomb Dome"—these are ways of acknowledging the lasting consequences of a nation's acts.

The phenomenon of conscience-wracked American tourists is so well known that it has fostered a local joke: En route to the A-Bomb Museum, a penitent American couple asks their taxi driver, well, how is Hiroshima *doing* these days, after all it's been through? "Things are terrible in Hiroshima," the driver says. "Now the Carp are dead." Envisioning the elegant, ancient carp that cruise through so many ceremonial pools in Japan, the Americans share a despairing thought: Even the carp! How long the damage goes on!

The joke, of course, is that the Carp are Hiroshima's beloved baseball team, who are "dead" and are in the midst of a rotten season.

In the years since the atomic bombing of Japan, many other parts of Asia, from Korea to Cambodia, have been disfigured by war. But in Hiroshima and most of these other one-time battle sites, the damage seems to have lasted longer in memory and imagination than in the landscape itself.

The first time I visited Hiroshima I wondered why it felt vaguely different from other big Japanese cities. Then I realized, with horror, that it was unusual because of its broad, open vistas, where the normal urban clutter had been blasted away. Yet even this subliminal reminder of wartime devas-

tation was misleading. Tokyo had been just as thoroughly pulverized as Hiroshima during the war. But Tokyo rebuilt itself so quickly and densely that nearly all signs of combat were erased. Hiroshima deliberately preserved its ghostly empty spaces, as part of its remembrance of the bomb.

I am not a scientist, and I can't pretend to offer any judgment about the hidden environmental costs that warfare may exact. For all I know the Japanese or Korean landscape could still be blighted in some subtle way by what happened decades ago during war. But I have had a chance to see many parts of Asia where 20th Century wars were fought. To judge from the surface—by the way cities and countrysides now look and the way people live—it appears warfare is remarkable for how little long-term damage it has done. To be more precise about it, Asia's experience indicates that war can be less damaging than peace.

Naturally there are some places where the scars of war endure. Korea has a harsh, extremely mountainous terrain, which, in other parts of the world, might be softened with forests. But there are very few towering trees in South Korea. Most of the forests were cut down or blasted apart during the Korean War: Few trees are more than 35 years old.

Flying into or out of Hanoi's ramshackle airport travelers notice strange circular ponds, each 10 to 20 yards in diameter and clustered together like grapes. These are craters left from the American B-52 attacks 20 years ago. Along Vietnam's central coast, in Da Nang and Nha Trang, I have seen surprisingly large numbers of children whose limbs are missing or malformed. They are far too young to have been maimed in combat, and because there are so many of them, it is hard not to think, as the Vietnamese government contends, that Agent Orange is to blame.

Yet to see Vietnam itself, as compared to its Southeast Asian neighbors, one can not help but conclude that war is one of the less environmentally damaging activities that people undertake. There must be parts of Vietnam's central highlands that still show the effect of massive defoliation campaigns. But in the several weeks I have spent during the last two years riding through the country and flying over it, I didn't see them. In fact the approach to the Kuala Lumpur airport in Malaysia is pocked with craters much larger and more intrusive than the ones in Hanoi: These are the remnants not of war but of peacetime tin-mining booms.

The tropical forest zone of Southeast Asia stretches across the borders of Vietnam, Laos, and Myanmar, which have all been ravaged by war, and Thailand, Malaysia, and Indonesia, which have generally been at peace for years. From the air the dense green cover looks far more intact over the former war zones in Vietnam and Laos than anywhere else.

In the Malay Peninsula and Sumatra, the rainforests were "peacefully" devastated early in this century when British and Dutch planters converted them to rubber and palm-oil plantations. And in Hanoi last year, a Vietnamese bureaucrat told me, with no apparent irony, about the new environmental problem facing his country. During the years of war, he explained, the shelling and danger had kept people out of the forests. "Now they're moving right in," he said, noting that timber companies were cutting trees for export and settlers were clearing hillside fields using the system of slash-and-burn.

The miracle that is the Asian economic boom, rather than the turmoil of Asia's 20th Century wars, looks to me like the real challenge to the region's environment. Thailand was not even a combatant during World War II and, since then, has enjoyed peace while countries all around it have fought. Its stability is a major reason why Bangkok is bursting with industrial activity—and burdened with probably the worst air-pollution problem in Asia. (The Thai government recently ordered a shift to unleaded gasoline, in the wake of medical warnings that lead in the air was causing widespread brain damage among children.)

And in the 1960s, when Indonesia and Malaysia were fighting over border claims on the island of Borneo, they did relatively little damage to its vast wilderness. In the 1990s they peacefully compete to cut down and sell its forests.

Presumably the Asian economies will grow beyond this awkward stage toward a less overtly rapacious version of capitalism, as some Western societies have done. But for now, peace is

a bigger environmental problem than war.

The most poignant illustration comes from the Philippines, where during World War II, Philippine soldiers fought side-by-side with Americans. Some of the most lasting environmental damage there occurred when the fighting stopped. The U.S. Navy, girding for the inevitable invasion of Japan, had, in 1945, amassed a mighty fleet in the Philippines. Then, when the Japanese surrendered after the bombing of Hiroshima and Nagasaki, the Navy was left with thousands of tons of surplus shells. Rather than haul them back home, it dumped the ordnance in Philippine waters, especially in the Lingayen Gulf.

When the warships left, poor Philippine fishermen went diving to retrieve the bombs, hoping to dismantle them and recover the gunpowder. In the years since, these fishermen have used the powder for "dynamite fishing" —the practice of setting off underwater explosions, usually near coral reefs, and then collecting the dead fish that drift to the surface. When they have failed to dismantle the bombs properly the fishermen and their families have been killed. When they have succeeded, they have pulverized the reefs.

War is hell, but environmental damage is not the main reason.

ERIC A. FISCHER

ERIC A. FISCHER is the National Audubon Society's senior vice-president for science and sanctuaries. He recently addressed an international conference in Montreal on the environmental effects of war. He has a doctorate in zoology and formerly served as deputy director of the Smithsonian Tropical Research Institute in Panama.

FOR FAR TOO LONG we humans have deluded ourselves into thinking we are separate from the natural world. Perhaps it is because so many of us live in cities, or perhaps the human tendency to think in terms of

opposites—black and white, war and peace, man and nature—blinds us to the deep interconnection between us and our environment. Whatever its cause, this delusion has contributed greatly to the current global environmental crisis.

Nowhere is this more apparent than in our approach to war. When we make war on each other we also make war on nature. A great deal of environmental damage has been inflicted on the Earth in the name of national defense.

In war environmental concerns that otherwise would weigh heavily are reflexively given lower status. Military needs become paramount. During the Gulf War, President Bush waived the requirements of the National Environmental Policy Act, permitting the Pentagon to test new weapons and perform other activities in the United States without the normal, public environmental review. Provisions of the Endangered Species Act and other environmental laws were not waived, but the option was available.

The waiver was used twice during the war, once to test a fuel air explosive and once to change flight operations at an air base. Had the war not been so brief, more potentially damaging activities, such as desert training exercises, also could have escaped review. Desert ecosystems are notoriously fragile and slow to recover from disturbance. In the fifty years since World War II the tank tracks made by General Patton's troops while training in the deserts of southern California have changed little.

During the Gulf War many environmentalists hesitated to express concerns about the ecological effects of the war for fear of being labeled unpatriotic or being accused of putting concerns about wildlife ahead of concerns about people. Had our government been engaged in behavior of similar destructiveness in peacetime, the protests would have been vociferous. Certainly, opposition to ecologically damaging projects unrelated to the Gulf conflict continued during the war.

War and peace do not really differ in the kind or extent of destruction that they inflict on the environment. The main difference is in intent. The Gulf War did not generate the first large oil spills or oilfield fires. Even the radiation and climatic effects of nuclear war have peacetime precedents.

The total damage inflicted on the Earth by recent wars pales in comparison to that wrought by peacetime habitat destruction, which in itself is a kind of silent war that humans are waging against other inhabitants of the planet.

While peacetime devastation is usually committed in the name of progress, war damage often results from direct attempts to thwart the enemy. The ten million gallons of the Agent Orange defoliant sprayed on Vietnam in the 1960s was justified as a way to decrease the ability of the enemy to hide and live off the land. The Iraqis purportedly intended the great oil spill during the Gulf War to foul Saudi desalinization plants and foil beachhead landings by U.S. Marines.

The inferno of oilfield fires in Kuwait, the petroleum-steeped coastal marshes and mangroves in the Persian Gulf, and the crushing of the desert's protective crust by armored vehicles show that the environmental effects of military confrontation often last well beyond the end of the conflict. And major unanticipated effects may occur in the aftermath of war, long after the last shots are fired.

The "Just Cause" invasion of Panama by U.S. troops in 1989 led to increased deforestation. The conflict occurred at the beginning of the burning season, the rainless time of year during which farmers clear land and most logging occurs. The invasion and its aftermath further weakened an economy severely damaged by U.S. sanctions. For many citizens low employment prospects in the cities made clearing land for agriculture a compelling alternative to city life. And the new Panamanian government could not protect the forests: The fighting destroyed vehicles and other equipment previously used by wardens to patrol protected areas, and the government had no funds to replace them.

The result was devastating. According to the Panamanian natural resources agency, twice as much forest was lost in the 1990 burning season as in an average year during the 1980s —or during the 1991 season. (By this year both the economy and a protective infrastructure had been partially restored.)

To prevent environmental war damage in the future we cannot count solely on efforts to avoid armed conflicts. The causes of war run deep. It is a routine activity for both human and animal societies. Some coral colonies attack neighbors that impinge too closely. Social primates engage in raids and border disputes with neighboring groups. And ants and termites attack other colonies with the help of soldiers whose bodies are specially adapted for fighting and defense.

Anthropologists who wish to study peaceful human societies have a hard time finding them. Fewer than one society out of ten can be classified as peaceful, and the modern state is particularly bellicose. In this century there has been only one year of global peace, and on average, three wars are occurring at any given time. No matter how much we may want peace, we can expect many more wars before we find other ways to solve conflicts.

What can we do to guard against environmental damage from the wars that will inevitably occur? For nuclear war the probable outcomes are so horrific that prevention is the only feasible approach. Since the bombing of Hiroshima and Nagasaki, the behavior of states with nuclear weapons has exhibited a restraint that would be admirable if it were applied to conventional conflicts.

Dealing with the environmental impact must become an integral part of any war planning. Failure to do so may lead to additional loss of life from such problems as lack of food and potable water. Environmental organizations and government agencies should bring their expertise to bear.

We also must enhance environmental protection during the conflict itself. Like other human social activities, war has accepted rules and conventions. We need to find ways to make environmental protection a stronger part of the laws of war. The United States could take a significant first step by declaring support for provisions in the 1977 Protocols to the Geneva Conventions that outlaw wartime activities having severe environmental effects. However, these protocols are vague and purely admonitory, and stronger measures are needed.

Nations must hold each other accountable for the environmental devastation they inflict in war. Environmental damage should no more be accepted as a cost of waging war than pollution should be accepted as a cost of doing business.

Indeed, only when we begin to give these ideas their rightful place in the international ethos will we finally be moving toward making peace with nature.

The Environment of Tomorrow

Martin W. Holdgate

Martin W. Holdgate is director general of The World Conservation Union (IUCN) in Gland, Switzerland. This article is adapted from a lecture given for the David Davies Memorial Institute of International Studies.

Earth is the scene of constant change. The summits of the Jura Mountains in Switzerland, more than 1,500 meters above present sea level, are formed of limestone laid down as soft sediments in the bed of a warm and shallow sea about 175 million years ago. There were once forests in Antarctica and dinosaurs and ice sheets in England. As Alfred, Lord Tennyson wrote, "There where the long street roars has been the stillness of the central sea."[1]

Such changes will always occur. Some of today's seas will be squeezed out of existence by the collision of continents to form new mountain ranges, as the Himalayas are being shaped by the collision of India and Asia. The Arctic ice may well expand again, providing the ultimate solution to the architectural problems of Europe and North America. Life forms will continue to evolve and drive their predecessors toward the extinction that is the ultimate fate of every species. On a longer time horizon, as astronomer Fred Hoyle has put it, "we shall certainly be roasted"[2] when the sun emerges from its present stable phase and expands to engulf and vaporize the Earth.

However, I am concerned with a more limited perspective. Throughout this article, I will use *tomorrow* to mean 40 years from now, in 2030, and the phrase *the day after tomorrow,* viewed through a haze of uncertainty and for that reason receiving less attention, as 40 years later, in 2070. I will address three simple questions:

• What will the world be like as a habitat for life if present trends continue?

• What are the implications of these changes for humanity and for the world of nature?

• What can we do about it if we would like tomorrow to be different?

Throughout history, people have adapted perforce to the cycles of the changing Earth and, where those changes have proceeded slowly, have probably hardly noticed them. Even natural catastrophes, such as massive floods and volcanic eruptions like that of Mount Vesuvius in 78 and 79 B.C., though they left scars on the body of civilization, had only a local impact on its progress. Such events doubtless caused our ancestors much suffering and social upheaval, but at least people were able to excuse themselves from responsibility, unless they chose to ascribe the disasters to the vengeful acts of a god irritated by human sin. In Java and Bali, the gods are still held responsible for volcanic eruptions. However, most of the major changes on the planet today are very much acts of humanity, and they result from the cumulative impact of two linked processes: the growth in human populations and the process that we call development.

Until the last 10,000 years, our species was a relatively uncommon animal, slowly increasing in numbers and extending its range to reach a total global population of about 500 million by 1000 A.D. The number of humans then doubled by around 1800, doubled again in the following century, and is now rising past 5.2 billion. This growth has been made possible by development, or the alteration of the Earth's environmental systems so that an increasing proportion of their nonliving resources and biological productivity serves human needs. As a result of development, some 40 percent of terrestrial plant productivity—the basic fixing of energy by the green mantle of the planet—is today used in one way or another by people. Changes in ecological systems have been an inevitable result of development. Because we eat cereals and our livestock are grazers, much of our land has been deforested and converted into pasture or that highly modified grassland we call cereal cropland. To expand croplands, people around the world have greatly altered patterns of water flow, sometimes successfully in stable irrigation systems but often unsuccessfully, leading to salt accumulation and soil sterilization—which has now damaged some 60 million hectares worldwide, an area equivalent to two-thirds of China's cropland.

As agricultural methods become more dependable, people not directly involved in the business of subsistence were able to create objects and ideas that enriched the total community. Such craftsmen applied increasing skill to the use of nonliving resources in

From *Environment*, Vol. 33, No. 6, July/August 1991, pp. 14-20, 40-42. First published as a lecture in 1990 by the David Davies Memorial Institute of International Studies. Reprinted by permission.

buildings, metal goods, and other artifacts and transformed areas of the physical environment with mines, quarries, and other structures. As people began to smelt metals, the movement of these elements through the biosphere inevitably increased, just as their agricultural and fuel-burning habits increased the fluxes of carbon, nitrogen, and sulfur. Development, by permitting the dominance of our species, has inevitably altered the world both by the nature and the scale of the transformations involved.

In recent years, some so-called environmentalists in developed countries have spoken of development as if the word were dirty. That is nonsense, as the World Conservation Strategy, prepared in 1980 by The World Conservation Union (IUCN) in partnership with the World Wildlife Fund and the United Nations Environment Programme, makes plain.[3] Development has been essential to the evolution and expansion of human civilizations, and more will be needed to help millions of people escape from today's poverty and squalor and to feed tomorrow's added billions. What we have to be concerned about is the nature and quality of development and the social and political structures needed to bring it about. It is clear from the destruction that has been the price of today's uneven and unsatisfactory development that the world cannot afford much more of the same. We need something different. To use today's catch-word, we need development that is sustainable—that is, it must not overcrop soil, pastures, forest, or fisheries or create products that spread from a beneficial activity, like industry, to blight other essential ones, like agriculture, the supply of drinking water, or the stability of the world's ecosystems.

When we demand that development should be sustainable, we must be clear about our meaning. We do not mean that the growth in a human activity, such as the cultivation of new land, must be capable of indefi-

Throughout history, people have adapted perforce to the cycles of the changing Earth and, where those changes have proceeded slowly, have probably hardly noticed them.

nite extrapolation—little such growth will be. Rather, we mean that the changes we make in our environment must not only improve the yield of a useful product to-

day but also go on supplying that product tomorrow, without side effects or unforeseen consequences that undermine other essential environmental functions. If overuse has such an impact, we may have to accept adjustment to a lower level of sustainable production. To quote the World Commission on Environment and Development, sustainable development means "meeting the needs of the present without jeopardizing the ability of future generations to meet their own needs."[4] For IUCN, conservation means preserving the world's natural resource base as the indispensable foundation for the future.

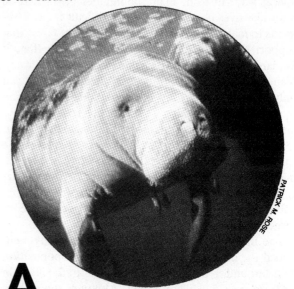

PATRICK M. ROSE

As we look to the possibilities of sustainable development, we would be wise to remind ourselves of three basic features of the Earth as a habitat. First, the Earth is a rather small planet with finite resources, and it receives a finite amount of energy from the sun. Its living and nonliving systems have interacted over time to create the habitats within which humanity evolved. Interactions between living and nonliving elements still operate in a fashion that regulates the overall environment in a manner analogous to the self-regulatory mechanism of a living being. James Lovelock pointed this out when he named the planetary organism Gaia, after the Greek goddess of the Earth. Hindu scholars had the same idea 2,000 years ago. There can be no sustainable development that does not preserve that essential planetary system on which all life depends.

Second, 70 percent of the Earth's surface is ocean, used only to a very limited extent by land animals. Although the plants in the sea fix about as much carbon annually as do those on land, humans only take from the sea about 100 to 120 million tonnes of assorted fishery products annually. Although I expect that our descendants will cultivate more marine plants and animals in shallow waters, most of the human future is going to depend on how we use the land.

Third, land will remain highly heterogeneous as a habitat. People living in different regions cannot expect to enjoy anything approaching environmental equality. It will always be easier to live comfortably off the produce of the temperate zones like western Europe than off the arid sands of the desert or the dry grasslands of the savannas. The broad pattern of life on Earth will prevail tomorrow as well as today, even though the detailed ecological pattern within those biomes will change. Biological diversity is and will remain concentrated in the tropics. Therefore, the nations of the world cannot be made environmentally equal. This would be true even if national frontiers had been drawn in an environmentally logical way. Instead, many boundaries bisect natural units like river basins and make it difficult for governments to manage in an integrated way the resources that support their populations. The implication is that, if humanity is to have a sustainable future on this diverse planet with many environmental inequalities, it can only be through a process of international cooperation that transcends anything we see today. To state that is immediately to pose an immense challenge if we are to seek success in as short a time as 40 years. And we have to contend with certain trends arising from the nonsustainable nature of the current development process that aggravate our problems and are likely to prove unstoppable on a 40-year time scale. I am particularly concerned with six of these trends.

The first is population growth. I have already mentioned the dramatic increase in the world's population from around 1 billion people in 1800, to 2 billion in 1900, to over 5.2 billion today. Although the rate of population growth has slowed somewhat from a peak of nearly 2 percent per year in 1970 to around 1.66 percent now, a cautious medium projection suggests that by 2030 there will be more than 8 billion people in the world and that our descendants will be lucky if, by 2070, stability has been achieved at around 10 billion. The trend is unstoppable because in many countries half the population is still under reproductive age. Even if these people only have two children per marriage, a near doubling is inevitable. Forty years from now, most of those people presently under 16 years of age will still be alive and their children will be between 10 and 35 years old—and some of them will themselves have children. Common sense tells us that not all people will limit their family size to two children per couple by 2030, and it will obviously require an immense change in attitude for this to happen by 2070. The total population achieved in a number of countries will depend on how many children survive, as well as on how many are born.

Today's population explosion is a tribute to the medical profession, which has greatly increased the average life expectancy in both developed and developing countries. Clearly, this advance could continue if the effectiveness of medicine is not undermined by malnutrition or social disorder and if new diseases do not appear suddenly to exploit the wonderful habitat represented by so much healthy human tissue. My own guess is that new diseases will indeed appear because the ecological niche is there for rapidly evolving viruses, which are likely to exploit the particular means of dispersal provided by human behavior—as AIDS has exploited the unique human habit of transferring blood from person to person, a parasite's dream. However, I doubt that new diseases will limit the growth of human populations, because I have great faith in the resilience of medicine.

On the other hand, malnutrition will limit population growth in areas where water supplies are inadequate, the soil is overexploited, and there are no new reserves of fertile land. Agricultural science, along with medicine, has been a major supporter of population growth. But in large areas of Africa and parts of India, per-capita food production has been declining recently even though total food production has been increasing. Countries with high population growth, limited rainfall, and a hot climate will be at risk until population increase is curbed by measures more humane than famine. It is important to look now at places where people are on a collision course with their environment because

Interactions between living and nonliving elements still operate in a fashion that regulates the overall environment in a manner analogous to the self-regulatory mechanism of a living being.

these are the places that need the most help. We have learned the lesson of the recurrent tragedy of famine from Ethiopia and parts of the Sahel. We have also learned that food and medical aid cannot do more than stave off such misery.

Moreover, people are understandably reluctant to sit still and starve. Now, as in the past, if populations really deplete their environmental resources regularly and over significant areas, they are likely to migrate, and finding a frontier here or a different government there is unlikely to deter them. As an increasing number of people come up against environmental limits, real threats to the peace and stability of nations may arise. The only alternative is to accelerate programs of sustainable development that meet human needs and, at the same time, provide the means and the incentive for stabilizing human numbers, country by country, in the areas of greatest risk. Clearly, these programs will require levels of international assistance beyond anything available today, and politicians in the countries concerned must recognize that population pressure is a genuine and urgent problem—something that a number of governments are presently reluctant to accept.

The second trend is deforestation, which, I suggest, is also unstoppable because of the need to feed more people, especially in the developing world. Moreover, I do not think it is honest to present all tropical deforestation as an environmental disaster. There are parts of the tropics—including areas of the Amazon basin—where conversion of forests to well-managed agriculture, particularly agroforestry, could be a perfectly acceptable pattern of development, though the governments concerned are fearful about the damage caused by present methods. Further deforestation in the tropical regions of South America, central Africa, and Southeast Asia is virtually certain. The governments concerned should steer the process in the direction of sustainable agricultural systems, while halting destructive deforestation of areas that are especially valuable as reservoirs of biological diversity, that are essential to regulate local climate, or that protect the land from erosion. Looking beyond 2030 to 2070, my prediction is that the great forests of Earth will by then be concentrated in the boreal zone, in the more rugged and uncultivable mountains of the tropics, and in tracts of sparsely populated tropical lowlands, where the soils are poor. I also foresee significant but relatively small patches of natural forest in the temperate zones and in the densely

> *If humanity is to have a sustainable future on this diverse planet with many environmental inequalities, it can only be through a process of international cooperation that transcends anything we see today.*

populated and heavily utilized tropical lowlands. We may not like that pattern, but it is the one that I suspect our descendants will have to live with.

Another trend that is unstoppable is desertification, or land degradation. I do not refer to the advance of the Saharan dunes but, rather, to erosive soil loss, salinization, and declining fertility as a result of poor or inadequate irrigation—which has, for example, affected some 60 percent of the cropland in Pakistan. However,

because of some good science in recent years, there are signs of a cure. In the Sahel, for example, food production has increased steadily over the past 20 years despite a lower average annual rainfall than in the preceding two decades. In Pakistan, measures to improve the quality of irrigation, to control salinization, and to plant crops that are resistant to salt are gaining ground. Accordingly, I do not expect the maps of the world to show more desert areas in 2030 than they do today, and by 2070 this phenomenon may have been brought under control or even reversed.

Fourth, I believe that the continuing loss of the planet's biological diversity is unavoidable. Before every conservationist starts jumping up and down and shouting, let me explain at once that I do not imply that destruction of biodiversity should continue at its present rate or that we should complacently accept this trend. We are, however, almost bound to lose a significant number of species from the Earth as a consequence of the development process impelled by the imperative of human need. Such loss is a logical necessity because, as I have said, it will be impossible to stop tropical deforestation. About 50 percent of the species believed to exist on Earth are insects and other small organisms living in the canopies of tropical rain forests. A significant number of these species are bound to disappear if the forests are destroyed. Islands, which are also significant reservoirs of unique species, are vulnerable to invasion by mainland life forms, which are increasingly spread around the world by today's unparalleled ease of transport.

The question we have to ask ourselves—and it is one that wildlife conservationists understandably shirk—is whether these losses will matter. My own answer is yes, they do matter, especially to those who can afford to care. But we must not oversell the disaster. The losses matter if you believe that species other than our own have a right to exist as a part of natural creation. They matter if you believe that the rich diversity of life maintains the equilibrium of Gaia, the planet, in ways we do not fully understand and that reducing diversity brings risks we do not comprehend. The losses also matter because there may be many species and genotypes in the wild that are or could be of considerable value to humanity and that will act as the genetic basis for evolution in response to future change. Despite all this, a sense of proportion is needed. The United Kingdom, for example, supports about 1,700 species of vascular plant, 25,000 invertebrates, 12 reptiles and amphibians, 54 freshwater fish, 50 mammals, and 200 birds. This is an extremely impoverished biota by world standards, yet I think we would regard it as one that is able to maintain essential ecological functions—even though people would soon grumble if their diet was based only on what they could grow or gather there. I am not arguing for impoverishing all of the world's ecosystems to this extent. It is always better to maintain a system with the widest possible range of ecological functions and the greatest practicable reservoir of genotypes. But I suggest that some loss of biological diversity does not inevitably bring collapse. The problem is that we do not know how much loss of what kinds of organism is tolerable. Figuring that out is one of the real challenges to science.

My final two unstoppable trends are quite different from the others. Trend five concerns pollution. Some kinds of damaging pollution are unstoppable over a 40-year time scale because they are already present in the environment and cannot be eliminated by 2030. For instance, polychlorinated biphenyls and other persistent chlorinated organic substances are widely dispersed in the ocean and in biological food chains, even though the most damaging kinds are no longer produced. Even if the production of chlorofluorocarbons (CFCs), which are responsible for depleting the ozone layer, ends by 2000, as many scientists wish, the persistence of these substances in the atmosphere is such that the ozone "hole" will not begin to fill in in less than 40 years, and the ozone layer will not be fully restored even by 2070.

The buildup of greenhouse gases is also most unlikely to be stopped by 2030; at best, the upward trend in carbon dioxide emissions will have been slowed, CFCs will no longer be released, and emissions of other greenhouse gases will be limited. Thus, the sixth and last unstoppable trend is climate change. Because of the many greenhouse gases already emitted, there is every reason to believe that by 2030 the world will be between 1° and 2° C warmer than it is now. Temperatures are likely to increase by more than this average in the higher latitudes of the Northern Hemisphere in winter, by slightly more than the global average in the temperate zones, and by slightly less than the global average in the tropics. Wet areas may get wetter, dry areas may get drier, and the force of storms may increase. By 2030, sea level may rise by 10 to 20 centimeters. As for the climate in 2070, much depends on how successful people are in curbing greenhouse-gas emissions. If they are not—if politicians drag their feet—then the world could be 2° to 4° C warmer, and sea level could rise 20 to 40 centimeters, posing severe risks to burgeoning human populations, especially those in the coastal zones of the tropics.

Because about half of the world's population lives in coastal zones, a 1-meter sea-level rise, which is greater than most experts consider likely, could spell disaster. Such a rise in Bangladesh would inundate nearly 15 percent of the country and displace some 10 million people.

In Guyana, 95 percent of the population and 90 percent of the agriculture are concentrated on a 3-mile-wide coastal strip that already stands near sea level and is protected only by slender barriers. The plight of fertile and densely populated deltas and low-lying coral atolls with most of their surface below the one-meter level is obvious.

Climate change also threatens the ways of life of many inland people. Each 1° C rise in temperature effectively displaces the limits of tolerance, or ranges, of species (including crops) by about 100 to 150 kilometers toward the poles or 150 meters vertically. Temperature changes may well prove less important than changes in

amounts of rainfall, but these figures do demonstrate the kinds of dislocation that could occur. The disruption of natural systems would be aggravated because the shifts in limits of tolerance could occur much faster than species of trees and other dominant plants would be capable of matching through natural dispersion. The spectacle of trees being "left behind" by their optimum habitats and, at the same time, ceasing to be able to maintain themselves by reproduction in considerable parts of their present range has been sketched out by many ecologists. The fact of the matter is, however, that scientists do not know how the world's living systems would respond to drastic change, and finding out is a major challenge that is only now being undertaken. There is little doubt that agriculturalists could help people grow new varieties of crops adaptable to the changed environment, but there would be considerable social stress nonetheless, especially if people had to change long-established dietary habits.

WORLD BANK—RAY WITLIN

Why are these trends unstoppable over such a long time scale as 40 years? All six trends have two features in common. First, all are caused by the innumerable small actions of a very large number of people, who do not see their actions as detrimental. Second, there is a substantial lag time between cause and effect for each trend. Let me prove my point by referring to one form of pollution I have not listed as unstoppable—acid rain. This phenomenon is stoppable because of two factors. First, the time lag between the emission of the sulfur as sulfur dioxide and its deposition as acid rain is quite short—only two or three days. Consequently, once emissions are reduced, acid depositions quickly decline, and the acidified ecosystems can begin to recover. Second, because the greatest proportion of the sulfur emissions come from a relatively small number of large point sources, specifically commercial power stations that burn coal, there is good reason to believe that by 2030 these sources can be cleaned up and acid rain will have been stopped.

However, the situation for greenhouse gases is quite different. CFCs enter the atmosphere from such products as leaky refrigerators, old-fashioned aerosol cans, industrial solvents, and the materials used to expand plastic insulating foam expand. The only solution here is to end CFC production and use in favor of less harmful substitutes. Even then, because of the 80- to 100-year residence time of CFCs in the atmosphere, the problems they cause will take decades to cure. Similarly, carbon dioxide, the principal greenhouse gas, is inevitably produced whenever forests are cleared or coal, oil, gas, or wood is burned in power stations, industrial heating plants, domestic stoves, or vehicles. Although carbon dioxide and other greenhouse gases, such as nitrous oxide and methane, which come from livestock, cultivation, fertilizers, and fuel combustion, do not have as long a residence time in the air as do the CFCs, the process of their absorption by green plants or by the oceans and freshwaters is not rapid. The greenhouse-gas problem is intractable because of the multiplicity of sources, the fundamental importance of fuel burning and agriculture in human life, and the significant lead time between curative action and restorative effect.

Neither the causes nor the potential impacts of the trends are evenly spread throughout the world. In developed countries like those of Western Europe, population growth is not a problem (even if we might prefer less crowded conditions) and forest areas are actually increasing (although the species being planted are not those that naturalists or environmentalists would welcome most and may not be the most tolerant of climate change). Desertification is not a problem either, although soil loss and improper irrigation techniques can be problematic, and losses of biological diversity, though unwelcome, are not as heavy as in other parts of the world. However, all of these trends are grave concerns in the developing world. Rapid population growth is correlated with poverty and is, at the same time, an impediment to the economic growth that might end poverty and provide people with the means and the incentive to limit the size of their families. Moreover, the pollutants that threaten to perturb global climate, destroy stratospheric ozone, and accumulate in living systems come overwhelmingly from the developed countries of the Northern Hemisphere, and their spread imposes extra burdens on the nations experiencing the worst of the other trends. Put crudely, the poor are in a mess, and the rich are making that mess worse.

The solution to these problems, which are growing increasingly graver, does not lie in science or in technology, but in politics. Of course we need more and better science to understand the planetary systems more fully and to manage our use of them more wisely; of course we need a better understanding of the factors that determine human behavior; of course we need better medicine, better agriculture, better forestry, more efficient use of energy, technology that produces essential products without generating intractable pollution, and products that are environmentally benign and that use scarce resources more economically. All of these are necessary, and much more as well. But the lesson of the recent decades is that scientific understanding and technological capacity are not the limiting factors. The blockage lies in human perception and the willingness of people to change how they behave at the individual, group, and national levels.

Many world leaders would reject a great deal of my argument, and some would say that it is a typical manifestation of the obstructive and negative thinking that prevails in developed countries, where fear of overpopulation, arguments against deforestation, demands for the conservation of biological diversity, and calls to halt the production and use of global pollutants are deliberate devices to hold back the development of the Third World and to maintain the present inequalities among nations. These leaders would argue that the real problem of the global environment is chronic poverty, which has been imposed and aggravated by the economic dominance of developed countries that grew rich by exploiting the raw materials of poorer countries and that continue to consume more than their fair share of those resources, to control commodity markets in favor of their own terms of trade, and to depress the growth of the Third World by issuing loans that benefit the lender far more than the recipient.

These arguments can be put eloquently and, indeed, at one level, are true. I am less sure that the environmental speeches of leaders of developed nations are dishonest and actually designed to supply justification for keeping the rich rich, the poor poor, and the latter dependent on the former. However, it does not matter

Each international agreement must be the culmination of a process thoroughly built from within each nation.

whether such a view is true or false. What matters is that it exists, and this argument will be heard a lot more as the United Nations prepares for the 1992 World Conference on Environment and Development. If we are seriously concerned about the future of the world as a habitat, with the quality of life for the people of tomorrow, with the peace of the world, and with the stability of the global economic system, we have to face this gap in perception and the suspicion that goes with it.

Perhaps the first thing to do is to let facts about the world environment speak for themselves. This means openly sharing all available scientific knowledge and letting the professional community in each country evaluate the significance of those facts for itself. This is

already happening to a considerable and increasing extent. There are internationally accepted compendia of environmental data, produced, notably, by the United Nations Environment Programme. There are many international bodies that bring together governments, nongovernmental organizations, and individual specialists from developing and industrialized countries. Some problems, such as climate change and ozone depletion, are accepted almost universally as serious and in need of concerted action. In both of these areas, the developing world wants and expects the developed countries to continue to share their knowledge; to help developing countries evaluate the unavoidable trends over the next 40 years; to take steps to reduce output of CFCs and greenhouse gases, which are already overtaxing the capacity of the environment and are leaving no room for development in the Third World; and to transfer alternative materials and technologies that will allow the Third World to develop without creating new environmental risks.

The costs of these four actions would all fall on the developed world. But is that unreasonable, when the causative agents of the ozone depletion and the greenhouse effect stem especially from the activities of developed nations? Unless the developed world shows itself willing to take these actions and shoulder a significant part of the costs, the credibility of its leaders as sincere advocates of action designed to benefit the whole world will be at stake. Any attempts by leaders of developed countries to sell the essential new technology at a profit or to transfer it on terms that add to the burden of debt that already hampers the capacity of the developing world to deal with other intense problems of sustainable development will be seen as proof of insincerity.

Such attempts will make it more difficult to persuade the leaders of the Third World to treat the issues of population growth, deforestation, and loss of biological diversity with equal seriousness. Similarly, if the developing countries are going to conserve their biological diversity, promote sustainable use of forests and rangelands, and experience positive trends in economic and social conditions that will favor population stability, they will need money, which today is largely concentrated in the developed north.

Convincing world leaders that they must talk and think as one community is essential, but I fear that it is the easiest of the tasks that lie ahead. Indeed, it is already half accomplished. For whatever suspicions and accusations they have, the leaders of the world do talk about the importance of the environment and have done so with increasing frequency since the first United Nations Conference on the Human Environment, held at Stockholm in 1972. Indeed, given the burgeoning number of international bodies, the almost incessant conferences, and the more than 100 international conventions, agreements, and action plans that now deal with the world environment, it might be deduced that action was well in hand. The problem is, as one of my colleagues put it recently, that "thunderstorms of rhetoric are followed by droughts of inaction." Words are cheap and action plans are easy to put on paper. Because such agreements create the cozy, inexpensive illusion of a problem solved, they are dangerous.

How do we turn the words into actions? Clearly, each international agreement must be the culmination of a process thoroughly built from within each nation so that, in signing the convention or plan, the leaders

Scientists do not know how the world's living systems would respond to drastic change, and finding out is a major challenge that is only now being undertaken.

concerned know that they express the will of their people and can commit themselves to putting the agreement into operation—either by acting within their own territory or by giving aid to other countries where the need is greatest. How can such a process be ensured? Within any particular country there are three components of certainty: a sound understanding of the problems and of the means to solve them; good organization

and administration; and the backing of a well-informed, committed public. None of these components can succeed without the others. Unfortunately, in many nations, environmental concerns have been grasped slowly and addressed half-heartedly through governmental measures that often lag behind public awareness and will. For example, most governments that have created ministries to address the environment have added them as new sectorial entities alongside the older establishments that deal with agriculture, energy, industry, defense, and local government. Many environmental ministries have been used primarily to govern national parks and nature reserves. However, if the environment is actually the fundamental resource base on which all development rests, then surely it is proper to see it as a country's "natural capital"—to use a term that is now widely used among economists. In other words, the environment is real national wealth, as

precious or more precious than the manmade wealth guarded by treasuries and analyzed by finance ministers in annual budgets.

Managing the environment as a national treasure would unite science, economics, and government in a new way. Science would be needed to understand both the potential of using the natural environmental capital for sustainable development and the sensitivity of the environment to various impacts; the techniques for doing this already exist. Economics is needed to place financial value on the resource and to convert the material income yielded by the environment into cash terms. Again, the techniques for more precise economic evaluation of environmental assets and for evaluating the true costs and benefits of alternative options for their development are now well advanced. Government is needed to define how the state's managerial system then proceeds to get the best value for its money out of those resources. Government must also ensure that individual sectorial departments do not undermine the whole system by the pursuit of narrow and often traditional policies.

However successful such national plans may be and however enlightened government administrations may become (and I would add that I know of no government in the world that has yet structured itself to deal properly with environmental issues), nothing will be accomplished without the support of the governed. This fact seems obvious in terms of the six unstoppable trends. For example, experience in China suggests that it is difficult for even an authoritarian and determined state to bring population growth rapidly under control. The power of governments to control pollution, desertification, deforestation and human-induced climate change is obviously limited, and it is expressed only by altering the behavior of individuals through education, training, incentives, help, and deterence of those individuals who nonetheless insist on seeking their own ends, regardless of wider interests. In the end, the problem and the solution are matters of perception.

Our ancestors viewed environmental disasters as "acts of God" and either tolerated them or sought to prevent them by appeasing affronted deities. Today, we are prone to regard such disasters as something done by someone else, which another "they," such as the authorities, must put right. Meanwhile, the chemicals we use, the gasoline we waste, the fumes we generate,

Real changes in direction are needed if we are going to build a world in which humanity lives in enduring harmony with nature.

the heat we let escape through poorly insulated walls and roofs, the litter we create, and the environmentally unfriendly pesticides we buy because they are cheap seem so trivial in proportion to the size of the problem that altering our behavior does not seem worth the effort—even if we can find alternatives to buy and places to send items for recycling. The perceived benefit is all too often of less value than the perceived cost. Yet, unless millions of people are helped, guided, and advised

to alter their behavior, the unstoppable trends will not stop and the day after tomorrow will be dark.

There is some indication that the 1992 United Nations Conference on Environment and Development could be a scene of conflict between the leaders of de-

veloped and developing countries, fueled by self-interest and greed for bigger slices of a finite cake. To avoid this disaster, all countries need to review and be prepared to adjust their strategies and policies so that the six unstoppable trends—and other apocalyptic horsemen—do not bring the roof down over our heads. We need to devote great effort to slowing or reversing the damaging trends, but real changes in direction are also needed if we are going to build a world in which humanity lives in enduring harmony with nature. Willingly, or perforce, many people face life-style changes, and many industries may face a smaller profit margin. But unless we begin to travel with hope we shall certainly never arrive, and unless we start quickly, we shall face regression rather than progress. Let us, therefore, begin to develop practical measures that will enlist the aid and inventiveness of all the world's peoples to make their contribution to tomorrow's world.

NOTES

1. *In Memoriam*, section 123, stanza 1.
2. F. Hoyle, *The Nature of the Universe* (New York: Harper, 1960).
3. International Union for Conservation of Nature and National Resources (IUCN), *World Conservation Strategy* (Gland, Switzerland: IUCN, 1990).
4. The World Commission on Environment and Development, *Our Common Future* (New York: Oxford University Press, 1987).

Global Change Ecology

Predicting the whole earth environmental consequences of human activities requires research that incorporates longer time frames and larger areas of study in an international context.

John J. Magnuson and Jennifer A. Drury

John J. Magnuson is professor of zoology and the director of the Center for Limnology at the University of Wisconsin–Madison. Jennifer A. Drury is a free-lance writer.

In the past, people influenced the environment on a local and regional scale. By the end of the Roman era, most forests around the Mediterranean were destroyed by human activity. However, with the Industrial Revolution our species began to alter the earth at a global level, and the rate of environmental degradation has grown at an accelerating pace since then. The intensity of human impact increases yearly as our population swells and per capita consumption continues to increase in the developing world. By the year 2000, global population is predicted to reach six billion, and eight billion by the year 2022.

In recent years, a growing environmental consciousness has created a great demand for scientific predictions of the global environmental consequences of human activities. In response, the new science of global change ecology is emerging, and scientists are making projections of probable environmental consequences of human activities. Predictions of impending radical environmental disruption have gained preeminence in much of the public debate, while a significant minority of scientists project much less serious consequences.

Indeed, the alarming projections coming from some scientists worldwide suggest that life on our planet may be altered significantly during the next century. For example, global increases in atmospheric carbon dioxide and other greenhouse gases may contribute to a general global warming that could reduce grain production in the Great Plains, lower freshwater reserves, and increase the demand for electricity. In addition, the hole in the ozone layer over Antarctica has increased markedly over the last decade, pesticides have rendered some drinking water unsafe, and the tropical rain forests are being cleared for agricultural purposes. Although the present rate of extinction is uncertain because the total number of species is not known, some scientists estimate that one out of every 1,000 species is lost each year. Such an extinction rate could be as much as 1,000 times faster than the rate apart from human influence.

These and other such predictions have a ring of truth to them, but in fact no one can be certain of the magnitude of the impending changes. So far, the methods of science have seldom been applied rigorously to whole earth ecology, because many of the processes of whole earth ecology occur over long time frames and expansive spatial dimensions. Computer projections are often used at present to complement the relatively limited data sets available for addressing the issues of global environmental change.

It is evident that traditional science cannot provide answers to a problem that encompasses every aspect of life on earth. Scientists from single disciplines working alone are not equipped to solve problems as complex as global change. How can answers be found?

A new approach

In this era of global change, science is challenged to adapt new perceptual modes. Analysis of microscopic components is clearly insufficient for understanding the dynamics of global change. New tools of macroscopic perception are required. Long-term ecological research is developing methods and institutions of mac-

From *The World & I,* April 1991, pp. 304-311. Reprinted with permission from *The World & I,* a publication of The Washington Times Corporation. Copyright © 1991.

COURTESY OF JOHN J. MAGNUSON

roscopic perception operating at longer time scales and on broader spatial scales.

This approach to research is necessary to understand important processes that occur very slowly or rarely—for instance, the disturbances caused by natural disasters such as hurricanes, or complex processes such as climate change that operate at a variety of scales in time or space. Without the information collected from long-term ecological research, observations of the present are likely to be misinterpreted and lost in the "invisible present." Many studies have shown that a conclusion reached after one or two years of observation is likely to be inaccurate or even opposite to that reached after five years of study.

One question is how many years are appropriate; and one consideration is the life span of the species being studied. A scientist cannot be certain of the long-term effects on an organism unless the organism is studied for

at least its entire life span. For some animals that may mean 20 years, for some trees, hundreds of years. Other factors to consider include cyclic patterns in climate and time lags between cause and effect that are longer than a year.

A natural and necessary complement to a lengthened time scale is a broadened spatial scale of research, to reveal "invisible places." Adjacent parts of the landscape influence each other. For example, a lake's water chemistry is influenced by the streams and land cover around the lake. Also, the water chemistry of lakes high in a landscape varies more from year to year than the chemistry of those low in the landscape.

The inherently complex interdependencies of the whole earth ecosystem dictate an interdisciplinary approach for global change science. Scientists from as many disciplines as appropriate to the question being answered must be included in the research. Geologists, zoologists, water chem-

■ Niwot Ridge is an alpine Long-Term Ecological Research (LTER) site in Colorado. Topics under investigation here include vertebrate population dynamics, decomposition processes, and nutrient cycling. The remote weather station shown is typical of many similar installations collecting climate change data at ecological research sites.

ists, and limnologists are only a few of the specialists who could be involved in a project. Eventually, complete global change studies should include social scientists as an integral part of the research team, because it is impossible to predict global change without assessing the effects of humans on the environment.

The new perceptual modes of long-term ecological research also dictate that a somewhat modified form of the scientific method be used. Traditionally, a scientist formulated a hypothesis (if ... then) based on current observations and then tested this question in a controlled setting. Long-term ecological research uses the same method but on different scales of time and space. Observations are not for a moment in time but for decades or centuries. In addition, the experimental setting is often the real world, not a laboratory. The hypotheses about long-term environmental responses are formalized models constructed to predict the future based upon past experiences. These models are finally tested in the real world, often over many years.

Coordinating the numerous scientists necessary for such lengthy periods requires formal or-

ganization. Without it, difficulties arise in keeping research on track. One example of a formal organization is the LTER program in the United States. (Franklin, et al, *Bioscience*, 40(7): 509-523.)

The LTER program

Long-Term Ecological Research (LTER) is a program funded by the U.S. National Science Foundation to understand a broad range of temporal and spatial scales. The program comprises

18 individual sites spread throughout the continental United States, Alaska, Puerto Rico, and Antarctica. Its operations are integrated nationally through a functioning network of collaborative research, planning, and data sharing. Jerry Franklin, Bloedel professor of Ecosystems Analysis, University of Washington, coordinates the network office at the University of Washington at Seattle. In addition, the LTER network collaborates with numerous federal agencies, such as the U.S. Environmental Protection Agency (EPA) and U.S. Forest Service, and a variety of state agencies.

Each LTER site was selected by a process of independent peer review. The quality of science proposed was the most important factor, but other considerations included the quality of the site, the availability of historical data sets, and the institutional capability to maintain and husband long-term research.

The sites range from arctic and alpine tundra to tropical rain forests, deserts, and dry grasslands, from coastal wetlands, estuaries, and unspoiled

Given the intrinsic complexity and interconnectedness of ecological processes, the study of global change ecology requires monitoring sites representative of different key types of ecosystems.
■ *Above left:* At the Kellogg Biological Station site in Michigan, studies include agricultural productivity and genetic transfer in soil microorganisms.
■ *Above right:* Extensively damaged by Hurricane Hugo in 1989, the Luquillo Experimental Forest LTER site in Puerto Rico has studies ranging from the recovery of an endangered parrot species to the interactions of land and streams.
■ *Right:* The U.S. LTER program includes sites ranging from Antarctica (not shown on the map) to Alaska.

Observations are not for a moment in time but for decades and centuries.

wilderness to grazing lands and cornfields. Comparative analyses among this diversity of landscapes will add more generality to our findings than results from studies of single sites.

An example of a LTER research site is the North Temperate Lakes site in northern Wisconsin. An hourglass-shaped lake has been divided in half by a barrier. One-half of the lake was acidified with sulfuric acid to assess the direct and indirect effects of lake acidification from acid rain. The other half was left as a reference. Other study lakes in the vicinity are being observed as additional references.

The study aims to assess both the direct and indirect effects of lake acidification from acid rain. Such studies can provide valuable information on how whole ecosystems, such as a lake, respond to human-caused stresses that push ecosystems beyond the boundaries of their natural variability.

The Invisible Present

A lake near Madison, Wisconsin, Lake Mendota, has been the site of a simple, yet extremely valuable, exercise in long-term ecological research. Last year the lake was ice covered for fewer days than the year before. Could this be a sign of global warming?

A process as complex as global warming is impossible to accurately recognize from one observation. How can it be observed?

Dale M. Robertson completed his 1989 Ph.D. thesis at the

University of Wisconsin–Madison on the duration of ice cover on Lake Mendota. Robertson's results shed some light on the possibility of global climate warming and the benefits of long-term ecological research.

The project also illustrates a dominant concept in long-term research. It is impossible to directly sense gradual change as it occurs over decades, because these changes become hidden in large year-to-year variations and time lags between cause and effect longer than a year. Thus these changes and the processes that cause them are "invisible" at the present time. We have referred to this idea as the "invisible present" (J.J. Magnuson, *BioScience*, 40(7):495-501).

Lake Mendota is unique because its ice cover has been documented continuously since 1855. When Robertson graphed the number of days the lake was frozen each winter for 132 years, he was able to see trends that are impossible to notice on a short time scale.

Robertson's analysis is important in many ways. If data from only one year are analyzed, the only information yielded is the number of days the lake was frozen. After 10 years, the large year-to-year variations can be observed. The range in duration of ice cover was from about 50 days to 120 days, and in 1983 Lake Mendota was frozen for 40 days less than the other nine years.

After 50 years, the data reveal that ice cover was unusually short not only in 1983 but in

most other El Niño years. El Niño refers to a warm current that sometimes appears off the coast of Peru around the end of December. The best explanation for these parallel occurences is that both El Niño and the short ice-cover years at Lake Mendota are caused by periodic shifts in the location of major atmospheric circulation patterns around the globe.

However, the most important observation comes from an analysis of the entire 132 years. It is here that the warming trend at the end of the "little ice age" in the late 1800s becomes visible for the first time. The future duration of ice cover on Lake Mendota with a doubling of greenhouse gases can be predicted by using global climate models and a model for lake ice based on the Lake Mendota record. This scenario, when tested against future changes in ice cover, will help us alter or gain confidence in our models of earth system behavior.

Robertson's ice cover analysis illustrates the importance of long-term observations in uncovering the invisible present. It is impossible to determine the effects of complex global processes unless studies are continued for several decades or longer. Many of the answers are undetectable in the present. Studies such as Robertson's ice cover analysis are sure to become more numerous as long-term research becomes more standard in science. This kind of science will reduce the uncertainty about the causes and mechanisms of global change in earth systems. Increased understanding will help people plan for and adapt to such changes and, eventually, to reverse the less desirable ones.

Invisible Places

A space-age technology is emerging as a primary tool for global change ecology. Remote sensing, the collection of data using a measuring device distant from the object observed, is revolutionizing the way scientists view the world in which we live. Whereas the microscope awakened scientists to the world of microorganisms and structures until then invisible to the human eye, remote sensing can be thought of as a macroscope, awakening scientists to a view of the world and its many landscapes and seascapes previously invisible to the human eye.

"It's a new tool and new

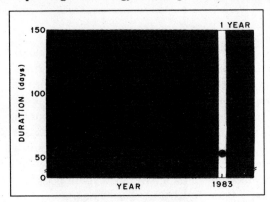

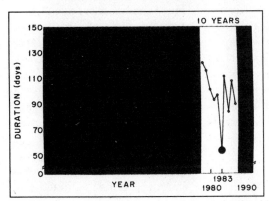

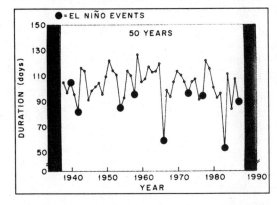

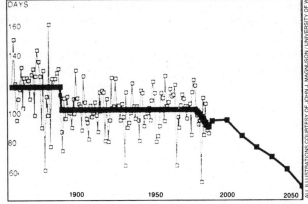

ALL ILLUSTRATIONS COURTESY OF JOHN J. MAGNUSON / UNIVERSITY OF WISCONSIN, MADISON

Revealing the "invisible present" requires putting specific ecological data into historical context. With its continuous record of ice cover duration extending back to 1855, Lake Mendota near Madison, Wisconsin, has become an important source of global climate change data, as shown by the sequence of four graphs (*counterclockwise from top left*).

■ The unusually warm winter of 1983, with its short duration of ice cover, stands alone in the one-year record.
■ A 10-year sequence reveals some of the yearly variability of climate.
■ A 50-year sequence reveals a cycle that seems to be synchronized with the cycles of the El Niño current near South America.
■ The full 132-year record reveals a significant warming at about 1890. This is highlighted by the mean ice cover line (the thick, black line) with its distinct break at about 1890. The

line from 1990 to 2050 is a projection of ice cover under the assumption of global warming caused by a doubling of the carbon dioxide in the atmosphere. Model testing can be done by comparing records of the actual yearly ice cover with the model's predictions.

■ *Top right:* Lake Mendota. Each winter, the expansion of ice on warm days causes the formation of pressure ridges that trace an irregular path across the lake.

To understand the lake at our own doorstep we must make it visible in the context of a larger spatial view.

scale that we've never been able to observe the earth from before," said Barbara Benson, data manager for the University of Wisconsin's North Temperate Lake LTER site.

Most remote sensing data consist of satellite imagery, although they can include aerial photography and acoustics (sonar soundings of the oceans and lakes). As a satellite orbits the earth, it digitally records the reflected radiation from the earth's surface onto magnetic tapes that a computer will later read. Each type of satellite has sensors that are sensitive to different portions of the electromagnetic spectrum. For example, some portions of the electromagnetic spectrum provide information on temperature, others on greenness or the amount of vegetation sensed from infrared reflections. These bands of the spectrum can be combined to give more detailed information about the earth's surface.

Existing satellites have a frequency for sampling the earth's surface ranging from twice a day to twice a month, determined by how often the different satellites pass over a particular site. For example, Advanced Very High Resolution Radiometer (AVHRR; owned by the National Oceanic and Atmospheric Administration within the U.S. Department of Commerce) passes over a site twice a day; Système Pour l'Observation de la Terre (SPOT; owned by the French Centre National d'Etudes Spatiales corporation and subsidized by the

French government) passes over a site every four days. Landsat, owned by Earth Resources Observation Satellite Company and subsidized by the U.S. government, passes over a site every 16 days.

The greatest strength of remote sensing is its ability to sample large areas at relatively fine-grained spatial resolution. A satellite image of the earth is not a solid mass of color but is made up of thousands of colored dots called *pixels*. Each pixel represents a particular piece of the earth's surface, with the smallest commercially available pixel representing a piece of land only 10 meters square.

Inevitably, the elements of a landscape interact. So, to understand a place, that is to make it visible, we must put it into the context of the larger surrounding landscape in which the "invisible place" is embedded. For example, the water in a lake will be affected by changes in land use and forest dieback, phenomena that cannot easily be perceived by on-site lake studies but that could be readily monitored by satellite remote sensing.

Context and detail are complementary sorts of information. Achieving larger spatial context requires relinquishing detail (high resolution).

Global change ecology requires putting the places we know at our restricted human scale of perception into regional and global contexts. Unless we do, we cannot be certain that, for

example, the apparent changes in climate in the time series of Lake Mendota ice records are unique to that particular lake or are a signal of changes occurring in the Midwest. Analysis of ice records on larger lakes across the Midwest, using the daily satellite data, will allow us to determine whether such changes are unique to one place or regional in extent.

Perhaps more importantly, the lake at our doorstep is changing, and we would like to know why so that we can either plan for and adapt to those changes or attempt to influence their course. The changes we see through the "invisible present" may be caused by the changes occurring either immediately around the lake or in the region, or by changes occurring around the world through burning of fossil fuels and deforestation. To understand the lake at our doorstep we must make it visible in the context of a larger spatial view that various macroscopes can help provide.

Satellite views alone cannot provide understanding of earth systems at a global scale. Networks of research sites, such as the Long-Term Ecological Network in the United States, need to develop and link in communication, data sharing, and interactive research. International organizations play an important role in assisting with the formulation of programs whose focus is beyond the needs of individual countries. One developing network is the International Geosphere Biosphere Programme (IGBP) headquartered in Stockholm. Plans are afoot to link the U.S. LTER network and other networks around the world with the IGBP to share data and results in pursuit of long-term ecological research.

The future

The natural progression for global change ecology lies in establishing networks to exchange data and information. The long-term vision is to link up 14 separate ecological-research networks to share data and results around the world.

However, it may be decades before scientists fully grasp the complexity of global processes. Global change ecology is still in its infancy, full of uncertainty. In addition the realities of economics and politics make even theoretically simple research difficult and time consuming.

While an expectant humanity demands immediate answers on the future effects of human activities on the biosphere, global change ecology must be a voice of moderation. The invisible present and the invisible place will remain so, so long as we persist in examining the world with our traditional tools of microscopic perception. The tools of macroscopic perception are being put into place, and they are beginning to bring the present and place of global ecology into focus so fact, not fiction, can guide the development of future global environmental policies.

The World's Population: People and Hunger

One of the greatest setbacks on the road to development of more stable and sensible population policies came about as a result of well-meaning and well-intentioned, but incorrect, population growth projections made in the late 1960s and early 1970s. The world was in for a "population explosion" the experts told us back then, and they predicted that one-quarter of the world's population would starve to death between 1973 and 1983. These population projections were wrong for the same reasons that similar predictions made by an English clergyman named Malthus in the 1790s were wrong: they were based on extrapolating the future from the trends of the past and such extrapolations are rarely correct. Shortly after the publication of the heralded works *The Population Bomb* (Paul Ehrlich, 1975) and *Limits to Growth* (D. H. Meadows et al., 1974), the growth rate of the world's population began to decline slightly. There was no cause and effect relationship at work here; the decline in growth was simply the process of the demographic transition at work, a process in which declining population growth tends to accompany increasing levels of economic development. Unfortunately, since the alarming predictions of massive famine and starvation did not come to pass, the world began to relax a little. Population growth was viewed by some as good rather than bad, and human ingenuity, it was said, could cope with increasing populations that meant, after all, increasing markets for manufactured goods and a larger labor force to produce those goods. Indeed, a counter-argument to the "population explosion" thesis began to gain favor by the late 1970s and some theorists actually began to set forth the notion that gradual population growth not only does not harm the environment but accelerates human progress. This "cornucopian" thesis notwithstanding, two facts remain: population growth in biological systems must be limited by available resources; and the availability of the Earth's resources is not infinite. That neither Malthus nor the alarmists of twenty years ago were correct in their predictions does not invalidate their basic premise: the Earth is a closed system and increasing numbers of people put increasing strain on that system. The "population bomb" may have failed to go off on time—but it is bound to go off eventually.

Consider the following: in Third World countries, high and growing rural population densities have forced the use of increasingly marginal farmland once considered to be too steep, too dry or too wet, too sterile, or too far from market for efficient agricultural use. Farming this land damages soil and watershed systems, creates deforestation problems, and adds relatively little to total food production. In the more developed world, farmers also have been driven—usually by market forces—to farm more marginal lands and to rely more on environmentally harmful farming methods utilizing high levels of agricultural chemicals such as pesticides and artificial fertilizers. These chemicals create hazards for all life and rob the soil of its natural ability to renew itself. The increased demand for food production has also created an increase in the use of previous groundwater reserves for irrigation purposes, depleting those reserves beyond their natural capacity to recharge and creating the potential for once-fertile farmland and grazing land to be transformed into desert. The continued demand for larger production levels also contributes to a soil erosion problem that has reached alarming proportions in all agricultural areas of the world, be they high or low on the scale of economic development. The need to increase the food supply and its consequent effects on the agricultural environment is not the only result of continued population growth. For industrialists, the larger market creates an almost irresistible temptation to accelerate production, requiring the use of more marginal resources and resulting in the destruction of more fragile ecological systems, particularly in the tropics. For consumers, the increased demand for products means increased competition for scarce resources, driving up the cost of those resources until only the wealthiest can afford what our grandfathers would have viewed as an adequate standard of living.

The articles selected for this second unit all relate, in one way or another, to both the theory and reality of population growth and to the consequences of increased human populations for environmental systems. In the first article of this unit, Nafis Sadik, the executive director of the United Nations Population Fund, outlines the current world population trends and suggests strategies for curbing population growth rates. In "World Population Continues to Rise," Sadik claims that the coming decade will be crucial in deciding the speed of population growth for much of the next century. At least one billion persons will be added to the global population in the 1990s, according to Sadik, with over 90 percent of this increase occurring in the Third World countries of Latin America, Africa, and South and East Asia. Such massive growth will continue unchecked except by food shortages and environmental disaster unless the developed nations make heavy investments in human resources and provide a firm base for economic development that will lower growth rates. In "Sheer Numbers," well-known writer Garrett Hardin elaborates on the theme of Sadik's article: population growth trends of the present decade indicate the probable trends of the next century. Hardin attributes much of the failure of environmental scientists and others to seriously consider population growth as an area for investigation and action to the chronic nature of the problem. Population growth is something that humans have almost always had to deal with. Therefore, it tends to be much more difficult to stir up human interest in this perpetual problem than in more immediate and dramatic problems such as war. Something could be done to solve the population dilemma, according to Hardin, if people could get beyond the politics and perceptions of the past. Werner Fornos, president of the Population Institute, echoes Hardin's viewpoint in "Population Politics." Fornos notes that UN demographers have revised upward their estimate of the most probable point at which human population will level off. The larger numbers suggested by the revised estimates are, in most geographic or regional instances, incapable of being supported by existing environmental and food production systems. Both the knowledge and the technology are present to stabilize world population well below the UN estimates. Unfortunately, political and ideological problems get in the way of solutions. In the unit's next selection, Lester Brown of the Worldwatch Institute also examines the gap between food and population. Brown argues that the economic policy of continued growth is a failure and that agricultural production cannot keep pace with an expanding human population. The imbalance between people and food is likely to widen, says Brown, and this is particularly the case for Third World areas already stressed by too many people and not enough good farm land. The title of Brown's article, "Feeding Six Billion," is symbolic of the unpleasant scenario of a badly unbalanced population/food ratio in a very short period of time. Brown's thesis puts him squarely in the camp of the biologists: the only solution to the world's population and food problem is workable population control. In the final selection of the unit, journalist François Monier discusses the population/food problem in the context of global climate change. Monier claims that

global warming is an inevitable consequence of human activities that release carbon dioxide and other greenhouse gases into the atmosphere. This global warming will produce significant changes in the ability of certain areas to produce food and a growing world population. The advantage in the world of the future will be with those countries that know how to adapt to a changing climate—in other words, to make sophisticated plans, change legislation and education systems quickly, and introduce the greatest possible flexibility to their use of resources.

It is clear from most of the selections in this unit that the global environment is being stressed by population growth and that more people means more stress and more poverty. It should also be clear that we can no longer afford to permit the unplanned and unchecked growth of the planet's dominant species. No closed environmental system can long sustain the kind of pressures that a population of more than five billion persons places upon it. Continuation of those pressures will wipe out the tremendous strides in human well-being that have been made over the last few centuries, and will assure that the environmental problems of the present will pale in significance beside those of the future.

Looking Ahead: Challenge Questions

Why do demographers anticipate that so much of the future growth of the world's population will occur in those sections of the world that would be classified as "Third World" or underdeveloped nations?

Why are problems of population growth dealt with differently by institutional and political systems than problems such as international security or defense? Are there ways to create a more responsive decision-making process to deal with population problems?

What is meant by "zero population growth" or ZPG, and how can countries achieve ZPG in the face of political and ideological systems that are directed toward growth objectives?

How do environments differ in their ability to sustain high levels of agricultural productivity, and how do these differences help to explain the current environmental conditions in many of the world's developing nations?

What kinds of changes in food production may be expected to take place in a world with warmer temperatures than the present? Are there strategies that countries can develop to maintain their agricultural productivity, even in the face of drastically changing climates?

World Population Continues to RISE

N A F I S   S A D I K

Nafis Sadik is executive director of the United Nations Population Fund, 220 East 42nd Street, New York, New York 10017. This article is adapted from the Fund's *The State of World Population 1990*.

The executive director of the United Nations Population Fund outlines the current world population trends and suggests strategies for curbing population growth rates.

The 1990s will be a critical decade. The choices of the next 10 years will decide the speed of population growth for much of the next century; they will decide whether world population triples or merely doubles before it finally stops growing; they will decide whether the pace of damage to the environment speeds up or slows down.

The world's population, now 5.3 billion, is increasing by three people every second — about a quarter of a million every day. Between 90 and 100 million people — roughly equivalent to the population of Eastern Europe or Central America — will be added every year during the 1990s; a billion people — a whole extra China — over the decade.

No less than 95% of the global population growth over the next 35 years will be in the developing countries of Africa, Asia, and Latin America.

It has been more than 20 years since the population growth rate of developing countries reached its peak in 1965-70. But it will be during only the last five years of this century that the additions to total numbers in developing countries will reach their maximum. This 35-year lag is a powerful demonstration of the steamroller momentum of population growth.

Racing to provide services to fast-growing populations is like running up the down escalator: You have to run very fast indeed to maintain upward motion. So far, all the effort put into social programs has not been quite enough to move upward in numerical terms. The absolute total of human deprivation has actually increased, and unless there is a massive increase in family planning and other social spending, the future will be no better.

Population Trends

Southern Asia, with almost a quarter of the current total world population, will account for 31% of the total increase between now and the end of the century; Africa, with 12% of the world's population today, will account for 23% of the increase. By contrast, eastern Asia, which has another 25% of the current world population, will account for only 17% of the total increase.

Reprinted with permission from *The Futurist*, March/April 1991, pp. 9-14. *The Futurist*, published by the World Future Society, 4916 Saint Elmo Avenue, Bethesda, Maryland 20814.

Similarly, the developed countries — Europe (including the Soviet Union), North America, and Japan, which represent 23% of the current world population — will account for only 6% of the increase. The remaining 15% of the world's population, living in developing countries, will produce 23% of the increase.

By and large, the increases will be in the poorest countries — those by definition least equipped to meet the needs of the new arrivals and invest in the future.

Because of the world's skewed growth patterns, the balance of numbers will shift radically. In 1950, Europe and North America constituted 22% of the world's population. In 2025, they will make up less than 9%. Africa, only 9% of the world population in 1950, will account for just under a fifth of the 2025 total. India will overtake China as the world's most populous country by the year 2030.

Toward the end of the twenty-first century, a number of countries seem set to face severe problems if populations grow as projected. Nigeria could have some 500 million citizens — as many as the whole African continent had around 1982. This would represent more than 10 people for every hectare of arable land. Modern France, with better soils and less erosion, has only three people per hectare. Bangladesh's 116 million inhabitants would grow to 324 million, with density on its arable land more than twice as high as in the Netherlands today. This does not take into account any land that may be lost to sea-level rises caused by global warming.

It should be emphasized that these are not the most-pessimistic projections. On the contrary, they assume steadily declining fertility during most of the next 100 years.

Food

Between 1979-81 and 1986-87, cereal production per person actually declined in 51 developing countries and rose in only 43. The total number of malnourished people increased from 460 million to 512 million and is projected to exceed 532 million by the end of the century.

Developing countries as a whole have suffered a serious decline in food self-sufficiency. Their cereal imports in 1969-71 were only 20 million tons. By 1983-85, they had risen to 69 million tons and are projected to total 112 million tons by the end of the century. These deficits have so far been met by corresponding surpluses in the industrialized countries — of which the overwhelming bulk comes from North America.

World food security now depends shakily on the performance of North American farmers. Following the drought-hit U.S. harvest of 1988, world cereal stocks dropped from 451 million tons in 1986-87 to only 290 million tons in 1989, down from a safe 24% of annual consumption to the danger level of 17%.

Poverty

The world produces enough food to feed everyone today — yet malnutrition affects as many as 500 million people. The problem is poverty and the ability to earn a livelihood. The total numbers of the poor have grown over the past two decades to around one billion now.

Absolute poverty has shown a dogged tendency to rise in numerical terms. The poorest fifth of the population still dispose of only 4% of the world's wealth, while the richest dispose of 58%. Economic recession, rising debt burdens, and mistaken priorities have reduced social spending in many countries.

But population growth at over 2% annually has also slowed social progress. So much additional investment has been required to increase the quantity of health, education, and other services to meet the needs of increased populations that the quality of service has suffered.

In many sectors, the proportion of deprived people has declined. But this is a reduced proportion of a higher total population swelled by rapid growth. As a result, the total numbers of deprived people have grown.

The growth of incomes may be affected by population growth. On a regional basis, there is an inverse relationship between population growth and growth of per capita income. There is a lag of 15–20 years between the peak of population growth and the peak growth in the labor force. Already there are severe problems in absorbing new entrants to the labor force in regions such as Africa or South Asia. Yet, in numerical terms, the highest rates of labor-force growth in developing countries lie ahead, in the years 2010–2020.

The labor force in developing countries will grow from around 1.76 billion today to more than 3.1 billion in 2025. Every year, 38 million new jobs will be needed, without counting jobs required to wipe out existing underemployment, estimated at 40% in many developing countries. Complicating the issue will be the spread of new, labor-saving technologies.

The land still provides the livelihood of almost 60% of the population of developing countries. But most of the best and most-accessible land is already in use, and what is left is either less fertile or harder to clear and work. The area available per person actually declined at the rate of 1.9% a year during the 1980s.

Urban and Education Issues

In recent decades, urban growth in developing countries has been even more rapid than overall population growth. Town populations are expanding at 3.6% a year — four and a half times faster than in industrialized countries and 60% faster than rural areas. Rural migrants swell the total, but an increasing share of this growth now comes from natural growth within the cities themselves.

The speed of growth has outpaced the ability of local and national government to ● provide adequate services. The number of urban households without safe water increased from 138 million in 1970 to 215 million in 1988. Over the same period, households without adequate sanitation ballooned from 98 million to 340 million.

The total number of children out of school grew from 284 million in 1970 to 293 million in 1985 and is

FAO PHOTO BY F. MATTIOLI

projected to rise further to 315 million by the end of the century. Also between 1970 and 1985: The total number of illiterates rose from 742 million to 889 million, and the total number of people without safe sanitation increased from about a billion to 1.75 billion.

A woman in Lesotho shovels earth while carrying her baby on her back. Ninety percent of the labor force here is women. In many cultures, women do much of the labor while rearing children. A key to success in reducing overpopulation, according to author Sadik, lies in reaching women in developing countries.

Today, the situation looks less promising. Progress in reducing birth rates has been slower than expected. According to the latest U.N. projections, the world has overshot the marker points of the 1984 "most likely" medium projection and is now on course for an eventual total that will be closer to 11 billion than to 10 billion.

In 15 countries — 13 of them in Africa — birthrates actually rose between 1960-65 and 1980-85. In another 23 nations, the birthrate fell by less than 2%.

If fertility reductions continue to be slower than projected, the mark could be missed yet again. In that case, the world could be headed toward an eventual total of up to 14 billion people.

Why should we be worried about this? At present, the human race numbers "only" 5.3 billion, of which about a billion live in poverty. Can the earth meet even modest aspirations for the "bottom billion," let alone those of the better-off and their descendants, without irreparable damage to its life-support systems?

Eating Away at the Earth

These increasing numbers are eating away at the earth itself. The combination of fast population growth and poverty in developing countries has begun to make permanent changes to the environment. During the 1990s, these changes will reach critical levels. They include continued urban growth, degradation of land and water resources, massive deforestation, and buildup of greenhouse gases.

Many of these changes are now inevitable because they were not foreseen early enough, or because action was not taken to forestall them. Our options in the present generation are narrower because of the decisions of our predecessors. Our range of choice, as individuals or as nations, is narrower, and the choices are harder.

The 1990s will decide whether the choices for our children narrow yet further — or open up. We know

more about population — and interactions among population, resources, and the environment — than any previous generation. We have the basis for action. Failure to use it decisively will ensure only that the problems become much more severe and much more intractable, the choices harder and their price higher.

At the start of the 1990s, the choice must be to act decisively to slow population growth, attack poverty, and protect the environment. The alternative is to hand on to our children a poisoned inheritance.

Danger Signals

Just a few years ago, in 1984, it seemed as if the rate of population growth was slowing everywhere except Africa and parts of South Asia. The world's population seemed set to stabilize at around 10.2 billion toward the end of the next century.

Already, our impact has been sufficient to degrade the soils of millions of hectares, to threaten the rain forests and the thousands of species they harbor, to thin the ozone layer, and to initiate a global warming whose full consequences cannot yet be calculated. The impact has greatly increased since 1950.

By far the largest share of resources used, and waste created, is currently the responsibility of the "top billion" people, those in industrialized countries. These are the countries overwhelmingly responsible for damage to the ozone layer and acidification, as well as for roughly two-thirds of global warming.

However, in developing countries, the combination of poverty and population growth among the "bottom billion" is damaging the environment in several of the most sensitive areas, notably through deforestation and land degradation. Deforestation is a prime cause of increased levels of carbon dioxide, one of the principal greenhouse gases responsible for global warming. Rice paddies and domestic cattle — food suppliers for 2 billion people in developing countries — are also major producers of methane, another of the greenhouse gases.

Developing countries are also doing their best to increase their share of industrial production and consumption. Their share of industrial pollution is rising and will continue to rise.

At any level of development, larger numbers of people consume more resources and produce more waste. The quality of human life is inseparable from the quality of the environment. It is increasingly clear that both are inseparable from the question of human numbers and concentrations.

A Case for Change

Redressing the balance demands action in three major areas:

1. A shift to cleaner technologies, energy efficiency, and resource conservation by all countries is necessary, especially for the richer quarter of the world's population.

Carbon-dioxide emissions will be hardest to bring under control. If the atmospheric concentration of carbon dioxide is to be stabilized, cuts of 50% to 80% in emissions may be required by the middle of the next century. These will be difficult to achieve even with the most-concentrated efforts.

Four major lines of action will produce the greatest impact, especially if they are pursued in parallel. The first is improved efficiency in energy use. The second is a shift from fossil fuels, which currently account for 78% of the world's energy use, to renewable sources such as wind, geothermal, and solar thermal. The third is halting deforestation. The fourth is slowing population growth.

There are no technological solutions in sight for methane emissions from irrigated fields and livestock. They have both expanded in response to growing rural populations and to meet expanding world demands for cereals and meat. The irrigated area has grown by about 1.9% a year since 1970, slightly faster than world population. Livestock and irrigation will both continue to expand in line with populations in developing countries. Reducing population growth is the only viable strategy to reduce the growth in methane emissions from these sources.

2. A direct and all-out attack on poverty itself will be required.

3. Reductions are needed in overall rates of population growth. Reducing population growth, especially in the countries with the highest rates of growth, will be a crucial part of any strategy of sustainable development.

Reducing the rate of population growth will help extend the options for future generations: It will be easier to provide higher quality and universal education, health care, shelter, and an adequate diet; to invest in employment and economic development; and to limit the overall level of environmental damage.

What Needs to Be Done?

Immediate action to widen options and improve the quality of life, especially for women, will do much to secure population goals. It will also widen the options and improve the quality of life of future generations.

Education is often the means to a new vision of options. It encourages a sense of control over personal destiny and the possibility of choices beyond accepted tradition. For women, it offers a view of sources of status beyond childbearing. Because of this, education — especially for girls — has a strong impact on the health of the family and on its chosen size.

Women assume the burden of childcare along with their other tasks. They are in charge of nutrition, hygiene, food, and water. As a result, the effect of women's education on child survival is very marked.

Education's impact on fertility and use of family planning is equally strong. Women with seven or more years of education tend to marry an average of almost four years later than those who have had none.

Yet, there remains a great deal to be done in women's education, even to bring it level with men's. Women make up almost two-thirds

A literacy class for mothers at a social center in Chad. Achieving literacy for women gives them access to new information, which in turn gives them more choices regarding children and family planning. Literacy programs are one of the main components of international organizations' outreach to developing countries to stem the world's population growth rates.

FAO PHOTO BY A. GIROD

of the illiterate adults in developing countries. The importance of literacy programs for adult women goes far beyond reading and writing: It also allows access to practical information on such matters as preventive health care and family planning, which are often part of the programs themselves.

Sustained improvements in health care give people a sense of control over their lives. With adequate health care, parents develop the sense that they have some choice over their children's survival. Parents' feeling of control over their lives is extended by modern family planning; they can protect the health of both mother and child by preventing or postponing childbirth. Preventive measures that the family itself can apply assist the process.

Support for Family-Planning Efforts

Political support from the highest levels in the state is essential in making family planning both widely available and widely used. Political backing helps to legitimize family planning, to desensitize it, and to place it in the forum of public debate. It helps win over traditional leaders or counter their hostility. It also helps to ensure that funding and staffing for family planning are stable and protected against damaging budget cuts or the competing demands of rival departments.

Support must extend far beyond the national leadership before programs take off. It may be necessary to involve a wide range of religious and traditional leaders in discussions before introducing population policies and programs on a wide scale. If these leaders feel that they have been sidestepped, their opposition may become entrenched. If they are consulted and involved, on the other hand, they may often turn into allies. In Indonesia, for example, Muslim religious leaders were consulted at national and local levels; they not only withdrew their opposition, but have added their voices to the government's call for family planning.

Four main barriers block the way to easy access to family planning. The most obvious is geographical:

FAO PHOTO BY F. BOTTS

An outdoor class in a rural primary school in Swaziland. The Swazi government, assisted by international organizations, delivers advice on family planning through maternal and child-health centers.

How long do people have to travel to get supplies, and how long do they have to wait for service when they get there?

The second barrier is financial: While many surveys show that people are willing to pay moderate amounts for family-planning supplies, most poor people have a fairly low price threshold. Costs of more than 1% of income are likely to prove a deterrent.

Culture and communication are a third barrier: opposition from the peer group, husband, or mother-in-law; shyness about discussing contraception or undergoing gynecological examination; language difficulties; or unsympathetic clinic staff.

A fourth barrier is the methods available: There is no such thing as the perfect contraceptive. Most people who need one can find a method suited to their needs — if one is available. However, if high contraception use is to be achieved, suitable services must be not only available, but accessible to all who need them.

Suitable services mean high-quality services. In the long run, the quantity of continuing users will depend on the quality of the service.

Service providers do not need to be highly educated, but they should be sympathetic, well informed, and committed to their work. The service must be reliable, so that users can count on supplies when they need them. Good counseling is one of the most important aspects of quality. Family planning is loaded with emotional, social, and sometimes religious values. It is vulnerable to poor information, rumor, and outright superstition. Along with reliable supplies and a good system for referral in problem cases, good counseling can make a big difference to continuation rates.

Two other channels are useful in broadening the base and increasing the appeal of family planning: community-based distribution and social marketing.

Community-based distribution (CBD) programs use members of the community — housewives, leaders, or members of local groups — to distribute contraceptives. Older married mothers who are themselves contraceptive users have proved the best candidates.

Maturity, tact, perseverance, and enthusiasm are essential requirements for the good distributor.

After many years of relying only on clinic and health workers to deliver services, family-planning programs are discovering the uses of the marketplace. The private sector provides contraceptives to more than half the users in many developing countries.

Social-marketing programs reduce the cost to the user and increase sales by subsidizing supplies. These two aspects — the integration of suppliers with regular health services through training and the subsidization of supplies — are felt to combine the ease of access of the market with the sense of social responsibility of service programs.

The potential of community-based distribution and social marketing has not been exploited in most countries. Out of 93 countries studied for one survey, only 37 had a CBD or social-marketing project. There is clearly a considerable potential for expansion as an essential complement to integrated health services.

The technology of contraception is usually thought of in terms of safety and reliability. But it should also be seen as another important aspect of improving access and choice in family-planning programs.

Currently, the most popular method worldwide is sterilization, with around 119 million women and 45 million men in 1987. Next in popularity was the intrauterine device (IUD), with 84 million users, followed by the pill, with 67 million. The pattern of use differs considerably from one country to another and between developing countries and developed. Sterilization is by far the most common method in developing countries, with 45% of users, though only one-quarter of these were male. The IUD comes next in popularity, with 23%. In the North, sterilization accounts for only 14% of users and IUDs for 8%.

Users balance all the advantages and disadvantages they are aware of before deciding on a method —

Research must continue for the ideal method of contraception: cheap, totally effective, risk-free, reversible, without undesirable side effects, and simple enough for use without medical provision or supervision.

No such method is yet on the horizon. But research continues to push forward the frontiers, under conditions that have become more and more difficult.

The business of developing new contraceptives has changed radically. Tighter controls on testing and rising risks of costly lawsuits have made drug companies wary. The leading role has been assumed by the World Health Organization (WHO) and by nonprofit organizations such as the Population Council and Family Health International. But real spending on contraceptive research and development has not increased.

Some promising new candidates have been developed. Norplant, already approved for use in several countries, is probably closest to wide dissemination. It consists of six tiny rods containing the progestin hormone levonorgestrol. These are implanted under the skin of a woman's inside arm. Norplant, particularly suitable for women who have completed their families, prevents pregnancies for five years before it needs replacing. A two-rod version providing protection for three years is being developed. Norplant is highly effective, and unlike the injectable Depo-Provera, its contraceptive effect ends soon after it is removed. The drawback for some situations is that it requires a physician to insert and remove. The cost, at $2.80 per year of protection, is more than the pill at $1.95, but less than Depo-Provera at $4 per year.

Other long-acting hormonal methods may be introduced during the 1990s. They include biodegradable implants, providing 18 months of protection, and injectable microspheres lasting between one and six months. Vaginal rings containing levonorgestrol, which can be inserted and removed by the woman, are also being tested. And the Population Council is researching a male contraceptive vaccine.

— **Nafis Sadik**

or no method. If they do not like the available alternatives, they will simply drop out and use no method at all, or revert to less reliable traditional methods. One recent study in East Java found that, among women who were not given the method they preferred, 85% had discontinued use within one year. Where women were given the method they wanted, the dropout rate was only 25%.

Diversity, then, is the key to providing options. Diverse channels of distribution create the widest possible access to contraception. Diverse technology offers the widest possible choice of methods. The combination maximizes use.

Developing Human Resources

Investment in human resources provides a firm base for rapid economic development and could have a significant impact on the environmental crisis. It is essential for global security. But in the past, it has often commanded a lower priority than industry, agriculture, or military expenditure.

It is time for a new scale of priorities: There is no other sphere of development where investment can make such a large contribution both to the options and to the quality of life, both in the present and in the future. Whatever the future returns, investment is needed now.

SHEER NUMBERS

Can Environmentalists Grasp The Nettle Of Population?

An Unexpurgated Look at Our Shrinking World

DR. GARRETT HARDIN

DR. GARRETT HARDIN *is a population buff.*

Afunny thing happened on the way to Earth Day 1990: population got clobbered.

On the first Earth Day in 1970 there probably wasn't a single celebration that did not include some strong statements about the threat of population growth. In the 1990 ceremonies, population was scarcely mentioned. So what happened between 1970 and 1990?

Did world population decline during that period? Hardly. On the contrary it increased by a whopping 47 percent (from 3.632 billion to 5.321 billion). While the rate of increase *did* decrease (from two percent per year down to 1.8), in absolute numbers of people added each year you get a different story: In 1970 some 73 million people were added; but in 1990 an estimated 96 million will be added. That's 32 percent more. The arithmetic puzzle is easily explained: a slightly smaller rate is now operating on a much larger base.

The deeper puzzle, though, is the change in public attitude. The differences can be illustrated by a comparison of two very effective activists, Paul Ehrlich and Frances Moore Lappé, both renowned authors and public speakers with divergent styles. Ehrlich, the author of *The Population Bomb* (1968), is often called a "doomsayer." The more optimistic Lappé, author of *Diet for a Small Planet* (1971), laughingly acknowledges that she is known as "the Julia Child of the Soybean Circuit." As stereotypes go, Ehrlich is a liberal, 1960s model: he urges international cooperation in tackling environmental problems. Lappé is liberal only in her call for individual freedom and individual virtue to solve community problems through market forces.

The index of Frances Moore Lappé's latest book, *Rediscovering America's Values* (Ballantine, 1989), reveals that she sees no more role for population explanations than did Adam Smith when he wrote *The Wealth of Nations*. Missing from Lappé's index are "Overpopulation," "Migration," "Immigration," and "Carrying Capacity." Since Lappé is so much concerned with the welfare of individuals, it is not surprising to note that the subject of "Poverty" receives 34 page-citations in her index. In twenty years, while world popula-

tion increased by 47 percent, population as the perceived cause of anything—even poverty (if Lappé is a true mirror)—went down to zero. Quite a change.

Lappé has correctly sensed the mood of our times. The political assault on population theory during the Reagan years was immensely successful. Reagan-type conservatism disapproved of trying to limit growth, whether of Gross National Product, individual wealth or population. The growth of government regulation was especially disapproved of: witness the savings and loan disaster. To offset these disapprovals, the growth of individual freedom was highly praised—unless it happened to be the freedom for women to control their own child-bearing.

By the time Denis Hayes, the organizer of Earth Day 1970, began to pass the hat to finance the 1990 event, it was clear that no one wanted to risk offending the powers that be by funding anything that touched on population. If Earth Day 1990 touched population at all, it was only with a ten-foot pole. One wonders what role population will play in Earth Day 2010?

"Population buffs" are sure that long before 2010 arrives population will once

From *E Magazine*, Vol. 1, No. 6, November/December 1990, pp. 40-47. Reprinted with permission from *E*, the Environmental Magazine, P. O. Box 6667, Syracuse, NY 13217. (800) 825-0061.

more be seen as a truly massive global problem. So what interferes with publicizing the issue now?

First, population suffers from being a chronic problem rather than a critical one, and the media have a hard time stirring up interest in chronic problems. Consider the situations that news editors had to wrestle with during the past year, and the juxtapositions that called for journalistic decisions:

Yesterday, the Berlin Wall came down, *and world population increased by a quarter of a million.*

Yesterday, Lithuania declared its independence from the U.S.S.R., *and world population increased by a quarter of a million.*

Yesterday, Iraq invaded Kuwait, *and world population increased by a quarter of a million.*

What we call "news" consists of crises—sharply focused occurrences that are easy to report. Chronic, time-extended happenings don't have much of a chance when competing for time or space in the evening broadcast or the morning newspaper. The world didn't pay much attention to global population growth when it reached 100,000 per day; we don't pay much attention now that it has mounted to 263,000 per day. Will we pay any more attention when it rises to a million per day? Let's face it: an increase in population just isn't news.

But population per se is not important: it is the *consequences* of overpopulation that cause the trouble. Yet many intelligent people fail to see the connection. Take pollution: it is hard to see how anyone could maintain that population size has nothing to do with pollution levels, but some people hold this position. Political philosopher Mark Sagoff, at the beginning of the Reagan regime, said: "Pollution results not from our numbers, but from our lifestyles and our rate of consumption."

Scientifically, this is an indefensible position. Long ago Paul Ehrlich introduced the public to the very simple "Impact Equation" that governs all relations between human beings and their environment:

Impact per person × *Number of persons* = *Total Environmental Impact.*

Would Sagoff really maintain that five billion people produce no more pollution than, say, five people? Do five billion people exhaust natural resources no faster than just a few? Surely there is no way that five billion people can live so simple a life that they make no more demands on the environment than a mere handful of men and women. While this is an absurd example, one would suspect that Sagoff's equally absurd assertion reveals a fear of any mention of population. We need to know why population has such a poor public image.

Poor public relations are generally deserved. Population buffs have, in the past, said a lot of foolish things. What they say is often easily interpreted to mean that population is the *only* cause of society's ills. (That's the reverse of Sagoff's error, and it's just as bad.) Everything that happens in this world has many co-acting causes, and our language should mirror that reality. Population may be a causal factor—one among many—but it is never the *sole* cause. The suffering in many parts of the developing world is caused neither by overpopulation alone nor by bad government alone. Is there anything we can do to improve another country's government? Or is there any way we can persuade others to reduce their population growth? These are not easy questions! And there are no easy answers.

The "Impact Equation" tells us that the environment is affected by population numbers as well as by the "lifestyle" of the people. Organizations and magazines devoted to nature preservation and the environment should treat both, yet they generally avoid the population issue. Why?

There's a good strategic reason for this silence. The more issues reformers take on simultaneously, the less success they have. If one of a dozen causes they support affronts someone, that person generally withholds support for their whole program. Subtract all the people who oppose any particular cause and pretty soon reformers have only a tiny army behind them—they have bitten off more than they can chew. By ignoring population, environmental organizations and magazines have avoided arguments and have probably maximized their support. But the day may come when most people recognize the truth of the "Impact Equation." At that point, evasion will no longer pay. Perhaps we are close to that day now.

The dilemma facing environmentalists can be illustrated by looking at the automobile pollution problem. Since cars are never perfect we should always be able to reduce the pollution they produce by engineering them better. During the past ten years a great deal of progress has been achieved in automotive engineering, and we should be proud of this progress. But as the population of people increases further, the population of automobiles will increase too. We'll have to improve their engineering still more to keep the total pollution from getting worse. But there are inherent limits to engineering, and successive improvements become progressively more difficult—after all, the efficiency of a heat engine cannot exceed 100 percent.

On the other hand, biological reproduction has no inherent numerical limits—population growth is almost always stopped only by external causes (starvation and war, for example). Alternatively it *could* be stopped by internal, psychological causes, like adherence to community ideals ("Stop with two!"). Naturally, we

---

"Whenever a growing population is involved, it is not possible to cure a shortage by increasing the supply."

---

don't want external controls; unfortunately, few people are pushing hard for internal ones. Whether we like it or not, so long as we steer clear of all controls, pollution will grow and environmental wealth will diminish.

Before much can be done about population more people are going to need to recognize the fundamental wisdom of the "Impact Equation." The easy thing is to focus our attention on foreign countries. The hard thing is to focus on our own country; in our own communities we risk turning neighbors into enemies. Trying to reform ourselves takes much more courage than does preaching to people on the other side of the world.

Local action calls for ingenuity. Take the community of Santa Barbara, California and its water problem. The shortage of water is already an acute problem in southern California (and the shortage of safe water will soon be a problem for much of the United States). Over the past century rainfall in Santa Barbara has ranged from just five inches per year to 45 inches. Planning must therefore allow for a ninefold variability. We've tried to iron out the annual variation with reservoirs and lakes, but dams are expensive and we can build only so many. Ultimately the water problem has to be attacked in two ways, simultaneously:

Population approach: control the in-migration of new residents (whether from other states or other nations).

Environmental approach: lower the per capita use of water.

The second approach has the greater appeal to most environmentalists—just think of the water that is wasted when someone washes their car at the curb. Think of the tremendous amounts of water needed to keep a golf course green in a region where there is no rain during the hottest months. Isn't there something immoral about maintaining such luxuries? Quite simply, if people didn't demand luxuries, basic resources would be available to a larger population.

Population buffs take exception to all this. If luxuries were completely outlawed, they say, the community would one day still have to say "No" to a further increase in the community's population. Why not say "No" now, then, while we still have some luxuries left? Aren't luxuries for a few better than luxuries for none? (The battle begins!)

The variability of resources makes such choices even more painful. About once a generation, coastal California has several very wet years in a row. Reservoirs fill up and the overflow of water runs off into the ocean. At this point developers and builders say: "Look at all that wasted water! Give us the building permits and we'll bring more people here, thus putting an end to the waste."

Indeed, many people benefit directly by increases in population: architects, real estate developers, landowners and banks, for instance. Let's be honest: the pressure to bring more people into the community comes from many sources, and it has more to do with making money than it does with kindness and sharing.

Unfortunately, after a few years of above-average rainfall the dry years come. What then? If building has expanded to use all the water in the "good years," the population will have grown to the point where the hardship felt by each resident will be much greater.

We need to borrow wisdom from the engineers. Every properly engineered system includes safety factors. A bridge is always built to bear a bit more than the heaviest possible load. That bit more is called the "safety factor." We don't regard the extra steel in a bridge as a form of waste. Similarly, elevators are designed with cables that are stronger than they need to be: again, a safety factor. Would we want to save money by eliminating this "waste"?

Returning to matters of population and the environment, should not our socially engineered systems also include "safety factors"? If so, the target size for population should be determined by the poor years—not by the best years, or even by the average ones. The truly mature citizen recognizes that times will get worse, sooner or later.

For most of human history, population growth has been very close to zero, with the death rate and birth rate nearly equal and very high. Then about three hundred years ago the death rate started falling in

Childfree By Choice

According to Zero Population Growth (ZPG), one out of seven couples in the U.S. chooses not to have biological children. Moreover, the proportion of women opting to have one child or none is the highest it's been since the Baby Boom. Some ten percent of American men of reproductive age have had vasectomies and another 400,000 undergo the procedure each year. These statistics would be heartening if not for the grim reality that overshadows them: world population is likely to double within the next fifty years if steps are not taken to bring birth rates down sharply. Though the birth rate in the U.S. is currently 1.9 children per woman, population continues to rise in sheer numbers because the Baby Boomers are now in their child-producing years, spawning an "echo boom" of their own. Additionally, nearly 13 million teenagers become parents each year.

Since each new birth places additional strains on an already-depleted planet, we are indeed called upon as concerned citizens to rethink practically every decision we make with an eye toward social responsibility. It seems only fitting to examine our attitudes toward procreation. In fact, the choice to forego producing offspring may be the most environ-mentally-significant choice one can make.

Clinical psychologists living in Berkeley, California, Charlie King and Olga Grinstead have been married several years and are in their mid-thirties. The couple incorporated their commitment not to have children into their wedding vows. "Part of it was concern about population growth, and part of it was a lifestyle decision," says Olga. "I also had a feeling that people have kids unconsciously, out of a sense of duty or obligation." Charlie agrees, adding, "There are enough kids out there needing homes. It seems wasteful to bring more into the world." The couple also points out that children born in an affluent country such as the U.S. create more of a burden, environmentally speaking, than do children born elsewhere. In fact, Americans produce three-quarters of the world's carbon monoxide and nearly half of the world's nitrogen oxide emissions, while accounting for only five percent of the world's population. The typical American consumes some 531 times what a typical Ethiopian citizen would consume, according to ZPG. In addition to draining world resources, "Kids are very good for using up surplus income," notes Olga. "That money could be given to a worthy cause. There's the idea in our culture that you ought to be living right up to the level of your earning capacity. To *not* consume up to capacity is a radical idea."

Dean, 47, lives in suburban Connecticut and is involved with a woman in a committed long-term relationship. Neither he nor his partner wanted to have children, and in 1988 Dean had a vasectomy. "Having the vasectomy was one of the best things I've ever done," he says. "It's freed me mentally and emotionally from worrying about the possibility of an accidental pregnancy." James, single and in his early thirties, lives in Lawrenceville, Georgia. He has decided, independent of marriage, not to father children. "A lot of couples do not consider when they're planning to have a child that it becomes a fulltime job and a fulltime expense. People seem to think that having children will bond their marriage together, and that's the worst reason for having a child. I believe there are enough children in the world already that need love. I would consider adoption," he says.

Such individuals are clearly the exception when it comes to attitudes about procreation. "There's no question we live in a pro-natalist society," says Diane Sherman, ZPG's director of communications. "People think that if you choose to have no children, you're a selfish yuppie. In fact, in most cases it is the child-free couples who have given more thought to

Europe. The birth rate fell too, but only after a time lag. In the interim, population grew. When the birth rate fell too, population growth slowed. When the two rates become equal again—hopefully at a low level this time—population will stop growing. In a few European countries this may already have happened. The change from a condition of high-birth-rate-plus-high-death-rate to one of low-birth-rate-plus-low-death-rate is called the "Demographic Transition."

During this transition the income of the average European grew. Did Europeans get richer because they were having smaller families? Or did they have smaller families because they were richer? Both views can be plausibly argued; perhaps both are partly true.

The important issue is devising a population policy is this: can the population of a country be lowered by gifts from the outside which make its citizens richer? Those who believe what has been called the "Benign Demographic Transition Theory" urge us to make poor people

rich in order to persuade them to have smaller families. Is this theory true?

Forty years ago many people thought so. But new knowledge has raised serious doubts. Surveys by Rose E. Frisch of Harvard University uniformly show

their birth rate, but only after a lag of one or two generations. Unfortunately, in two generations a rapidly growing country like Guatemala can swell to four times its present size; the relief of such a country's desperate need cannot wait for

---

> "In rich countries the demand per person must be reduced; in poor countries, where per person demand can hardly be lowered further, the size of the population must somehow be reduced.

---

that when poor women are better fed they become more—not less— capable of producing children.

Supporters of the Benign theory have argued that better nourished women would become more rational too, and would in fact refrain from producing more children. But again, studies made in Guatemala, Turkey and other countries indicate the opposite. Perhaps better fed populations may *eventually* lower

two generations. Michael Teitelbaum and many other demographers have come to the conclusion that the Benign Demographic Transition Theory simply isn't true.

What a pity! If the Benign theory were true we rich Americans could, by making personal sacrifices, solve population problems in poor countries. We could stop eating meat and produce more plant food to feed the starving elsewhere. American families could give up their second car and invest the savings in food and housing for the poor.

Shortages cannot be successfully attacked until the logic of the situation is understood. I may feel that I have a shortage of money, but objective bystanders (particularly if they are very poor) may feel that I am just too greedy. In simple logic every "shortage" of a resource is equally a "longage" of desire. Why, then, do we hear so much of "shortages" and so little of "longages"? Probably because the demand to cure a shortage is an egotistical one, and we're pretty good at bending language to suit our egos. By contrast, admitting that we may have a longage of desire implies the disturbing possibility that the trouble is within us.

Put another way, he who promises to cure our shortages can make money by selling us things; but he who says he will cure our longages threatens to change our inner nature. Sales representatives by the tens of millions make a good living. Ministers and philosophers, who try to reduce our demands, have a harder time living well, materially speaking.

Population buffs fall into the second category: it's no wonder we pay so little attention to them. Despite all claims to the contrary, we live in a limited world. No matter how clever we are, as our population increases we will not have more abundant wildernesses, cleaner air,

---

their position." ZPG has recently published a report titled "Planning the Ideal Family: the Small Family Option" which, while acknowledging that decisions about childbearing are and should be voluntary choices, provides information and encouragement to those considering having few or no children. Planned Parenthood, a nationwide family planning organization, takes a similar approach. "The right to make decisions about reproduction is an inherent individual right," says David Andrews, the group's executive vice-president, "but people should be less alone in making those decisions."

Most people choosing to remain child-free experience some form of disapproval from those around them. Says ZPG's executive director, Susan Weber, "Many times, personal desires and good judgement take a back seat to what others think is the right thing to do." Pressure from friends and family, notions of procreation as an affirmation of one's masculinity or femininity, and feelings of obligation to carry on a family name or produce grandchildren all contribute to a view of childbearing as a mandatory experience. Kathleen Newland, in a paper written for the Worldwatch Institute entitled "Women and Population Growth: Choices Beyond Childbearing," notes how narrow

assumptions about motherhood serve to close off options for women. "For most women, purpose and accomplishment have been defined largely—and sometimes exclusively—in terms of motherhood...A sound policy must aim to expand women's choices, on the assumption that women are no more naturally inclined to limit themselves to motherhood than men are inclined to limit themselves to fatherhood." The ZPG report concurs, noting that "persistent pressure to reproduce comes down hardest on women. For centuries, the public has adhered to the notion of a 'maternal instinct.' In fact, research conducted by anthropologists suggests that motherhood is a learned, rather than inherent response."

In her book, *Whatever Happened to Ecology?*, environmental writer Stephanie Mills discusses her own decision not to have children. Says Mills, "Unless I assume the power and responsibility to do things in my life according to ecological values, I can hardly expect the world around me to change. Although civilization affords little support, even less spiritual direction, and vitually no role models for women who choose not to be mothers, there does seem to be an ecological demand for them." And for men who choose not to be fathers as well.

— *Leslie Pardue*

purer water or less traffic. Conventional wisdom says that one cures shortages by increasing supply. But whenever a growing population is involved, it is not possible to cure a shortage by increasing the supply.

Since altering the supply won't reduce the need, we must work on the demand aspect of the "Impact Equation." In rich countries this means that the demand per person must be reduced; in poor countries, where per person demand can hardly be lowered further, it is the size of the population that must somehow be reduced.

Environmentalists who want to save wilderness should define these amenities as part of the safety factors of society. Such a policy could be called selfish. But if we are concerned with what we hand down to our children, we must seek to protect these amenities from being destroyed by the present generation. We must face up to the problem of lowering the growth rate of most societies. We would also be wise to decrease the per capita resource use in the Western world especially.

Now we see good reasons why most environmentalists have for so long steered clear of population questions. Dimly or clearly, they have realized that tackling the problem of population control would be painful, and that proposing truly effective measures would result in accusations of selfishness, provincialism, bigotry and even racism (the all-purpose curse of our generation). But, since increasing the supply becomes increasingly more diffi-

Organizations Working On Population Issues

National Audubon Society— Population Program
Government Relations Office
801 Pennsylvania Avenue, SE
Washington, DC 20003
(202)547-9009

Planned Parenthood Federation of America
801 Seventh Avenue
New York, NY 10019
(212)541-7800

Population Crisis Committee
1120 19 Street, NW
Suite 550
Washington, DC 20036
(202)659-1833

Population—Environment Balance
1325 G Street, NW
Suite 1003
Washington, DC 20005
(202)879-3000

Population Institute
110 Maryland Avenue, NE
Washington, DC 20002
(202)544-3300

Sierra Club—International Population Program
Attn: Nancy Wallace
408 C Street, NE
Washington, DC 20002
(202)547-1141

Zero Population Growth
1400 16 Street, NW
Suite 320
Washington, DC 20036
(202)332-2200

cult, and since doing so leads only to greater demands, we human beings must somehow find the means and the will to tackle the demand side of the population equation. This necessity cannot be evaded. It is understandable that environmentalists should have for so long avoided population questions; but is it excusable for them to continue to do so? Surely not, if they are interested in the welfare of their own children—and of the children of all humanity.

My prediction? After we have succeeded in jettisoning the partial taboo imposed on population during the past decade, and after we have made enough people aware of the fundamental truth of the Impact Equation, we will then be in a position to do something about population problems. If we do, the celebrations that take place on Earth Day 2010 should be quite different from the ones that took place this year. The topics of population and environment just might be honestly joined.

Population
Politics

*N*o challenge is more urgent

than stabilizing world population. The knowledge

and technology exist to reach that objective.

Only the will is missing.

Werner Fornos

WERNER FORNOS is the president of the Population Institute, a Washington, D.C.-based nonprofit organization dedicated to a more equitable balance among the world's population, resources, and environment. Fornos formerly headed George Washington University's Global Population Information Program. He has served as a consultant on family-planning programs for the Agency for International Development, the American Public Health Association, and Westinghouse Health Systems and has worked on family-planning projects for Tunisia, Pakistan, Bangladesh, Mexico, Turkey, the Philippines, Indonesia, China, Sri Lanka, and Kenya.

United Nations demographers have revised upward their estimate of the most probable point at which human population will level off. The latest projection is 11 billion, a full billion higher than the 1989 estimate. And U.N. projections, based on government figures, tend to be conservative. Without considerably more support for family planning, the world's population will likely stabilize closer to 14 billion in the next century. Humanity is growing much too fast for its own good.

In 1830, 1 billion people inhabited the earth. A century passed before the population reached 2 billion. Thirty years later, in 1960, it hit 3 billion; 15 years later, 4 billion; and by 1986—only 11 years later—5 billion. Global population now stands at 5.4 billion, and the 6 billion mark could be reached in 1995.

Within a generation, 3 billion young people, virtually all of them in the developing world, will reach their reproductive years. Since most of these potential parents are already alive, their threat to the planet is not conjecture. Their numbers portend environmental disaster, hunger and starvation, epidemics, monumental unemployment, and civil strife and disorder.

As we watch the evening news and the tragedy of Ethiopians herded into food camps, we should keep in mind that the average Ethiopian woman has seven children during her lifetime. A Population Institute study concluded that had Ethiopia launched a family-planning program in the mid-1960s, and had it been half as successful as many begun at that time, the number of births prevented would have equaled the number of Ethiopians dependent on relief in the last famine. Despite losing millions of people to famine, drought, and war in the 1980s, Ethiopia's population of 48.3 million is projected to double in 23 years.

Consider the following:

☐ In 1989, for the third year in a row, the world consumed more food than it produced. The gap between production and consumption can be measured in declining grain reserves, supplies set aside for future shortfalls. In 1986, the reserves topped 459 metric tons, enough to feed the world for 101 days. Since then, consumption has outpaced production and the U.S. Department of Agriculture estimates that world food reserves would last only about 30 days.

☐ Even more revealing is declining per capita food production, the yardstick of humanity's ability to feed itself. This drop, a result of weak investments in agriculture combined with spiraling population, is common throughout the developing world. In Africa, the continent where population growth is fastest, per capita food production is in the sharpest decline.

The surge of human expansion throws dangerously
out of kilter the intricate ecological balance that sustains life.

☐ More than 90 percent of world population growth occurs in the developing world—the nations that are least able to feed, educate, employ, and otherwise provide for their people. More than 1 billion people already live in absolute poverty. Nearly 3 billion people have inadequate sanitation, some 1.5 billion lack primary health care, and more than 1.7 billion don't have access to a safe source of water. One in three children under the age of five is seriously malnourished. The Third World, where unemployment often hits 30 percent or more, must produce 800 million new jobs in the next decade to accommodate its burgeoning numbers.

By diluting modest gains among an ever-increasing pool of citizens, runaway population growth strangles hopes for economic progress in the Third World. It floods already crowded schools and job markets with the young. And where family-planning programs are unavailable, it needlessly exposes millions of women to the physical hardships of unwanted pregnancy and childbirth.

The industrialized world will not be immune from the fallout. And as worthwhile causes line up for foreign aid and humanitarian assistance, the United States will have to set priorities. With rapid population growth causing or exacerbating so many other global problems, no challenge is more urgent than stabilizing world population. We have the knowledge and technology, by no means perfect but certainly adequate, to reach that objective in a rational, humane manner within our lifetime. All that is lacking is political will, the commitment of leadership.

Population and the Environment

For all its destabilizing impact on Third World economies, societies, and political institutions, rapid population growth is perhaps even more devastating to the global environment. The continuing surge of human expansion throws dangerously out of kilter the intricate ecological balance that sustains life.

In her 1990 *State of World Population* report, Nafis Sadik, executive director of the United Nations Population Fund, notes that rapid population growth in poor countries has already "begun to make permanent changes to the environment" that will reach critical levels in this decade. "At the start of the 1990s, the choice must be to act decisively to stop population growth, attack poverty, and protect the environment. The alternative is to hand on to our children a poisoned inheritance."

Part of the problem is the developing world's swift urbanization. By the year 2000, 18 of the 20 largest cities will be in the Third World. Mexico City epitomizes the massive pollution associated with rising megacities. Its densely crowded ghettos, snarled traffic, faulty or nonexistent sewage and waste-disposal systems, and sprawling, inefficient industries all contribute to the end product: a polluted, water-short, smog-ridden city whose population doubles every 20 years.

Urbanization relates to a second key factor in the ecological transformation brought about by population growth: the rapid destruction of the world's forests. Enough timberland to cover 40 Californias will have disappeared within the final two decades of this century.

Three main causes propel the wide-scale razing of trees and scrubwood: the push to convert land for human settlement or agriculture, the desire to harvest forests as a quick cash crop, and the reliance on wood as a primary fuel source. Surging population growth exacerbates all three. More people demand more fuel and food, driving the conversion of forest to farmland. And by overwhelming economic development plans, population growth deters the shift to other fuels.

The Third World as a whole—both industries and families—still relies on wood for half its energy needs.

HUNGER ZONES

In the shaded areas, devoting all farmland to subsistence agriculture would still not feed the population.

Families are especially dependent, relying on wood for 70 percent of their heating, cooking, and other fuel needs. With fuel prices soaring due to the Iraqi crisis, the percentage could climb even higher. As cities grow, their inhabitants look far afield for wood, creating vast rings of wasteland around many urban centers.

In Africa, deforestation is dramatic, and the taking of scrubwood has been particularly severe in northern parts of the continent. In 1900, trees and brush covered 40 percent of Ethiopia; today, as a result of population growth and the need for fuel, only 4 percent of Ethiopia's forests remain.

Once deprived of vegetation cover, the sandy soil of the African Sahel is especially vulnerable to erosion. Consequently, its vast deserts expand six miles a year. Studies by the U.S. National Research Council and the Congressional Research Service have linked population growth in the Sahel to soil degradation, food shortages, and decreasing agricultural productivity. Worldwide, by the year 2000, deserts will consume an additional area one and a half times the size of the United States.

In Nepal, population pressures are driving people to cultivate marginal land. Farmers carefully tend slopes in the Himalayas so steep that the soil is promptly washed away in the next seasonal monsoon. Nepal's major "export" is its precious topsoil, although the country earns not a dime of foreign exchange for its sacrifice. In fact, 25 billion tons of arable topsoil vanish from the world's cropland every year.

While we can only estimate the absolute limits of the earth's sustainability, we know that by the year 2000, acute fuel shortages will affect 350 million people. Sixty-five countries that depend on subsistence farming may be unable to feed their people, ten more than now.

Food: The Limits of Production

Since the end of the Second World War, food production has nearly tripled, fostering the mistaken belief that technology will always outpace demographic expansion. But Norman Borlaug, who won the 1970 Nobel Peace Prize for pioneering the Green Revolution, vehemently disagrees. Borlaug says the Green Revolution "bought us some time" but that global agriculture production can't double in the next 40 years to match population growth. He believes the only hope to feed the world four decades hence is "by coming to grips with the population monster."

Like Borlaug, most scientists and economists have come to realize the limits of the Green Revolution, as agricultural productivity slows even in nations that pioneered it. Grain harvests in India, Mexico, and most of the rest of the developing world leveled off in the 1980s. In 1988, India again began importing food, and harvests are now shrinking in many Third World countries. Africa was once a grain exporter, but for 20 years its food production has grown only half as fast as its

By the year 2000, deserts will consume an additional area one and a half times the size of the United States.

population. Africa's food production increases 1.2 percent annually, its population 3.1 percent. Nearly 200 million Africans—almost a third of the population—are hungry and malnourished.

Third World agricultural yields are suffering because these nations are coming up hard against limits inherent in nonsustainable agriculture. Many of the earlier spectacular productivity gains were achieved by exploiting land in ways that yielded quick but short-lived results. Putting marginal lands into production boosted yields temporarily but eventually eroded soil and robbed land of its vitality. Chemical fertilizers and pesticides also helped in the near term but damaged the long-term productivity of the land.

Water supplies are similarly finite. Irrigation systems that tapped into limited groundwater supplies allowed farmers to water crops—until the water ran out. One need look no farther than the Great Plains to see the limits that water shortages impose on farming. The problem is just as serious in other parts of the world. As water tables drop in the major agricultural regions of the United States, China, and elsewhere, many acres will have to be withdrawn from irrigation. By the year 2000, freshwater shortages in many developing coun-

tries will have increased by 35 percent.

Many nations, both developing and industrialized, realize the drawbacks of an approach that seeks only to raise production. Under the Food Security Act of 1985, the United States is withdrawing 11 percent of its cropland from production because it is susceptible to erosion and should never have been farmed. In recent years, Mexico has lost one out of every ten acres it had in grain production either because the land was overworked or because encroaching population took it for other uses.

The same pressures are reducing cropland in virtually every nation. China, the world's second largest food producer, and the Soviet Union have each lost 13 percent of their grain-producing cropland over the past dozen years to erosion, soil degradation, and nonagricultural uses. The rate is just as startling in India, where the government estimates that overcultivation is deteriorating more than 6 million acres annually. At the same time, overgrazing and the taking of fuelwood are converting 2.5 million acres of Indian pastures and cropland and 3.7 million acres of forest into wasteland every year.

The net result is that the world must feed a growing population with less cropland. Global agriculture is literally losing ground to population growth.

At the close of the twentieth century, we find two worlds rapidly diverging. While the Northern Hemisphere implements extensive efforts to control or mitigate pollution, the Third World is pressing to industrialize and can't afford the often costly procedures needed to minimize the pollution that results.

Where population growth is stable, economic growth is possible, and money to protect the environment is available. In the affluent, industrialized world, nations are making some progress in efforts to undo earlier damage to the environment. In Europe and North America, urban air may not be clean, but its quality is often better than it was five years ago. Efforts are under way to clean up bodies of water from the Mediterranean Sea to Lake Erie, and the water in many streams and rivers is cleaner than it was in the 1970s.

There is, however, less call for optimism in the developing world. When population doubles each 20 or 30 years, there is rarely enough economic vitality to feed and clothe people, much less to implement farsighted environmental plans. In this world, surging numbers of people combined with soaring debts thwart economic progress. Here, there are few signs that deforestation, pollution, and other unsound environmental practices will abate in the near future.

The Risk of Being Wrong

The woes faced by the developing world are not its alone. Ecological disasters have little respect for national boundaries. Once unleashed by neglect in one region, they spread, disrupting ecological balances until they alter the entire global environment. For example, most experts agree that the fuel and consumer demands of rapidly proliferating humanity contribute to depleting the ozone layer, the earth's shield against the sun's ultraviolet rays. And by the end of this century, the air we breathe will contain a third more carbon dioxide than now, thereby contributing to global warming. And we know that these problems are exacerbated by a global population too big for its resource base.

The industrialized nations also have a great stake in the struggle to feed the human family. The humanitarian imperative is clear enough, but those who require a less charitable incentive need only consider the political implications of widening hunger in the Third World. Many developing nations, including important U.S. allies, have long walked the edge of hunger, dependent on foreign subsidies to see them through. In Egypt, food subsidies eat up 14 percent of the national budget. Over the past 12 years, the United States has provided Egypt with wheat and corn worth more than $3 billion. Every third loaf of bread baked in Egypt is made from wheat and corn ground in the United States.

Such dependence is expensive to the United States and perilous to the recipient nation. The uncertainty of food supplies can easily explode into violence and

IT IS BETTER TO HAVE ONE CHILD ONLY 只生一个

unrest when shortages or price hikes trigger panic or anger. Egypt, Sudan, Tunisia, and Morocco are but a few of many nations rocked by bloody food riots in recent years. It is in the interest of the United States to encourage agricultural self-sufficiency and stability in its developing neighbors. It is no answer to point out that the United States and Europe normally have food surpluses. There has yet to be developed a perfect system for distributing extra food to the world's poor while retaining economic incentives for U.S. and European farmers to continue to grow surpluses.

These crises call for industrialized nations to assist their Third World neighbors in slowing population growth. Toward that end, representatives of 79 countries met in Amsterdam in November 1989 to discuss international population initiatives for the next century. They concluded that global commitment—or the lack thereof—to vastly widening the availability of family planning in the coming decade will determine the ultimate size of the world's population. Failure to make family planning significantly more accessible would be humankind's ultimate blunder.

Indeed, when the Duke of Edinburgh delivered the 1990 Rafael M. Salas Memorial Lecture to the United Nations, he warned that the "fuse of the population bomb has already been ignited, and the consequences . . . for the future of the world will be a great deal more devastating than any nuclear holocaust." He stressed that "common prudence would therefore suggest that the risk of being wrong"—the risk of overestimating the earth's carrying capacity—"is not worth taking."

The United States: The Problem or the Solution?

The risk of being wrong is precisely the irresponsible game the U.S. government has played for six years. In the mid-1980s, the Reagan administration concluded that population growth isn't a factor in development, a message James Buckley, leader of the U.S. delegation to the 1984 International Conference on Population in Mexico City, delivered to that forum. The conferees listened in stunned disbelief.

For nearly two decades, the United States had led the

chorus telling the world's poorest countries the exact opposite: population growth had to be balanced with natural resources before sustained economic progress could be achieved. At the 1974 World Population Conference in Bucharest, U.S. delegates and representatives of other industrialized nations had implored developing countries to lower their fertility rates. At that time, many Third World and Eastern bloc delegates had suspected that the very idea of reducing population growth was a conspiracy by former colonial masters to suppress emerging nations, ensure the exploitation of their resources by the industrialized world, and limit the strength of Third World armies. These nations equated population with power, prestige, and prosperity.

Ten years later, at the International Population Conference in Mexico City, those misconceptions had all but vanished. Population growth had declined from 15 to 50 percent in countries as diverse politically, culturally, and geographically as Colombia and Sri Lanka, Thailand and Costa Rica, Indonesia and Zimbabwe, South Korea and Mexico. Virtually every decline was predicated on a leadership commitment to a voluntary reduction of population growth and a strong national family-planning program.

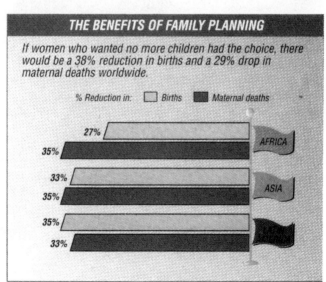

THE BENEFITS OF FAMILY PLANNING

If women who wanted no more children had the choice, there would be a 38% reduction in births and a 29% drop in maternal deaths worldwide.

% Reduction in: ☐ Births   ■ Maternal deaths

27% — 35%   AFRICA
33% — 35%   ASIA
35% — 33%

CHART SOURCE: FAO   PHOTO: UNITED NATIONS/JOHN ISAAC

2. WORLD'S POPULATION

Every day, Romanians have 3,000 abortions.

Meanwhile, pro-life groups in the United States oppose funding family-planning efforts there.

Almost every African country had been unreceptive to birth-control programs in 1974, but by 1984 most nations on the fastest-growing continent had launched family-planning projects or were contemplating launching them. Today 91 percent of the world lives in countries that directly support access to contraception. When the United States changed its mind, the developing world had already committed itself to programs aimed at lowering fertility or maternal and child mortality and morbidity. Simultaneous progress in development in such countries as Indonesia, South Korea, and Thailand was already evident.

The abrupt U.S. turnaround on international population policy in 1984 was an effort to placate domestic abortion opponents. Until then, the United States had been the primary donor to both the United Nations Population Fund (UNFPA) and the International Planned Parenthood Federation (IPPF). These were, respectively, the largest multilateral agency and the largest private voluntary organization providing family-planning services to developing countries. With funds other than those contributed by the U.S. government, IPPF affiliates provided voluntary abortion services in 11 nations. UNFPA has never funded or engaged in abortion activities. In any case, the 1973 Helms amendment to the foreign assistance act prohibits both organizations from using U.S. funds for abortions.

Unable to impose their agenda in their own country, extremists in the Reagan administration and in Congress chose to victimize poor women in the developing world with the 1985 Kemp-Kasten amendment to the foreign assistance act. The amendment decreed that the United States would not contribute to "any organization or program" that "supports or participates in the management of a program of coercive abortion or involuntary sterilization."

Since the approval of Kemp-Kasten, the U.S. government has refused to contribute to the UNFPA on grounds that the organization "supports or participates in the management" of the Chinese program. UNFPA was providing support to China's family-planning program, which allegedly forced women to have abortions to lower the fertility rate. Leading Chinese officials have repeatedly denied that the nation's population program is coercive. Furthermore, probes by the U.S. Agency for International Development (AID) failed to uncover evidence that UNFPA funds were used to promote or perform abortions, coercive or otherwise. In fact, UNFPA contributes only about $10 million to China's annual $1 billion family-planning program, hardly enough to make a difference in management.

While the United States provides bilateral population assistance to some 40 countries, the United Nations has population programs in 140 nations. And while U.S. population funds flow into other developing countries through private organizations, the fact remains that the United States has severely reduced its contribution to IPPF. By pulling out of U.N. population efforts, the United States has, in effect, turned its back on multilateral solutions to overpopulation. The multilateral approach is significant because no-strings-attached aid is better accepted by developing nations.

The Real Coercion

The U.S. withdrawal from the UNFPA is a classic Catch-22. Logically, the largest multilateral provider of family-planning services *should* provide population assistance to the most populous nation in the world. Moreover, the demand for abortion might actually increase in China if the UNFPA ceases its family-planning work there. Strangely, that is precisely the condition the U.S. government has established for resuming its contributions to UNFPA. And the Bush administration rigidly maintains the Reagan policy. In November 1989 Congress attempted to restore funds for UNFPA, with the stipulation that funds not be spent in China. Bush vetoed the entire foreign aid bill because the UNFPA would have received funds.

Even Romania has become a pawn in the U.S. battle over abortion, not because Romania wants to continue its current rate of some 100,000 abortions per month, but rather because it wants to prevent them.

Under Nicolae Ceausescu, Romania had outlawed

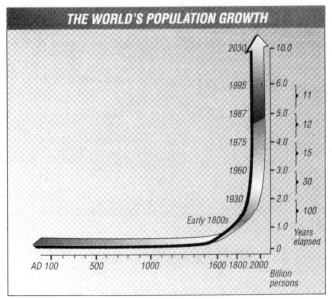

THE WORLD'S POPULATION GROWTH

CHART SOURCE: FAO

The Save the Children Federation advises this village family-planning council in rural Bangladesh.

PHOTO: UNITED NATIONS/PHILIP TEUSCHER

family planning. The dictator believed that a high fertility rate was key to building a stable and prosperous nation. After his fall, demands for abortion deluged Romanian hospitals, reaching 3,000 a day. UNFPA and IPPF moved in with emergency family-planning programs.

In July 1990, Rep. Bill Lehman (D-Fla.) and Rep. John Porter (R-Ill.), the latter a strong Bush supporter, sought to ensure that $1 million would go to UNFPA and IPPF for nonabortion family-planning efforts in Romania. But abortion foes Rep. Christopher Smith (R-N.J.) and Rep. Henry Hyde (R-Ill.) pushed through an amendment disallowing funds for Romania from going through UNFPA and IPPF. Smith insisted that four private groups could do the job.

That might have been acceptable had it been true; it was not. None of the four agencies Smith named had programs in Romania. None had plans to institute programs there. Two of the agencies even recommended UNFPA and IPPF.

When President Bush threatened to veto the Romanian aid bill if UNFPA and IPPF were the executing agencies, Rep. Richard Durbin (D-Ill.) described the position of pro-life groups and the president as "not only naive but morally and logically inconsistent." Although Durbin usually votes for anti-abortion legislation, he charged that "because this administration and some members [of Congress] have nothing short of a vendetta against two organizations who cannot use this money for abortion, they would deny the funds to

Romania. Abortions will continue in that country at the rate of 3,000 a day while so-called pro-life groups fight over the efforts to bring family planning to Romania."

But Romania is only one among many pawns in the U.S. ideological war over abortion. The Bush administration's bizarre stance on this issue means that more than 100 nations are denied U.S. population assistance through the UNFPA.

Fortunately, other UNFPA contributors reject this illogic. No nations, even those most closely allied with the United States, have followed the U.S. example. Germany, Canada, the United Kingdom, Japan, and the Scandinavian countries have actually increased their UNFPA donations. They understand the devastating consequences of rapid population growth and are determined to fill some of the void until the U.S. government regains some semblance of sanity.

Concerns about coercive population measures are legitimate, but if the high fertility rates in the Southern Hemisphere don't slow soon, coercion to one degree or another will be the tragic rule, not the exception. Country after country will conclude that they simply cannot allow human numbers to double every two or three decades. They will find their only option is to institute draconian laws to limit family size. There can be little security or stability in countries that resort to such drastic actions. And what security or stability can there be in a world with 20 or more such countries?

The next time Congress debates the question of coercion, it would do well to keep in mind the results of the World Fertility Survey, a mammoth study conducted during the 1970s by the International Statistical Institute, headquartered in the Hague. Half a billion women in the developing world want to limit the size of their families but can't effectively do so because they lack access to birth control or can't afford it. In 60 percent of the developing world, half the population is actually denied birth control because the demand for family planning far exceeds the services that providers can now supply. According to the United Nations, 75 percent of Latin American women not planning their families would like to do so, 43 percent of Asian women, and 27 percent of African women. In Africa, the percentage would be much higher were it not for the fact that 90 percent of the women in 10 African countries haven't even heard of contraception.

In a world with a variety of effective and inexpensive contraceptive methods, this is shameful. Denying a vital need that can be easily provided amounts to a singularly repugnant form of coercion.

The Highest Priority

Ten years ago, it was reasonable to expect that with a concerted effort the earth's population would level off at 8 billion. But even the U.N.'s estimate of 11 billion will be overshot unless population stabilization becomes a top priority on the international agenda.

An international population strategy must aim to lower the fertility rate to 2.4 children per woman, the level required for stabilizing population at present death rates. The world can achieve that goal by the year 2000 if the proportion of couples in developing countries using contraception rises from the current 40 percent to 72 percent, according to the World Bank.

In view of the vast unmet demand for family planning, efforts must be accelerated to establish, fund, and implement effective national population programs and to increase funding for population assistance programs. Population and family-planning assistance from all sources—multilateral, bilateral, private, and host-government allocations—totals approximately $4.5 billion, according to the U.N. The experts meeting in Amsterdam concluded that the total must double to $9 billion by the end of this century to provide a hope of holding down global population. The United States and other donors must take the lead by at least doubling their contributions.

Even so, the nations of the Third World can never make significant strides toward self-sufficiency until they also shrug off crushing foreign debts. But Indonesia and Brazil, the fifth and sixth most populous countries in the world, are cutting and selling wood from their ecologically vital rain forests to help pay the interest on current debts. One way to accomplish two objectives at once is to tie debt forgiveness to sound and farsighted policies to halt and reverse needless destruction of the environment. These would help our Third World allies advance their economies while writing off loans U.S. banks can't realistically hope to collect.

Similarly creative foreign-debt programs should be instituted to encourage Third World governments to expand educational and employment opportunities for women. Studies have repeatedly documented the relationship between declining birth rates and women's rising economic and social opportunities. Debt forgiveness should be used to extend at least an eighth-grade education to all women and to attack discriminatory barriers in education and employment that encourage women to stay home and bear children rather than enter the workforce.

The industrialized nations should jointly establish an Urban Incentive Fund, to be administered by UNFPA, to assist developing nations committed to alleviating the flow of migrants into the Third World's megacities. The Urban Incentive Fund could provide low-interest loans that would enable developing countries to establish well-planned "magnet cities" outside of congested urban centers to accommodate rural migrants. The loans could also finance transportation, housing, water, and sewage projects to ensure that the magnet cities do not repeat the environmental travesties of the primary urban conglomerates.

A major step for the United States would be to enact a bill like one that Rep. Peter Kostmayer (D-Pa.) and Rep. Claudine Schneider (R-R.I.) introduced in 1990, which would have authorized $500 million for population programs overseas. The measure would have earmarked $60 million or 16 percent of the total authorization, whichever is less, for the UNFPA. The no-strings-attached multilateral approach of the UNFPA has won the trust and confidence of the entire developing world. For instance, UNFPA provides population assistance to 140 developing countries, while the United States provides bilateral aid to 40. Only through a coordinated, concerted effort encompassing multilateral and private voluntary organizations, as well as bilateral programs, will population growth be significantly slowed.

Time and again, Congress has demonstrated its comprehension of the consequences of overpopulation and its dedication to bringing population into a better balance with global resources. Indeed, Congress showed its concern when it voted to resume the U.S. contribution to the UNFPA before President Bush's veto scotched it. For the sake of poor people throughout the world who are virtually forced to have babies they do not want, regardless of the risk to the lives and health of both the children and their mothers, the United States must not evade its responsibility.

We could demonstrate our resolve not only by increasing population assistance but also by consolidating the 48 separate entities within AID's Office of Popula-

During the 40 years of the Cold War,

there was always money to end lives but seldom enough to improve the quality of life.

tion into a centralized organization. The Office of Population operated much more efficiently and effectively in the 1970s, when it was a centralized unit. During that decade, fertility declined more dramatically in several countries receiving U.S. population assistance than it has since. In addition, the president could appoint an ambassador-at-large for population matters, a career diplomat committed to the idea that reining in population growth is essential to a nation's development.

During the 40 years of the Cold War, somehow there was always money to end lives but seldom enough to improve the quality of life. Even given the high priority placed on reducing the national debt, the United States can certainly afford more than the current $300 million for humanitarian family-planning aid in a foreign assistance budget totalling $15.5 billion. While such aid is frequently overlooked or ignored, there is no better hope for global stability and security than establishing universally available, voluntary family-planning services, slowing the rate of population growth, and reducing the destruction of the world's resources.

FEEDING SIX BILLION

*Population growth is exceeding farmers' ability to keep up.
How do we reach a balance between food and people?*

LESTER R. BROWN

Lester R. Brown is president of the Worldwatch Institute and former administrator of the U.S. Department of Agriculture's International Agricultural Development Service.

ur oldest enemy, hunger, is again at the door. We're exploring the outer reaches of the solar system, reaping the benefits of the computer revolution, and working wonders in medicine, but our ingenuity can't seem to resolve this age-old problem. With more hungry people in the world today than when this decade began, there's little to celebrate on the food front as we enter the nineties.

Between 1986 and 1988, drought-damaged harvests in key producing countries dropped world grain stocks to one of their

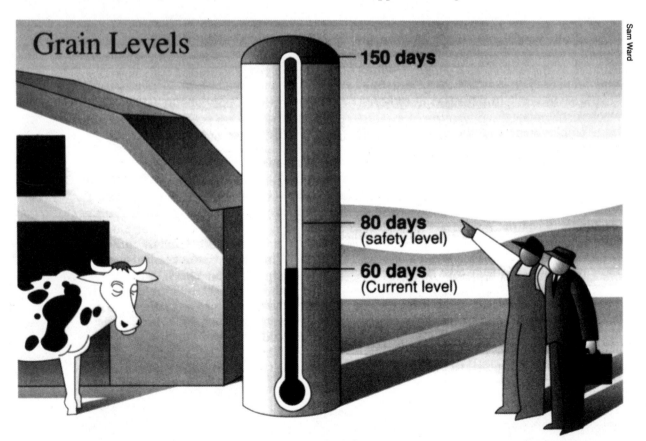

lowest levels in decades—little more than is necessary to fill the "pipeline" from field to table. As a result, prices increased by nearly half.

With higher prices and better weather in 1989, it was widely assumed that production would surge upward and stocks would be rebuilt. But this is not happening.

According to the July estimate of the U.S. Department of Agriculture, this year's harvest will be 13 million tons below the projected consumption of 1,684 million tons. The world's depleted grain stocks will not be rebuilt this year. If stocks cannot be replenished even with good weather, when can they be?

Rebuilding stocks depends on pushing production above consumption, but this is becoming more difficult. A worldwide scarcity of cropland and irrigation water, combined with a diminishing response to the use of additional chemical fertilizer, is slowing the growth in world food output. Meanwhile, 88 million mouths are added to the world's population each year. Now there's also evidence that the cumulative effects of environmental degradation are showing up at harvest time. Deforestation, for example, is leading to increased rainfall runoff and crop-destroying floods. Soil erosion is slowly undermining the productivity of one-third of the world's cropland. Air pollution and acid rain are damaging crops in industrial and developing countries alike. Data from experimental plots indicate that yields of some crops, such as soybeans, are reduced by the increased ultraviolet radiation associated with stratospheric ozone depletion. Global climate change in the form of hotter summers also may be taking a toll.

We knew that we couldn't continue to damage our life-support systems without eventually paying a heavy price, but how would we be affected? When would we know we were in trouble?

We may now have the answer. Food scarcity is emerging as the most profound and immediate consequence of global environmental degradation, and it is already affecting the welfare of hundreds of millions of people. The 1988 rise in world grain prices may have been a signal of trouble ahead.

The Environmental Wild Card

After a generation of record growth in world food output following World War II, it sometimes seemed that the rapid ascent in food production could continue indefinitely. Between 1950 and 1984, the world grain harvest expanded some 2.6 times, or nearly 3 percent per year, raising per-capita grain production by more than one-third. But, in the half decade since, output rose only 0.2 percent per year. This five-year period is obviously too short to show a trend, but it does suggest an unsettling slowdown in food output, one that is partly attributable to environmental degradation.

Among the environmental trends adversely affecting agriculture, soil erosion tops the list. As the demand for food has risen in recent decades, so have the pressures on the earth's soils. Soil erosion is accelerating as the world's farmers are pressed into plowing highly erodible land, and as traditional rotation systems that maintain soil stability break down.

Some one-third of the world's cropland is losing topsoil at a rate that undermines its future productivity. An estimated 24 billion tons of topsoil washes or blows off the land annually—roughly the amount on Australia's wheatland. Each year, the world's farmers must try to feed 88 million more people with 24 billion fewer tons of topsoil.

This loss is beginning to show up in diminished harvests. Studies undertaken in the U.S. Corn Belt conclude that the loss of one inch of topsoil reduces corn yields from 3 to 6 bushels per acre, or an average of 6 percent. Wheat yields follow a similar pattern.

Soil erosion and cropland loss in Third World nations is intimately linked with another form of environmental degradation: deforestation. As firewood becomes scarce, villagers begin to burn crop residues and animal dung for fuel, depriving the land of organic matter and nutrients. With less organic matter, the soil's ability to absorb and store moisture decreases, making the land more vulnerable to drought. Further, loss of organic matter increases runoff, thus reducing the percolation of rainfall into the subsoil and the recharge of aquifers.

Increased runoff, in turn, leads to flooding. This is now strikingly evident in the Indian subcontinent, where deforestation is destroying tree cover in the Himalayan watersheds. The area subject to annual flooding in India expanded from 47 million acres in 1960 to 124 million acres in 1984, an area larger than California. Accelerated runoff as a result of deforestation was evident in early

September 1988, when two-thirds of Bangladesh was inundated for several days. That flood, the worst on record, led to extensive crop damage.

Damage to crops from air pollution now is measurable in the automobile-centered societies of Western Europe and the United States and in the coal-burning economies of Eastern Europe and China. A U.S. Environmental Protection Agency study estimates

Table 1.

Year	Population	Increase by Decade	Average Annual Increase
	(billions)	(millions)	(millions)
1950	2.565	N. A.	N. A.
1960	3.050	485	49
1970	3.721	671	67
1980	4.477	756	76
1990	5.320	843	84
2000	6.241	921	92

World Population Growth by Decade, 1950-90 with Projections to 2000

Source: Francis Urban and Philip Rose. *World Population by Country and Region, 1950-86, and Projections to 2050.* U.S. Department of Agriculture.

that ground-level ozone spawned by fossil-fuel burning is reducing the U.S. corn, wheat, soybean and peanut harvests by at least 5 percent. Walter Heck, the U.S. Department of Agriculture representative on the EPA study panel, estimates that cutting ground-level ozone by half would cut crop losses by up to $5 billion.

All plants and animals are affected to some degree by the increased exposure to ultraviolet radiation resulting from depletion of the stratospheric ozone layer. Data gathered from experimental plots by professor Alan Teramura at the University of Maryland indicate that each 1 percent increase in ultraviolet radiation reduces soybean yields by a like amount. This suggests that the worldwide depletion in the ozone layer, roughly 3 percent over the last two decades, now may be reducing the output of soybeans, the world's leading protein crop. Unfortunately, no one is monitoring this loss to determine its precise dimensions.

The effect of hotter summers on world food output can be estimated from the computerized projections of global climate change. As global warming progresses, farm output could be cut sharply in North America and Central Asia, the regions of the earth likely to experience the greatest temperature rise. If the summer of 1988, which

reduced U.S. grain production below domestic consumption for the first time in modern history, is a glimpse of what summers will be like, then the days of the North American breadbasket could be numbered.

The Population Factor

Even while wrestling with the new uncertainties associated with hotter summers, farmers recently learned that they may have to feed more people than they had reckoned. The United Nations Population Fund announced this May that U.N. demographers have revised their earlier projections of world population upward, chiefly because of failed family planning efforts. Instead of leveling off at 10 billion, world population will settle at 14 billion. For a world that can't adequately feed 5.2 billion inhabitants today, this comes as sobering news.

Some progress has been made in scaling back population growth from an annual rate of close to 2 percent in 1970 to 1.7 percent today, but this decline has been so slow that the annual increment is actually going up. The world is now projected to add 921 million people during the nineties, the largest increment ever for a decade [see Table 1].

In many developing countries, soaring population now has a dual effect on food balance: it increases demand as it degrades the agricultural resource base. For instance, crowded cities and villages create a need for firewood that exceeds the sustainable yield of local forests. Deforestation is the outcome, which in turn increases rainfall runoff and soil erosion. Once started, this vicious cycle is hard to stop. As a result, the agricultural base for hundreds of millions of people is deteriorating on a scale whose consequences are fearful to imagine.

As a result of continuing rapid population growth and slower growth in world grain output, grain production per person has fallen sharply during the late eighties, interrupting the long-term gradual rise since mid-century. Thus far, this fall has been cushioned by consuming part of the record level of grain stocks on hand when the decline began. But, even after drawing down stocks, per-capita grain consumption fell 2 percent between 1987 and 1989.

Nowhere is the agricultural breakdown more evident than in Africa. With the fastest population growth of any continent on record, a combination of deforestation, overgrazing, soil erosion, and desertification has helped lower per-capita grain production some 20 percent from its peak in the late sixties. This fall has converted the continent into a grain-importing region, fueled the region's mounting external debt, and left millions of Africans hungry and physically weakened, drained of their vitality and productivity.

Sadly, there is nothing in sight to reverse this fall in African living standards. World Bank analysts assessing several scenarios for Africa's future label the one based on a simple extrapolation of recent trends "the nightmare scenario."

Nowhere to Grow

From the beginning of agriculture until mid-century, growth in the world's cultivated area more or less kept pace with that of population. After that point, the growth in cultivated area slowed to a crawl. After falling in the mid-eighties, it recovered somewhat as the United States returned to production cropland previously idled under farm commodity programs.

Each year, millions of acres of cropland are lost, either because the land is so severely eroded that it is not worth plowing anymore or because of its conversion to non-farm uses, such as construction of new homes, factories and highways. Losses are most pronounced in the densely populated, rapidly industrializing countries of east Asia, including Japan, South Korea, Taiwan and China. Beijing, in its near desperate efforts to save cropland, is publicly advocating cremation instead of burial. Despite this effort to reduce competition between the living and dead for precious land, and numerous other land-saving measures, nonfarm uses still claim more than a million acres of cropland per year.

Other densely populated countries that are suffering heavy losses include Egypt, Indonesia, India and Mexico. Both the Soviet Union and the United States are pulling back from their rapidly eroding land. A combination of abandonment of badly eroded cropland plus an increase in alternate-year fallow has reduced the Soviet grain area some 13 percent since 1977. Fearing a similar fate, the United States adopted a five-year plan beginning in 1986 to plant 40 million acres of rapidly eroding cropland to grass or trees.

Worldwide, the potential for expanding the cultivated area profitably is limited. A few

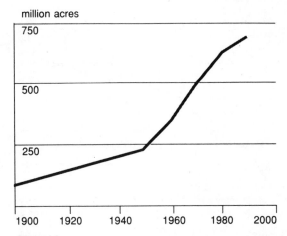

million acres

Figure 1.
World Irrigated Cropland, 1900–1989

countries, such as Brazil, will be able to add new cropland but, on balance, gains and losses for the nineties will offset each other, as they have during the eighties. Food for the 921 million people to be added during the nineties will have to come from raising land productivity.

The prospect for expanding the irrigated area is only slightly more promising. After growing slowly during the first half of this century, the irrigated area expanded from 232 million acres in 1950 to 615 million acres in 1980, increasing the irrigated area per person by 52 percent [see Figure 1]. Since 1980, however, the growth in world irrigated area has slowed dramatically, falling behind population growth.

Several countries, including this United States and China, actually are losing irrigated land as water tables fall and as water is diverted to nonfarm uses. With the net gain in irrigated land estimated at only 59 million acres during the eighties, the supply of irrigation water per person has shrunk by close to 8 percent. Although the cropland area per person has been falling steadily for decades, this is the first decade in which both cropland area and irrigation water per person have declined.

The Necessary Ingredient

From mid-century on, the increasing use of fertilizer has been the engine powering the growth in world food output. Between 1950 and 1989, world fertilizer use climbed from a meager 14 million tons to an estimated 146 million tons. If, for whatever reason, fertilizer use were abruptly discontinued, world food output would plummet by an estimated 40 percent.

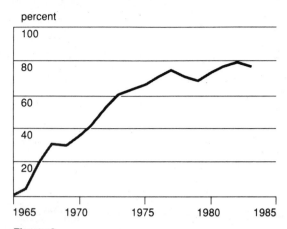

percent

**Figure 2.
Share of Wheat Land Planted to High-Yielding Varieties in India, 1965–1983**

The contribution of irrigation and high-yielding varieties to food output derives heavily from their ability to boost the effectiveness of fertilizer. Investments in irrigation would yield low returns if not for heavy amounts of fertilizer. Hybrid corn, which was initially developed in the United States and has since spread throughout the world, was a winner because it responded readily to heavy applications of nitrogen fertilizer. Third World farmers were attracted to the high-yielding dwarf wheats and rices that brought the Green Revolution, precisely because they were so responsive to fertilizer. Before these varieties came along, fertilizer use on traditional cereal varieties in the Third World was limited.

Rapid growth in fertilizer use depends on rapid growth in irrigation and in the spread of high-yielding seeds. Like the growth in irrigation, the adoption of high-yielding crop varieties follows an S-shaped trend. Over time, adoption by farmers increases slowly at first as a few innovative farmers plant new varieties and then more rapidly as large numbers of farmers see their advantage. Eventually, adoption slows and levels off as the new strains are planted on all suitable land.

The adoption curve for high-yielding wheats in India, which helped to more than double the country's wheat harvest between 1965 and 1972, illustrates this well [see Figure 2]. The spread of these new strains is unlikely to reach all wheatland because much of what's left is semi-arid land, where the new varieties do not fare well. Graphing the adoption of high-yielding corn in the United States, high-yielding rices in Indonesia, or high-yielding wheats in Mexico shows the same S-shaped growth curve.

Once the new fertilizer-responsive varieties are planted on all suitable land, growth in fertilizer use slows.

Such a trend is now apparent in some major food-producing countries. After multiplying several times over between 1950 and 1980, fertilizer use in the United States has leveled off during the eighties. Within the Soviet Union, heavy subsidies have led to the overuse of fertilizer. Recent agricultural reforms, which are keyed to the adoption of world market prices, are expected to actually reduce fertilizer use in 1989. Unlike China in the years following its economic reforms of the mid-seventies, the Soviet Union can't anticipate a huge increase in its grain output because the country is already on the upper part of the S-shaped curve for fertilizer responsiveness.

Many developing countries also are experiencing diminishing returns. Analyzing recent crop trends in Indonesia, agricultural economists Duane Chapman and Randy Barker of Cornell University note that "while 1 pound of fertilizer nutrients probably led to a yield increase of 10 pounds of unmilled rice in 1972, this ratio has fallen to about 1 to 5, at present."

Like the growth of any biological process in a finite environment, the rise in grain yield per acre will eventually conform to the S-shaped curve. So, too, will the response to any input contributing to grain yield, such as fertilizer. Not surprisingly, the fertilizer-use curve appears to be conforming in textbook fashion to the S-shaped growth curve so familiar to biologists [see Figure 3]. Future fertilizer use will expand as varieties with an even greater yield potential are developed, but the output gains are likely to be modest compared with the quantum jump that came from the initial adoption of the high-yielding varieties.

The ultimate constraints on the rise of crop yields will be imposed by the upper limit of photosynthetic efficiency, a limit set by the basic laws of physics and chemistry. As the genetic potential of high-yielding varieties approaches this limit, their response to the use of additional fertilizer diminishes. Advances in plant breeding, including those using biotechnology, can hasten the rise of yields toward the photosynthetic limit, but there is little hope of altering the basic mechanics of photosynthesis.

The Social Fallout

As grain supplies tighten in more and more countries, world prices will rise. A foreshadowing of this development occurred between July 1987 and July 1988, when world grain prices jumped by roughly one-half. They have stayed at that level through July of this year. For many heavily indebted Third World countries, rising grain prices are combining with falling personal incomes to pose a policy dilemma. Price hikes are needed to stimulate output and encourage additional investment by farmers, but the world's poor can't cope easily. Perhaps a billion or more of the world's people already spend 70 percent of their income on food. For many in this group, a dramatic rise in the cost of grain is life threatening.

The social effect of rising grain prices is much greater in developing countries than in industrial ones. In the United States, for example, a $1 loaf of bread contains roughly 5 cents worth of wheat. If the price of wheat were to double, the price of the loaf would increase only to $1.05. However, in developing countries, where wheat is purchased in the market and ground into flour at home, a doubling of retail grain prices translates into a doubling of bread prices. A food-price rise that is merely annoying to the world's affluent can drive consumption below the survival level among the poor.

Even before the recent price increases, the social effects of agricultural adversity were becoming evident throughout Africa and Latin America. A World Bank study, using data through 1986, reported that "both the proportion and the total number of Africans with deficient diets have climbed and will continue to rise unless special action is taken."

In Africa, the number of "food insecure" people, defined by the Bank as those not having enough food for normal health and physical activity, now totals more than 100 million. Some 14.7 million Ethiopians, nearly one-third of the country, are undernourished. Nigeria is close behind, with 13.7 million undernourished people. The Bank summarized the findings of its study by noting that "Africa's food situation is not only serious, it is deteriorating."

Perhaps the most tragic consequence of the deteriorating food situation in Africa and Latin America is the alarming increase in nutrition-related child mortality. In Madagascar, where soil erosion rates exceed even

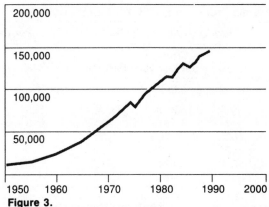

1,000 metric tons

Figure 3.
World Fertilizer Use, 1950–1989

those of Ethiopia and where the deteriorating food situation remarkably parallels that of Africa as a whole, infant deaths climbed by nearly one-fifth in the first six years of this decade.

Throughout Africa, the increase in malnutrition among children is putting additional pressure on already strained medical facilities. A survey at the Lusaka University Teaching Hospital in Zambia showed that of 433 children admitted in one week in late 1987, more than 100 died, most of them because their malnutrition had reached an irreversible stage.

UNICEF reports that in Nigeria between 120,000 and 150,000 children die each year of malnutrition. In its *State of the World's Children, 1988,* UNICEF declares that "for approximately one-sixth of mankind, the march of human progress has now become a retreat and, in many nations, development is being thrown into reverse. After decades of steady economic advance, large areas of the world are sliding backward into poverty."

In *The Global State of Hunger and Malnutrition, 1988,* the U.N.'s World Food Council reports that the number of malnourished preschoolers in Peru increased from 42 percent to 68 percent between 1980 and 1983. Infant deaths have risen in Brazil during the eighties. The council summarized: "Earlier progress in fighting hunger, malnutrition and poverty has come to a halt or is being reversed in many parts of the world."

An Unpleasant Scenario

The world enters the nineties not only with a low level of grain in reserve, but with little confidence that the "carryover" stocks can be rebuilt quickly. Sketching out the consequences of a poor harvest when stocks are

Table 2.

	World Grain Production			Per Person	
Year	Total	Change per Decade	Percent	Total	Change per Decade
	(million metric tons)			(pounds)	(percent)
1950	631	N. A.	N. A.	542	N. A.
1960	847	+216	+34	613	+13
1970	1,103	+256	+30	653	+ 6
1980	1,442	+339	+31	710	+ 9
1990	1,688[1]	+246	+17	699	- 2
2000	1,846[2]	+158	+ 9	653	- 7

[1] Assumes 1 percent increase over 1989 world harvest of 1,671 metric tons as estimated by USDA in June 1989. [2] Assumes no appreciable gains or losses in world grain area and a rate-of-yield increase for world grain between 1990 and 2000 that will equal the increase in Japan's rice yield between 1969-71 and 1986-88.

down begins to sound like the social equivalent of science fiction. If the United States were to experience a drought-reduced harvest similar to that of 1988 before stocks are rebuilt, its grain exports would slow to a trickle. By September of that fateful year, the more than 100 countries that import U.S. grain would be competing for meager exportable supplies from Argentina, Australia and France. Fierce competition among importers could double or triple grain prices, driving them far beyond any level previously experienced.

By November, the extent of starvation, food riots and political instability in the Third World would force governments in affluent industrial societies to consider tapping the only remaining food reserve of any size—the 450 million tons of grain fed to livestock. If they decided to restrict livestock feeding and use the grain saved for food relief, governments would have to devise a mechanism for doing so. Would they impose a meat tax to discourage consumption, or would they ration livestock products, much as meat was rationed in many countries during World War II?

While the most immediate consequences of a disastrous harvest would force millions of the world's poor to the brink of starvation, the international monetary system also would be in jeopardy. Debt-ridden Third World governments desperately trying to import enough high-priced grain to avoid widespread starvation would have little foreign exchange for debt payments. Whether the major international banks could withstand such a wholesale forfeiture of payments is problematic.

Looking at the Nineties

Projecting future food production trends was once a simple matter, but as yields in many countries approach the upper bend on the S-shaped growth curve, simple linear extrapolations of the recent past become irrelevant. The grain output of several countries, including China, Indonesia, Mexico and the Soviet Union, has shown little or no growth since 1984. In addition, land degradation and hotter summers—the former difficult to measure and the latter impossible to project with precision—will shape future production trends.

In one respect, however, projections are simpler now than in the past. Since cultivated area is not likely to change appreciably during the nineties, assessing the production prospect becomes solely a matter of estimating how fast land productivity will rise. Historically, the rise in world cropland productivity, as measured by grain yield per acre, peaked during the sixties when it climbed 27 percent during the decade. It rose only 21 percent during the seventies and an estimated 19 percent in the eighties. It seems likely, given the emerging constraints outlined earlier, that the growth in world cropland productivity will slow further during the nineties. Perhaps a harbinger of things to come, it increased only 2.6 percent in the last half decade.

Even though projecting food production trends is now a complex undertaking, we do have a model to help us. The recent experience of Japan, where grain yields started rising several decades before those in other countries, offers insight as to how rapidly land productivity might rise for the rest of the world. The world grain yield today, taking into account the wide range of growing conditions, appears to be roughly where Japan's grain yields were in 1970. For example, this year's rice yield in China is 1.5 tons of milled rice per acre, exactly the same as Japan's rice yield in 1970.

Since 1970, Japan's rice yield per acre has risen an average of 0.9 percent per year, scarcely half the 1.7 percent annual growth in world population. If the world can raise grain yields during the nineties at that rate, then grain output will increase by 158 million tons, for an overall gain of 9 percent [See Table 2].

With world population expected to increase by more than 921 million, or 17 percent, per-capita grain production would fall

7 percent during the nineties. If the world cannot do any better over the next decade than Japan has over the last two decades, then a steady deterioration of diets for many and starvation for some seems inevitable.

The key question for the nineties is whether the world will even be able to match the Japanese. Despite the powerful incentive of a domestic price support for their rice pegged at four times the world market, Japanese farmers have run out of agronomic options to achieve major gains in productivity. Farmers in the rest of the world, who are not as skilled, literate or scientifically oriented as are those in Japan, will find it difficult to do better.

Biting the Bullet

A deterioration in diet and an increase in hunger for part of humanity is no longer a matter of conjecture. It is a matter of record. Evidence that the world is in trouble on the food front is mounting. In Africa and Latin America, both the absolute number of people and the share of population that is hungry is increasing. Infant mortality, the most sensitive indicator of a society's nutritional state, appears to be rising in dozens of countries.

If the Japanese agricultural record provides a reasonable sense of what the world can expect during the nineties, and if the world continues with business-as-usual policies in agriculture and family planning, a food emergency within a matter of years may be inevitable. Soaring grain prices and ensuing food riots could both destabilize national governments and threaten the integrity of the international monetary system.

Barring any dramatic technological breakthroughs on the food front, the gap between population growth and food production will widen. In all-too-many countries, the opportunity to slow population growth with the time bought on the food front by the Green Revolution has been wasted. To be sure, there will be further gains in output from the Green Revolution, but they are not likely to match the impressive jumps registered from the mid-sixties to the mid-eighties.

The world needs to continue to strengthen agriculture in every way possible. A massive international effort is needed to protect soil, conserve water and restore the productivity of degraded land. But the Japanese experience suggests that even doing

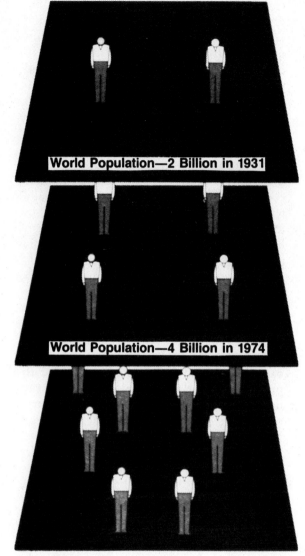

World Population—8 Billion in 2018

everything feasible on the food side of the food/population equation is not enough.

Avoiding a life-threatening food situation during the nineties may depend on quickly slowing world population growth to bring it in line with food output. The only reasonable goal will be to try and cut it in half by the end of the century, essentially doing what Japan did in the fifties and what China did in the seventies.

Reaching that goal will be extraordinarily difficult, perhaps more demanding than anything the international community has ever undertaken. Cutting population growth is not merely a matter of providing family planning services, although this is obviously essential. It depends on raising public understanding of the relationship between family size today and the quality of human existence tomorrow. Unless these goals are given top priority in national capitals, the effort will fail. Braking population growth is unlikely without much greater investment in education, particularly for Third World women, and in health care and other social improvements that facilitate the shift to small families.

Nor is a family planning effort of this magnitude likely to succeed if the international community does not effectively address the issue of Third World debt. The economic and social progress that normally leads to smaller families is now missing in many debt-ridden countries. Unless debt can be reduced to the point where economic progress resumes, the needed decline in fertility may not materialize.

For the United States, the obvious first step is to restore its funding of the United Nations Population Fund and International Planned Parenthood Federation, which was canceled several years ago in response to pressures from the political right. Given the link between population growth and environmental degradation, President George Bush cannot credibly call himself an environmentalist if funding is not resumed. In the process, the United States could reinvigorate the international family planning effort.

Future international political stability and global economic security may depend on reining in population growth more than any other single factor. Since national security depends on global economic security and international political stability, it may need redefinition. Uncontrolled population growth may now be a far greater threat to future political stability in the world than the East-West ideological conflict that has dominated world affairs over the past generation.

The best forum for getting all these issues on the table would be an emergency U.N. conference of the world's national political leaders. Such a gathering would permit a review of the shifting food and population balance and an examination of the projections so that people everywhere could better understand the consequences of continuing on the current demographic trajectory.

The time may have come for world leaders to issue a call to action. It may now be appropriate for the United Nations' secretary general, the president of the World Bank, and national political leaders to urge couples everywhere to stop at two surviving children. Difficult and harsh though this may seem, bringing population and food into balance by lowering birthrates is surely preferable to doing so inadvertently by allowing death rates to rise.

FORECAST: FAMINE?

Will a warmer planet yield less food?

Françoise Monier

Françoise Monier is a writer for L'Express.

We cannot escape a change of climate. The hypothesis of a general increase in the temperature of the earth's atmosphere became a reality several years ago, and we now know our children will live on a planet transformed by the consequences of the greenhouse effect.

Thousands of scientists share this view, and more and more of them are working on the meaning of this upheaval. Every prospective study — of soil, water, species, population migrations or international relations — must be revised to take this increase in global temperature into account.

The most basic of all problems that may result concerns food. With temperatures a few degrees higher and perhaps much less rainfall, how will we feed our planet?

Before answering, researchers have first had to prove to the international community that an increase in global temperature is no myth. Proof is essential because, to limit the phenomenon, our way of life, methods of production and agricultural techniques will have to change. The idea that the combustion of fossil fuels — coal and oil — and of wood increases the proportion of certain gases, particularly carbon dioxide (CO_2) in the atmosphere emerged a long time ago. But only about 20 years ago was it forecast that the accumulation of CO_2 in the atmosphere would contribute to increasing

global temperature. It was only discovered 10 years ago that other gases—methane, chlorofluorocarbons, nitrogen oxides—have the same effect. They act as the glass in a greenhouse: letting sunlight in but blocking the infrared rays reflected into space by the earth.

Almost a generation passed before the warnings of a few scientists became a worldwide preoccupation. Two events helped make people aware of it. First came the discovery of a hole in the ozone layer over the Antarctic. A team of British and American physicists and meteorologists confirmed its existence on 20 October 1986, when mission chief Susan Solomon announced live from the American McMurdo base that, for several weeks of the Antarctic spring, there is a hole in the ozone layer which protects the earth against certain dangerous solar rays. For the experts, this "puncture" in the ozone layer, though not directly connected to climate warming, nevertheless foreshadows the related problems of increasing global temperatures.

Then came the summer of 1988: exceptionally hot and dry in the USA, Canada and —although less publicised—in China, too. World food reserves dropped from 101 to 54 days. The warmer climate was not the only cause of alarm; there were also poor harvests in the Soviet Union, monsoons that arrived late in India, and locusts that attacked grazing lands in Africa. But governments appealed to the experts for an answer. The latter already know that, over the past 100 years, the average global temperature has in-

Reprinted with permission from *CERES*, September/October 1990, pp. 15-19.

77

creased by 0.7 °C, and theoretical calculations forecast an increase of 1.5 °C–4.5 °C within the next 50 years.

Up to this year, the scientific community was divided on the consequences of a warmer atmosphere. Pessimists shared the view of climatologist James E. Hansen, director of the Goddard Institute, who claimed that the drought that hit the Midwestern U.S. in 1988 was a demonstration of the predicted climatic revolution. Sceptics, like Richard Lindzen, Professor of meteorology at the Massachusetts Institute of Technology (MIT), considered it a natural phenomenon, caused by a variation in the sun's activity. Some optimists were even encouraged by the fact that CO_2 favours plant growth, and that for a few years certain European forests had prospered.

It was thus necessary to reconsider the statistical data, comparisons with climates of the past, and forecasting models. In late 1988, the United Nations requested reports from the Intergovernmental Panel on Climate Change (IPCC), and the first of them were submitted recently. Experts discussed them behind closed doors in late May and early June, and were to deliver their final texts in Stockholm in August.

One group, which met in London, worked on the phenomenon as such. Its members were sufficiently confident of their results to make their initial conclusions public. For the first time, the scientific community clearly stated that gases emitted by industrial activities are definitely responsible for an abnormal warming of the atmosphere. They blamed not only CO_2, methane and chlorofluorocarbons, but water vapour, which also accentuates temperature increase.

This text could be considered a watershed in the ongoing debate.

If we continue living, consuming and producing as we do now — Scenario A of the IPCC experts — the average global temperature increase will be 0.3°C per decade (ranging from a minimum of 0.2°C to a maximum of 0.5°C) for the coming century. Thus, in the year 2025, average temperatures will have risen by 1°C, and by 3°C in 2090. "By more than in the past 10 000 years", states the report. Other scenarios, in which gas emissions diminish, forecast lower increases. But the experts focused their attention on Scenario A and attempted to refine their statements.

Thus, the warming will be greater on land areas than over the oceans, and more pronounced in the northern hemisphere. Although it is difficult to forecast regional variations at this stage, the report announces that the warming in Southern Europe and in central North America will be more accentuated, with less rainfall and more evaporation. Finally, the level of the oceans will rise by 6 cm a decade for a century (from a minimum of 3 cm to a maximum of 10 cm), because water volumes will expand, and glaciers will melt.

Naturally, the IPCC admits numerous factors such as the role of the oceans, are still not well known. But these dark areas do not prevent the IPCC from reflecting on the most crucial question: the impact of global temperatures on life. Water, oceans, ecosystems, forests and agriculture — all those elements that contribute to the biological activity of the earth—will be affected.

The survival of mankind is therefore directly concerned.

A preliminary response to this question was given by the second IPCC research group which met in Moscow in the first week of June, presided over by Youri Izrael of the Soviet Union. A global forecast had never before been made on the evolution of total biomass. Hydrologists, oceanographers, forestry experts and agronomists are not used to working on such vast scales. They have very limited means for defining a situation, and from which they can follow changes. Remote-sensing satellites can provide basic data. But teams capable of analysing their figures and photographs have only just begun their task. A new discipline, agrometeorology, has been created to deal with these issues.

The Izrael group's first observation was that warming will have major effects on agriculture — good and bad. It is impossible to say whether the overall effect will be positive or negative, given the vast array of different situations. The authors of the report are cautious, and believe food resources will be sufficient to meet demand — even for Scenario A. But the cost will be higher than at present. This could be disastrous for developing countries, which already have problems feeding their inhabitants without plunging further in debt.

Some cereals, such as wheat and rice, may benefit from higher CO_2 levels. But this is not the case for maize, sorghum or millet. In the laboratory, certain crops make better use of the water generated when the atmosphere is heavy with CO_2. But we cannot be sure similar results will be obtained in the field. The northern areas of the northern hemisphere may gain from the change: it may be possible to grow wheat in Siberia, Scandinavia, and northern Canada. Forests may flourish 300 km north of present limits. It may only be a question of adapting varieties to the new situation.

Unfortunately, all is not good news: it isn't certain that northern soils are suitable for cereals. The same question arises for rainfall, which could occur as violent storms — much less useful for such soils. If this happens, erosion will occur sooner, particularly if, as foreshadowed by some forecasts, the differences between dry and rainy seasons are greater. Agronomists are concerned about the move-

ment of parasites. "The increase in global temperatures could extend both the habitat of harmful insects and crop diseases, presently restricted to tropical countries, to sub-tropical and temperate regions", warns the Moscow report.

It is commonly assumed that agriculture can adapt to change. This is true of areas where climate normally varies and farmers benefit from advice and technical assistance. But it is much less true for areas with low crop yields, such as developing countries. In either case, there will be a price. For instance, if the productivity of maize and soya farms decreases in the USA, animal feed and meat prices will rise. There are rumours, however, that world agricultural production of some crops could increase by an average 10 per cent.

Experts are also considering the green-

The balance of certain gases in the Earth's atmosphere, particularly carbon dioxide, is very delicate and appears to have a direct effect on the global temperature. The loss of the biomass, in the form of timber, can have a major impact on how fast the planet's temperature rises. As this temperature increases, certain grains and foodstuffs may become difficult or impossible to grow—potentially causing food shortages. (Photo credit: United Nations)

house effect on ecosystems, because the success of cropping systems depends on maintaining biological diversity. The risk is that diversity could be seriously reduced. All types of ecosystems could be affected, because the ideal climate of each ecological niche will have migrated several hundred kilometres. In the northern forests, biomass could be reduced 20 per cent. Part of the Eurasian tundra zones could disappear. The productivity of the semi-arid areas of the Mediterranean could drop drastically and the desert could spread. The worst affected would be mountain or insular reserves, and regions where ancient plant and animal communities subsist, protected up to now against disturbance: genetically-impoverished species, those whose survival depends on special physical conditions, and species that only live in limited areas, or reproduce slowly.

There would be competition between plants growing on the boundaries of ecosystems. "Invasions" of destructive species would multiply, as we have seen in Florida with *Melalenca quinquenervia* bamboo. Other invasions would follow: parasites attracted by the presence of dead wood and weakened plants. Scientists are forming groups all over the world to measure possible migrations of insects and pathogens.

But water remains the critical factor for agriculture. The study of changes in hydrological systems has only started, while rainfall, with its seasonal distribution, daytime variations and evapotranspiration, can modify vegetation more than temperature change.

A group of threatened nations has already reported to the United Nations. They are the 14 island states threatened by the rising level of the oceans, accounting for 700 000 inhabitants, in the Pacific and Indian Oceans, the Caribbean and Mediterranean. The United Nations Environment Programme estimates that 70 swampy regions, dune areas and estuaries are also threatened. The rising level of the ocean has been observed in the Maldives and the Guyana coast and does not occur as a gentle phenomenon. It starts with storms and small tidal waves. Potable water becomes scarce because salt water from the sea penetrates the aquifers close to shores; and erosion accelerates.

Often, coastal areas formed by alluvial material are extremely fertile, as in Southeast Asia where rice fields have developed on former swamp areas, and small aquaculture units produce shrimp, fish or alevins to stock neighbouring bays. Invasion of these areas by the sea affects the food security of part of Asia and Africa.

About 10 years ago, it was thought the process would be slower and populations would adapt. But a study by the National Centre for Atmospheric Research (NCAR) shows that an unexpected meteorological event — drought, floods, a tidal wave — completely upsets social and economic life. The study focuses on a problem neglected to now: the lack, both physical and legal, of available land areas. This constitutes a warning, as does the presence of more and more numerous groups of "ecological refugees". Climate becomes a component of international rivalry. This can already be seen in cases where several countries share the resources of the same river.

Warming will be greater on land than over oceans, and more pronounced in the north. . . .

Steps to be taken were discussed at a conference on climate, organized by UNEP and the World Meteorological Organization (WMO) in Geneva in 1989. The idea of an agreement on restricting greenhouse gases is gaining ground. But it will be much more difficult to ratify than the Montreal Agreement on the limitation of CFCs. In this case, as was the case for CFCs, developing countries have fewer means to adapt to change, although they produce far less greenhouse gases — only 30 per cent CO_2 for 75 per cent of the population. They expect rich countries to provide technical and economic aid to adapt their production systems to new conditions. Forestry and agriculture will also be affected by a warmer climate. Rich countries have organizations that can create new varieties or cropping techniques to take advantage of the increase in temperature and minimize negative effects. The South, however, needs assistance.

It is difficult to organize aid when forecasts are imperfect. Present observations and models are global. At regional level, observer stations need to be distributed more rationally in each continent, and new remote-sensing satellites, teams capable of making models, and better qualified computer personnel are required. Developing countries should be associated more than they are now with this phase of the research, which is predominantly focused on northern countries. Initiatives such as those of the tree study group — formed by 36 researchers from the northern and southern Sahara to advance work on the adaptation of trees and shrubs of arid and semi-arid zones — should be systematically multiplied. The methane produced by swamps and rice fields, for instance, has not been quantified. The assimilation of CO_2 by cultivated plants as a function of temperature, humidity and sunlight should be the object of special programmes, so as to give an idea of the potential agricultural output of each region. Production levels can be completely altered by parasite attacks. Here again, available knowledge must be reconsidered.

New industrial and energy policies will be necessary. But it is easier for the Netherlands than for China to give up using coal in power stations and replace it with natural gas. The same problem applies to forests: Canada can give up exploiting some of its trees more easily than Thailand. And before encouraging tropical countries to set up forestry programmes, it would be advisable to de-termine the CO_2 balance in tropical forests and the precise impact of deforestation.

All this requires enormous investments. Particularly for water management. In most of the Third World, food security depends on improving irrigation. It will become more crucial with warming of the atmosphere. These heavy investments assume that crops with a higher added value will be grown to be profitable. It also means that soil protection measures must be applied.

The problems of increasing global temperature will not be solved with an agreement to limit emission of greenhouse gases. They require gigantic research efforts with emphasis on data collection, establishing regional scenarios, and acquiring in-depth knowledge of the role of key ecosystems.

Every field of biology will be involved in future research. But this is where expert training is most lacking.

International competition is likely to become stronger. There will be fights for access to food stocks, to get hold of new technology, to acquire the best soils. These will be won by those who know how to adapt, i.e. countries capable of making sophisticated plans, and changing their legislation and educational systems quickly, to introduce the greatest possible flexibility.

Energy: Present and Future Problems

There has been a tendency, particularly in the developed nations of the world, to view the present high standards of living as exclusively the benefit of a high technology society. As recently as the 1960s, noted scientists described the technical-industrial civilization of the future as being limited only by a lack of enough trained engineers and scientists to build and maintain it. This euphoria reached its climax in July of 1969 when American astronauts walked upon the surface of the moon, an accomplishment brought about solely by American technology—or so it was supposed. It cannot be denied that technology has been important in raising standards of living and in permitting moon landings. But how much of the growth in living standards and how many outstanding and dramatic feats of space exploration have been the result of technology alone? The answer is few—for in many of humankind's recent successes, the contributions of technology to growth have been less important than the availability of incredibly cheap energy resources, particularly petroleum.

As the world's supply of recoverable (inexpensive) petroleum dwindles and becomes more important as a tool (or, as the 1991 events in Kuwait have shown, even a weapon) in international affairs, it becomes increasingly clear that the energy dilemma is the most serious economic and environmental threat facing the Western world and its high standard of living. With the exception of the population problem, the coming fossil fuel energy scarcity is probably the most serious threat facing the rest of the world as well. The economic dimensions of the energy problem are rooted in the instabilities of monetary systems produced by and dependent upon inexpensive energy. The environmental dimensions of the problem are even more complex—ranging from the hazards posed by the development of such alternative sources as coal and nuclear power to the inability of Third World farmers to purchase necessary fertilizers produced from petroleum that has suddenly become very costly. The only answer to the problem of dwindling and geographically vulnerable inexpensive energy supplies is conservation. Conservation requires a massive readjustment of thinking away from the exuberant notion that technology can solve any problem.

The difficulty with conservation, of course, is a philosophical one that grows out of the still-prevailing optimism in high technology. Conservation is not as exciting as putting a man on the moon. Its tactical applications—caulking windows and insulating attics—are dog-paddle technologies to people accustomed to the crawlstroke. Does a solution to this problem entail finding answers to energy conservation in the form of technological fixes with which people are so enamored? The first offering in this unit suggests that technological fixes may well be the salvation of the world's energy consumers—but technological fixes of a very special type. In "Energy for the Next Century," Will Nixon, an associate editor of *E Magazine*, concludes that the development and elaboration of a technology based on such alternative energy sources as solar power, wind power, and geothermal energy could well take up the slack in the energy budget represented by a decrease in the availability of petroleum. Nixon points out, however, that no alternative energy sources will develop by themselves. Rather, government action is required to replace oil and nuclear power with the alternative, sustainable energy forms of the future and, unfortunately, the development of the U.S. National Energy Strategy pays far too little attention to alternative energy. In the second article, author William H. Miller also argues that the National Energy Strategy is less than adequate in its emphasis (or lack thereof) on alternative energy development. Miller's view is that, unlike education, drugs, urban poverty, or many other public policy issues, the solution to the U.S. energy problem is clear-cut and should be based on an *econvironergy* policy: conserve energy, develop renewable (alternate) sources of energy, and increase domestic exploration for oil and natural gas to reduce dependence on foreign energy sources. Conservation provides the basis for the third article in this section. In "The Negawatt Revolution," physicist Amory Lovins of the Rocky Mountain Institute claims that, using existing technology, the United States could save three-fourths of all the energy it currently uses. Most of this savings would come in the corporate and industrial systems of America, where such simple measures as switching from incandescent to fluorescent lighting or switching from older to new fluorescent systems can effect energy savings of 70 to 90 percent. What all consumers must realize, concludes Lovins, is that conservation is not only sensible environmentally but is, to put it simply, profitable. Profitability is also at the heart of the argument for alternative energy advanced in the next selection in this unit. In "Here Comes the Sun," Christopher Flavin and Nicholas Lenssen, both of the Worldwatch Institute, conclude that scientific and engineering breakthroughs now exist that make it practical (meaning profitable) to begin producing our electricity, heating our homes, and fueling our cars with renewable energy. Because the conventional wisdom among government experts suggests that we are struck with dependence on fossil fuels, Flavin and Lenssen contend, there has not been nearly as much government support for the continuing development of marketable alternative energy. A vigorous public commitment of research and development dollars could push renewable energy into the mainstream of energy use. The potential benefits of such a commitment to research are already being realized in one area of alternative energy development, according to a team of researchers from the Oak Ridge National Laboratory, authors of the final selection in the Energy unit. In "Energy Crops for Biofuels," Janet H. Cushman, Lynn L. Wright, and Kate Shaw describe research currently under way that aims to make cultivated trees and grasses an important source of fuel for transportation and electricity generation. While burning biomass (wood, for example) for fuel is an ancient practice, the modern approach to biofuels is reliant on the conversion of biomass to fuel before combustion. Even more important, perhaps, are the potential developments in genetic engineering that make the prospects of producing plants that can act as manufacturers of fuel, something more tangible than scientists of even a decade ago would have believed.

Answers to energy questions and issues are as diverse as the world's geography. But all the answers require a reorientation of thought and the action of committed groups of people who have the capacity to change the dominant direction of a culture. The transition—in terms of both energy resources and patterns of thought—has already begun. Whether it can be completed in time to prevent massive social, economic, and environmental upheaval is a question of the utmost importance.

Looking Ahead: Challenge Questions

What are some of the environmental indicators that suggest a critical need for people to shift to a postpetroleum energy system? How does the National Energy Strategy of the United States fit with the objective of a postpetroleum society?

What is the relationship between the U.S. reliance on foreign oil to supply domestic needs and the development of the National Energy Strategy? How can a public strategy for energy development link or balance both economic and environmental concerns?

What are some of the most available means of conserving electrical energy by developing more efficient systems? How prepared are producers and consumers of electrical energy to adopt these conservation measures?

What are the most important forms of alternate energy currently available, and how might political barriers to the development of those energy sources be removed?

How can the development of biofuels and the application of biotechnology to crop raising contribute new sources of energy to countries like the United States? Are there environmental restrictions or drawbacks to the development of biofuels?

ENERGY FOR THE NEXT CENTURY

All signs seem to indicate we're heading—however slowly—toward a post-petroleum society. The policies we craft in the next decade might well determine whether it will be like Mad Max or Ecotopia.

Will Nixon

WILL NIXON *is Associate Editor of* E Magazine.

Just when the peace movement had run low on placard slogans—"No Blood For Oil" having lost some of its octane as the Gulf War wound down—George Bush announced his long-awaited National Energy Strategy (NES) at lunchtime on February 20. Not since Jimmy Carter had a president decided that energy planning should involve more than checking the gas gauge before heading off on a long trip. Indeed, Carter had donned a cardigan sweater, turned the White House thermostat down to 68, and solemnly told the nation that his energy policy was "the moral equivalent of war." Cynics now quipped that George Bush had dropped the equivalency part. But the NES was much more than that. It admitted that the United States couldn't kick the foreign oil habit, which was the great dream of presidents in the 70s, but offered almost 100 ways to increase our own energy production. It recommended opening up Alaska's Arctic National Wildlife Refuge for oil and gas exploration, reviving the nuclear energy industry by eliminating some of those pesky public hearings, freeing the oil and gas industries from some cumbersome regulations, and burning more garbage for energy. As for encouraging us to use less energy to begin with, well, better luck next time. The NES virtually ignored better car mileage standards and mass transportation. It dropped tax breaks for renewable energy and federal lighting standards. And, perhaps most negligent of all, it never got around to the reason we need an energy policy in the first place—to stop the flood of carbon emissions into our atmosphere which cause global warming. "When people talk about the need for a national energy policy, they don't mean writing down the mistakes of the past 10 years, which is what the Bush Administration has done," said Alexandra Allen of Greenpeace.

The next morning, a handful of Greenpeacers went to the Senate offices where Admiral James Watkins, secretary of the Department of Energy (DOE) was due to appear before the Energy Committee. Senator Albert Gore of Tennessee had already promised a "battle royal" on Capitol Hill over Bush's energy plan. The Greenpeacers arrived at 7:30 AM, but the Washington media pack had already claimed the room. So they stood out front, holding a "Reality Check" for Bush and Watkins. It was made out for "Untold Billions" for "Oil Wars, Oil Spills, Global Warming and Nuclear Waste."

In the gray and white world of Washington, Greenpeace provides the most color, but they were hardly alone in their outrage at the President's energy plan. Only a year ago the DOE had suggested that its plan would be almost the reverse of what Bush finally announced. Deputy Secretary Henson Moore had said, "Energy efficiency and renewables are basically the cleanest, cheapest and safest means of meeting our nation's growing energy needs in the 1990s and beyond."

"Based on comments by the DOE last year," said Scott Denman of the Safe Energy Communications Council, "and based on George Bush's claims in his State of the Union address that his strategy would support conservation, efficiency and alternative fuels, what we've been given is a cruel hoax. It's not an energy strategy, it's an energy tragedy."

Out in San Francisco, Chris Calwell of the Natural Resources Defense Council (NRDC) said, "The themes that emerge are simple and familiar: drill more, nuke more, pay more, save less." His group released an analysis entitled: "Looking for Oil in All the Wrong Places," with some math to show just how wrongheaded the NES was. If we drilled the Arctic Refuge and the Outer Continental Shelf, as the NES proposed, we might produce a little more than six billion barrels of oil or its equivalent in natural gas, and these areas would be drained dry by the year 2020. But if we used our renewable energy supplies—the sun, wind, rivers that flow into hydroelectric dams, and plants and trees that can be turned into biomass fuels—and we made our world far more energy efficient, we would have the equivalent of 66 billion barrels of oil. As Robert Watson, the author of this report says, "We *will* have a post-petroleum society. The question is whether it will be like Mad Max or Ecotopia."

The NES hadn't always looked so grim. For more than a year, the DOE had held hearings across the country, patiently listening to an army of experts who hadn't found a sympathetic ear in the executive branch since Ronald Reagan took office and put the DOE under the sleepy care of dentist James Edwards. Admiral Watkins had made his name in Washington, DC by taking over Reagan's troubled AIDS commission and producing a good report. "He showed he has a good ear. He's receptive to contrary points of view," says Scott Denman, who spoke at two hearings late in 1989. "He heard from every energy expert in the country."

These experts had a remarkable story to tell: the United States, with what scrambled energy policies we do have, already saves an estimated $160 billion a year in energy costs compared with 1973. "The really interesting thing is that total energy consumption was virtually identical in 1973 and 1986," says

"There are all sorts of incentives for fossil fuels which, to be fair, should be eliminated or applied to energy efficiency and renewables."

John Morrill of the American Council for an Energy-Efficient Economy. "And yet our GNP grew by 40 percent during those years."

Until the 70s, conventional wisdom had insisted that energy growth was synonymous with economic growth, but changes wrought by the oil shocks of 1973 and 1979 tied that thinking in a knot. The Corporate Automobile Fuel Efficiency (CAFE) standards passed in 1975 pushed cars from 13 miles per gallon to 27.5 in 1986, and now save us five million barrels of oil a day. Utilities dropped oil for cheaper coal. Homes were weatherized. Some states took action, such as California, which required 1980 refrigerators to use 20 percent less electricity than the 1975 models, causing an entire industry to change. (By the mid-80s refrigerators, the biggest energy drains in most homes, had improved by 35 percent.) Not until 1985 when President Reagan

Bush's National Energy Strategy

President Bush's National Energy Strategy (NES) includes almost 100 proposals for increasing our domestic energy production. Among his major initiatives are:

• Offering 1.5 million acres of the Arctic National Wildlife Refuge for oil and gas exploration. The Department of the Interior also announced plans to lease thousands of square miles on the outer continental shelf for similar exploration, from the Beaufort and Chuckchi Seas off Alaska, to the Gulf of Mexico, to large tracts off Southern California and the Eastern Seaboard.

• Reviving nuclear power by shortening post-construction public hearings and by transferring the problem of permanent radioactive waste disposal to an independent corporation not subject to the public constraints on a governmental agency. The NES would also extend the life of

currently operating plants to 60 years from their current 40.
• Amending the Public Utility Holding Act to allow utilities to expand across state lines and to allow private power plants to compete with utilities. This change could spur the private development of nuclear power plants regardless of public utilities' desires.
• Reducing regulations on natural gas pipelines and eliminating Federal review of natural gas imports or exports.
• Requiring the owners of fleet vehicles, such as taxis or delivery vans, to switch to alternative fuels: alcohol, natural gas or electricity. By the year 2000, ninety percent of the new fleet vehicles must run on an alternative fuel.
• Continuing the present tax breaks for energy conservation.

An Alternate National Energy Strategy

Numerous groups have called upon the country to adopt energy policies that

would increase our energy savings—an approach which they believe will be faster, cleaner and cheaper than increasing our energy production. Among their major proposals are:

• Improving our energy efficiency by two to three percent a year, the rate we achieved in the late 70s and early 80s.

• Increasing automobile fuel efficiency standards to 45 mpg for cars and 35 mpg for light trucks by the year 2001.

• Raising the federal gasoline tax by 50 cents a gallon to help finance mass transit and other energy efficiency programs.

• Adopting "least cost" planning methods that include energy savings and environmental effects when making all of our energy decisions.

• Increasing the funds for federal research on energy conservation and renewable energy until they are at least equal with those for nuclear power.

• Creating new tax credits for renewable energy and a carbon tax on fossil fuels.

rolled back the CAFE standards and oil prices dropped, did our energy savings stall.

One after another people appeared before the DOE with ideas on how to get us back on the savings track. After all, we still have a ways to go—Japan and Germany use half as much energy per dollar of economic output as we do. "The message from elected officials, citizens groups and energy related businesses was conservation, conservation, conservation," Denman says.

"What we've been given is a cruel hoax. It's not an energy strategy, it's an energy tragedy."

The DOE listened. "Admiral Watkins looked at the issue, and he began to say the one thing that's loud and clear is that improved conservation and energy efficiency are where the most gains can be made and the government can do a lot more than it has," said Christopher Flavin of Worldwatch who testified at the first hearing in August 1989. In April 1990 the DOE released an interim compendium of what it had heard so far which was dominated by conservation and energy efficiency. But then the report left the DOE for the White House.

"President Bush knows as much about energy as anyone we've ever had in the White House," says one DOE official. What Bush knows, though, is producing energy, not saving it. His home state, Texas, is the worst energy hog in the country. It can't even fill up its own gas tanks, importing 48 thousand barrels of oil in 1989 on top of 736 thousand barrels from its own wells. "Texas has 42 percent fewer citizens than does California," reports a study by the consumer advocacy group, Public Citizen, "but it uses 37 percent more energy." California has a state energy department and Texas doesn't, which just shows what a little planning can do.

"The implicit assumption of the NES is that the market runs at top efficiency," says Michael Brower of the Union of Concerned Scientists (UCS). "Any conservation or renewable energy we have is the amount we should be getting because we have a free market. Our point is that the market is not really free. There are all sorts of incentives for fossil fuels which, to be fair, should be eliminated or applied to energy efficiency and renewables. And the free market doesn't take into account the cost to the environment." This spring the UCS, NRDC, and the Boston-based nonprofit environmental research group, the Tellus Institute, released an alternative national energy strategy to show what we can do by taking conservation and efficiency seriously. "Nobody has ever made this serious an effort to meet the skeptics on their own ground," Brower says.

More than a dozen states now give serious consideration to energy efficiency and the environmental costs of fossil fuels when planning out their needs. Rather than simply worrying about energy supply, they consider demand, and often find that the cost of reducing the demand is a lot cheaper than increasing supply. This new approach, labeled "demand side" or "least cost" planning, now dominates on the West Coast and in New England where states have to import fossil fuels. States that produce coal and oil, though, tend to stick to the "supply side" view.

"If every state took the least cost approach there would be no reason to build any new power plants— coal or nuclear—in this decade," says Michael Totten of the International Institute for Energy Conservation. And if we took this approach in all of our energy decisions, we could be using seven percent less energy by the year 2000, instead of 34 percent more, as we would on our present course. In the next century, after we've captured these savings, renewable energy sources such as windmills, photovoltaics that turn sunlight into electricity, and turbines that run on biomass fuels could be competitive with fossil fuels.

But the NES took the supply side view. And many observers laid the blame on John Sununu more than on George Bush. "Sununu has always been hostile to demand side measures," Robert Watson says. "We both think the other's solution is a drop in the bucket. He dismisses conservation out of hand as being irrelevant. We dismiss nuclear power as largely being irrelevant. But I think we have the better basis for dismissing it."

After President Bush's announcement, the NES became just another influential sheaf of paper on Capitol Hill where Congress was moving into full gear to make 1991 the year for energy. The President was doing his part in the Persian Gulf, said one Republican staffer, now Congress had to do its part at home. The legislators introduced some 80 bills on energy, including one by Senators Bennett Johnston (D-LA) and Malcolm Wallop (R-WY) that was an oil and gas industry dream. But Senator Richard Bryan (D-NV) reintroduced his bill to improve auto fuel efficiency standards by 40 percent to 40 miles per gallon by 2001. It had been defeated the previous year, but now he had 35 co-sponsors. In the House, Representative Barbara Boxer (D-CA) offered an even better bill to push the standards to 45 mpg. And Representative Philip Sharp (D-IN) came up with a bill to make the federal government a major champion of energy efficiency. The battle had indeed come home.

One More Final Showdown at the Arctic Refuge

The Exxon Valdez spill seemed to save the Arctic Refuge, but the Gulf War put it right back into play. To the oil and gas industry, the Refuge could provide one tenth of our country's future production. To environmentalists, the Refuge is the last untouched

> "If Detroit was really smart they'd go into the mass transit business–most of our roads will not be able to handle many more cars. And in many places there's just no more room to build new roads."

corner of North America with the full spectrum of arctic and subarctic ecosystems.

"It's the only one up there," says Mike Matz, of the Sierra Club office in Washington, DC. "The oil industry has the rest of the North Slope, and all of the offshore areas, so there's no reason they need it." In fact, nearby Prudhoe Bay has 27 billion barrels of oil of which nine billion can be recovered, but another field nearby has 15 to 20 billion barrels. The oil industry is simply waiting until it is more profitable to drill. The Department of the Interior has done seismic testing of the Arctic Refuge and estimated a one in five chance of finding oil. They then estimated that *if* oil is found the field could yield maybe 3.2 billion barrels.

"We use 18 million barrels a day, so that's about a 180 day supply," Matz says. And it would take a minimum of seven years to begin shipping oil from the Refuge. So why do the oil companies want it so badly? "The financial health of the oil companies depends not on production, but on how much oil they have access to," Matz says. A gusher in the Arctic Refuge would immediately improve any company's financial standing even if they didn't drill it for years. "So all of this shouting about needing it for our national energy security is bogus," Matz concludes. He believes that, after a close debate, the Senate will kill the idea.

Arctic Refuge or not, the United States remains an oil power on its last legs. The average Saudi oil well pumps 9,000 barrels a day, while ours pump 15. Our domestic reserves could easily be dry by 2020 at our present rate of consumption. But the world's known oil reserves jumped from 615 to 917 billion barrels between 1985 and 1990, mostly in the Persian Gulf where it's cheaper to drill. During the late 80s we stopped producing two million barrels a day— as much as we imported from Kuwait—because the price fell too low. "A lot of the stripper wells, ten barrels a day or less, were shut down because they were just not economical," Watson says. "But if we kept eking out that stuff from already existing fields we'd have a lot of little wells that could add up to something big. We'd prefer that to sinking new wells in Alaska. But we'd need incentives for oil recovery

No Blood For Oil

All of those working for a renewable energy future are hoping that the public disquiet over the Persian Gulf War will create the political momentum needed for change, from support for better car fuel mileage standards to more funding for solar power. Before the Iraqi invasion of Kuwait our energy debate had revolved around the impact of fossil fuels on global warming. War made the issue much more personal. In late January, Dave Kraft of the Illinois-based Nuclear Energy Information Service said, "Two years ago we were arguing from abstractions, but recently we've been arguing about preventing piles of body bags."

Such major environmental groups as the Natural Resources Defense Council, the Sierra Club and the National Audubon Society, want to capitalize on the public mood for change, but they avoided any real stand on the war itself. They are not in the business of foreign affairs, they say. Mike McCloskey, chairman of the Sierra Club, said that the group "doesn't have a position on this war—or past wars," though it did issue statements about "extending warfare to birds, shrimp and

sea life." McCloskey said he is "gravely concerned about the oil spills," but he added that he "wouldn't want to characterize the whole war" as an attack on Gulf ecology.

Some groups, though, saw the Persian Gulf War as *the* environmental crisis of the moment. They weren't about to keep quiet. Bill Walker, media director of Greenpeace's office in San Francisco, says Greenpeace "was the first major environmental group to overtly oppose the war. As environmentalists, we had to take a stand against the war, which was being fought over a resource, oil, which we knew to be toxic to the planet." Walker points out that Greenpeace, which had significant contingents at both major peace marches in Washington, DC last January, "was founded on principles of non-violence. We believe that the planet cannot support armed conflict. We don't know how you can call yourselves environmentalists and not oppose war."

At the Earth Island Institute, also in San Francisco, Gar Smith, editor of *Earth Island Journal*, said, "We suspended a lot of our ordinary work, and used our expertise to get information out." The group even shut its offices on

occasion to go to teach-ins and demonstrations. Both Greenpeace and Earth Island say they lost members and publication subscribers because of their stand against the war.

The Washington, DC-based Friends of the Earth (FOE) also took an early and firm stand against the war. FOE warned of the environmental consequences of the Gulf War in early January; its releases, predicting oil spills and well fires, later proved highly accurate. "Unfortunately, nobody paid any attention until it actually started occurring," said Dr. Brent Blackwelder, vice president of policy, who noted that FOE had "postponed or reduced in intensity" some of its other campaigns to concentrate on the war. "The war essentially laid waste to the entire Fertile Crescent," said Blackwelder. "We essentially poisoned the Persian Gulf and obliterated Kuwait to 'save' it."

Like many of the activist spokespersons, Blackwelder is critical of the richer, more established groups that were slow to respond to the war crisis. "A lot of the groups regarded it as too controversial," he said. "They didn't want to take a stand."

—*Jim Motavalli*

in these fields because things like injecting steam to melt the thick globby oil or sideways drilling add five or ten bucks for each barrel of oil recovered."

"The fact is, oil is too cheap in this country," he continues. "It's simultaneously one of the most valued and one of the least valued commodities we have. It's valuable enough to send half a million people over to the Middle East—and our citizens are probably the most precious resource we have—yet we're not willing to spend an extra buck per gallon, or to require the auto industry to increase its fuel economy."

Car Wars

"The biggest untapped oil field in America is riding 18 inches off the ground in our gas tanks," says Paul Allen, communications director of the NRDC. Allen didn't invent this wonderful soundbite, nor will he be the last to use it as Congress begins debating new CAFE standards in one of the major energy battles of the session. "CAFE was one of the best energy policies we ever had," Watson adds. Passed in 1975 when new cars averaged 13 mpg, these standards pushed the standards up to 27.5 mpg in 1985 when Reagan rolled them back to 26 mpg, basically because Ford and GM couldn't make the higher standard and wanted to avoid the fines.

New standards have been contemplated for a while, but the Persian Gulf War gave momentum to the issue. After all, two thirds of our oil goes to transportation. "It's become much more politically mainstream," says Deborah Bleviss, author of *The New Oil Crisis and Fuel Economy Technologies*. "It has a good chance in the House, but in the Senate it's much more iffy because of John Dingell, who single-handedly held up the Clean Air Act for years." Dingell, who runs the Energy and Commerce Committee, represents Dearborn, Michigan and the auto industry. A new CAFE law would cost Detroit billions and probably add $500 to the price of a new car.

Detroit's old counterattack against CAFE is safety: when the head-on collision comes, do you want to be in a 27.5 mpg tank or a 40 mpg sardine can? "The real irony is that they've never really cared about safety," Bleviss replies. As for the big car/little car scenario, nobody really knows, she adds. The only tests we have, crashing a speeding car into a flat wall, simulate the impact of a car hitting its own weight. And the land rovers and mini-vans that are selling so well these days do poorly in these tests. Smaller cars can be built more safely, but Bleviss says, "The question is, do we want to?"

"If Detroit was really smart they'd go into the mass transit business," Watson says. "Our roads just won't be able to handle many more cars. And in many places there's no more room to build new roads." Our cars are indeed crowding in on themselves like lemmings—each day Californians lose a total of fifty years in traffic delays. Many feel we need to forgo the old suburban ideal of a quarter acre for every home and a car in every garage, and adopt a clustered, European style of life. But Watson warns, "What we're talking about is changing the American Dream, and that isn't going to happen anytime soon. Our conception is still of your big tailfin car driving out to your suburban home on empty freeways. But we haven't seen that since 1965, if we ever really saw it at all."

Power to the Future

Some say our energy future will become a showdown between solar power and nuclear energy. They're related, of course, since the sun is a giant fusion reactor—but solar power leaves the problems with fusion 93 million miles out in space. But all the problems we've created with fossil fuels have revived the nuclear alternative. The NES calls for increasing our nuclear power capacity by 30 to 80 percent over the next three decades.

Dr. Jan Beyea of the National Audubon Society, sometimes cast as an apostate for not rejecting the nuclear idea out of hand, says, "When we talk about global warming, we mean that if we don't take steps now, in 50 years the die will be cast; there will be no hope of preventing global warming from then on." Our planet releases six billion tons of carbon into the atmosphere each year; Beyea and many other scientists believe we must cut back by 20 percent by the year 2010 to avoid the greenhouse crisis.

"We need energy conservation in the short run, but in the long run we have solar and nuclear," Beyea continues. "Both are problematic. The price of solar power during the day isn't bad, but at night it gets very expensive because of storage. Although I'm skeptical, Dr. Larry Lidsky at MIT says that an 'idiot proof' nuclear reactor can be built, so I think we should give him the dollars to see if that's true."

Unlike the nuclear proponents of the early 50s who promised energy "too cheap to meter" Beyea comes to his view out of pessimism. "The public has a choice," he says. "They can choose solar, or they can slip back into nuclear. I detest nuclear power, but I don't think one stone can be left unturned. This is no joke. This is the final environmental battle."

For now, though, nuclear power is one of the great white elephants of our industrial history. The United States has 111 plants which produce 25 percent of our electrical power, but nobody has ordered a new plant since 1978. The NES includes some band-aids for the industry, such as speeding the post-construction approval process, but one Congressional staffer doesn't take them seriously. "Even if they had a majority in Congress, which they don't, the industry still wouldn't get an order."

"You show me one utility that is really hot about nuclear, and I'll show you two dozen that wouldn't touch it with a ten foot pole," says NRDC's Watson. "When we've got reams of material that show the cost of saving electricity is one or two cents per kilowatt hour, why would anyone spend ten cents— 25 in the case of Seabrook—for electricity from nuclear power?" These days many utilities meet

> "The United States—so huge, so sunny, so windy, and so dotted with geothermal hotspots like those already heating homes in Boise, Idaho— has renewable energy supplies that dwarf our oil and gas reserves."

their needs with energy savings campaigns supplemented by new 250 megawatt combined cycle plants that use oil or natural gas. Like many, Watson believes that natural gas will be our transition fuel to a renewable energy future.

The United States—so huge, so sunny, so windy, and so dotted with geothermal hotspots like those already heating homes in Boise, Idaho—has renewable energy supplies that dwarf our oil and gas reserves. And renewable sources, mostly from hydroelectric dams and wood burning power plants, already produce eight percent of our energy compared to the seven percent from nuclear power. Back in the late 70s the DOE created great expectations, confidently predicting that by the year 2000 twenty percent of our power would come from the sun. But Ronald Reagan wasn't interested. He took down the solar panels Carter had installed on the White House roof. He slashed the DOE's research and development budget for renewable energy from $557 million in 1981 to $78 million in 1989. And in 1985 he let the 40 percent solar tax credit for homeowners expire. A $700 million solar industry, which had installed solar water heaters in a million homes, preventing the need for a 1,000-megawatt power plant, rapidly became a $70 million one. Scott Sklar of the Solar Energy Industries Association says, "We lost 35,000 people in 1985, which was more than the U.S. auto industry."

California alone refused to give up the dream. Today it has become the world's showcase for solar and wind power. The parabolic mirrors of Luz International in the Mojave Desert produce 90 percent of the solar electricity generated in the world today— 194 megawatts a day—or one percent of Southern California Edison's peak supply. The wind farms that stretch across the passes at Altamont, Tehachapi, and San Gorgonio, looking like crops of airplane propellers, can generate 1,500 megawatts. The first wind farm built in 1981 was quickly dubbed a "tax farm" because it seemed to produce more writeoffs than electricity. But Randall Swisher, executive director of the American Wind Energy Association, says those bad old days are long gone. "The cost of wind energy was 25 cents per kilowatt hour at the first wind farms in California in 1981, but it's between five and nine cents today, and we expect that to decline at least another 40 percent." And California is only the 14th windiest state in the country. Now that these technologies are proving themselves, proponents believe the federal government should get back in the business of supporting them with new tax breaks and research funds.

"If we get everything we think we can," Sklar says, referring to the energy legislation now pending in Congress, "solar and renewable energy could be producing 20 percent of our country's power by 2000."

"If we stop fooling ourselves that nuclear is only eight cents a kilowatt hour and coal four to six, instead of 12, 14, or 16 like it should be," says Robert Watson, referring to the environmental costs of those industries, "and we had the same level of effort going into solar and alternative technologies that we have going into the moribund nuclear industry, then we would have a very viable industry within ten years."

BALANCE SOUGHT:
ENERGY, ENVIRONMENT, ECONOMY,

William H. Miller

"It should be the goal of the U. S. by the year 2010 to . . . quadruple the energy provided by renewable energy resources. . . . Oil and coal use should be cut by at least 50%. . . . No new nuclear reactors should be built."

Public Citizen

"Those who seek to curtail environmentally sound domestic use and production of fossil fuels threaten to imperil our energy security and economic well-being. An effective national energy strategy must call for development of our domestic reserves both onshore and offshore. . . . Action is needed to revitalize nuclear power."

National Assn. of Manufacturers

There you have it. Two diametrically opposite prescriptions for the U. S.'s energy policy.

No wonder it took 20 contentious months for the Bush Administration, riven by internal disputes and inundated with similar conflicting recommendations from hundreds of other interest groups on scores of other policy specifics, to create its National Energy Strategy released in February. Chances are the energy debate will rage even longer as it shifts to the 102nd Congress.

Chances are, too, that the debate will spill over to the Presidential election campaign in 1992.

And properly so. Energy is a critical public-policy issue. The Gulf War was largely fought over it. And it remains vital to America's future standard of living and industrial competitiveness.

This isn't the first time, of course, that the U. S. has striven to plan its energy future. There was "Project Independence" in the Nixon and Ford Administrations, designed to wean the nation from dependence on foreign oil. That too was the aim of Jimmy Carter's "National Energy Plan," which featured an all-out effort to develop synthetic fuels through the Synthetic Fuels Corp.

Even the Reagan Administration had an energy policy, although it was never formally stated. Its focus: Develop the Strategic Petroleum Reserve (SPR) for emergency use; otherwise, let the free market reign.

" No wonder it took 20 contentious months for the Bush Administration . . . to create its National Energy Strategy. "

Project Independence and the Synthetic Fuels Corp. are long gone, victims of the glut of oil—the world's dominant fuel—that emerged in the 1980s. One of the reasons for the surplus, certainly, was the return of the free market in the U. S. With oil and natural-gas prices no longer controlled, exploration surged early in the decade. Prices fell markedly. Consumers enjoyed—and still do—the best of all possible worlds: cheap fuel, and plenty of it.

And significantly, when oil prices briefly doubled after Iraq invaded Kuwait last summer, no lines formed at gasoline stations as they did during the oil "shocks" of the 1970s. Again, credit goes to the absence of controls—as well as to the supply reassurance offered by SPR, increased production by other oil-producing countries, and the demand-stifling economic recession. In fact, the panic was so subdued that when the shooting war erupted between the United Nations coalition and Iraq in January, oil prices actually declined.

It could be argued that the laissez-faire policy has "worked." But in two major areas, it has been a conspicuous failure:

• It has caused a deepening of the U. S.'s dependence on foreign oil. Domestic oil demand, which had fallen to 15.7 million barrels a day (bbl/d) in 1985, climbed to 17 million last year. Meanwhile, domestic production has been declining; aging U. S. oil fields produced only 7.2 million bbl/d last year, down from 8.9 million in 1985. As a result, the U. S., the world's largest energy consumer, now imports more than 50% of its crude-oil and petroleum products. Alarmingly, fully one-third comes from the politically volatile Middle East.

The dependency is higher than it was even before the Nixon, Ford, and Carter efforts to bring it down. Analysts estimate that on its present course, the U. S. will need to import two-thirds of its oil by the year 2000.

The higher the percentage, obviously, the greater the impact on the nation's trade deficit, of which oil now accounts for a troubling 54%. Also, the higher the percentage, the more vulnerable the U. S. becomes to supply disruptions.

• The laissez-faire policy has failed to account for what economists call the "market externalities"—mainly, the cost of environmental cleanup. As Joel Darmstadter, senior fellow at Resources for the Future, a Washington think tank, puts it: "We've done inexcusably too little to factor the environment into our energy transactions. Compelling evidence exists, locally and globally, that energy does a lot to hurt the environment. That's not been reflected in energy prices."

Given these failings, the effort to develop a new and lasting energy consensus is essential. Adding even more urgency is the little-noted electricity shortage that many experts believe is inevitable later this decade. But the task, as policymakers are finding, is daunting. Disputes over policy specifics abound.

However, the job needn't be as difficult as it seems. Legislators wrestling with the countless prickly decisions might find it easier if they keep in mind what Mobil Oil Corp.

Chairman Alan Murray, who is also chairman of the American Petroleum Institute (API), calls the "Triple E: the Economy, the Environment, and Energy."

Congress needs to balance all three Es. In short, it needs to come up with what might be called an *econvironergy* policy.

It's not a matter of choosing any one of the "Es" over the other two. Many environmental zealots are so alarmed about mankind's destruction of the planet that they insist policy should tilt exclusively their way; similarly, many industry executives are so driven by the need for economic growth and the availability of cheap, abundant energy that they give only lip service to environmental peril. They confuse an ever-rising gross national product (GNP) and the amassing of material goods with the quality of life.

But either/or choices don't necessarily have to be made. For example, a decision in favor of opening the Arctic National Wildlife Refuge (ANWR) to oil drilling doesn't have to come at the expense of cutting

Although Japan and European countries rely heavily on nuclear power (below), safety fears and high costs have given it a bad name in the U. S.

AMERICAN PETROLEUM INSTITUTE

research dollars for alternative forms of energy. An intelligent econvironergy policy involves a wide variety of steps, and *everybody* can get a piece of the action.

Moreover, an examination of the positions of individual interest groups reveals substantial common ground. Consider the two groups cited at the beginning of this article. Public Citizen, a Ralph Nader group, predictably declares that "energy conservation and energy efficiency should be the cornerstone of the nation's energy policy." But the National Assn. of Manufacturers, despite its call for fossil-fuel development, also emphasizes that "the efficient use of energy should be a cornerstone of a national energy strategy." The two protagonists, it turns out, share the same cornerstone.

Meanwhile, API, the trade association of Big Oil, in the same policy statement sentence that asks for "removal of unnecessary impediments to domestic petroleum production," also urges the government to "encourage economically efficient energy conservation"— words that would please Public Citizen. And for its part, Public Citizen, for all its stridency for renewable fuels, declares that natural gas "can serve as a 'bridge fuel' for the next several decades." Natural gas is an, ugh, fossil fuel produced by those, ugh, API members.

And who are the biggest investors in solar energy, the darling technology of environmentalists? Two biggies of Big Oil and the API: Amoco Corp. and Mobil!

Position statements of environmental and industry groups also frequently agree in their advocacy of "least-cost" energy options—that is, those forms of energy with the lowest *total* economic, environmental, and societal cost. (Unfortunately, disagreement arises in the calculation of the values given the component costs of the total.)

Too, there's general acknowledgement, even among groups lobbying for individual technologies, that the nation should strive for a broad diversity of energy sources.

Finally, there's recognition—and not solely among Adam Smith-style free marketers—that heavy-handed government intervention sometimes can be counterproductive.

A thoughtful econvironergy policy that balances the three Es would build on these and other points of agreement. For starters, the policy should address conservation, an area given too short shrift by President Bush's February proposal.

For all its energy profligacy, the nation has made remarkable progress in energy efficiency. From 1979 to 1989 the amount of energy needed to produce a dollar of GNP declined 20%—from 25,000 to 20,000 Btu. Energy consumption per capita meanwhile fell nearly 7%—from 349,000 to 325,000 Btu. Industry has led the way, trimming its energy demand 17% (from 23 trillion Btu to 19 trillion Btu) during the same period.

Lately, though, the progress has been slowing. Clearly, the U. S. does not have an energy-conservation ethic. It should. Reminds Michael Shepard, director of the energy policy at the Rocky Mountain Institute, a Snowmass, Colo., resource-policy group: "It is much cheaper to save energy than to produce new sources."

Mr. Shepard believes the nation can slash its current energy demand by as much as two-thirds. Although that sort of cutback would carry an intolerable cost to the economy— one of the Es—if forced too quickly, it's obvious that enormous opportunities for better efficiency exist.

For example, the Electric Power Research Institute, Palo Alto, Calif., the research arm of the electric-utility industry, estimates that energy savings of 25% are possible in electricity usage alone. One way of achieving that, says the Energy Foundation, a San Franciso-based energy grant-making organization, is by changing utility regulation "to make energy efficiency profitable." The foundation reports that Pacific Gas & Electric Co., capitalizing on new regulatory incentives in California, projects it can save 37 billion kw-hr of electricity in the next 10 years—and save customers $3.5 billion in the process.

Further gains also are possible in industrial energy usage. "It's not true that all the easy stuff has been done," claims James L. Wolf, executive director of the Washington-based Alliance to Save Energy. Upset that many companies have diminished the role of their corporate energy managers, he cites the savings potential offered, for example, by installing energy-efficient lamps in plants and offices. He also points out that 36,000 miles of uninsulated pipe still exist in U. S. manufacturing facilities.

For the most significant savings, though, the nation must tap the transportation sector. It accounts for 10 million barrels of the U. S.'s daily energy consumption of 17 million barrels of oil. Most of that oil is burned by automobiles.

It would be folly, in a nation of vast distances, to suppress the role of the automobile in the U. S. But surely Americans can be enticed to drive less. Commuter carpooling and mass transit, to name two initiatives, should receive greater encouragement.

Dramatically greater savings, though, would come from a significant increase in the federal gasoline tax, which currently is far lower in the U. S. than in other industrialized countries. Such a boost has been advocated by many, even by some executives in the petroleum industry. (Conoco Inc. CEO Constantine Nicandros, notably, has urged a 50¢/gallon hike.) Raising the tax not only would curb wasteful driving, but also would be a quick and significant revenue-raiser for the ailing federal treasury.

But a gas-tax hike is likely to remain a non-starter. "It's just not politically doable," concludes the Alliance to Save Energy's Mr. Wolf, recalling the machinations Congress went through last year before approving a meaningless 5¢/gallon rise in the tax. The issue boils with regional conflicts. Moreover, the tax falls disproportionately on the poor. Were a large increase to be enacted, it would need to be accompanied with a rebate program for lower-income drivers.

Wiser would be a Btu tax or other broadbased levy on all forms of energy. Although opposed by manufacturers, such a tax not only would spur conservation across the board, but could be adjusted so that cleaner-burning fuels received lower rates.

Another way to achieve transportation-fuel savings would be to toughen the Congressionally imposed corporate average fuel economy (CAFE) standards for cars. They're credited with doubling the average miles per gallon of the nation's auto fleet (to 29). A further toughening is likely this year.

Critics, however, fret over the economic penalty of the standards. They point out that the increase has been achieved so far by downsizing, which has made cars less safe, and by use of lighter-weight materials and improvements in engine efficiency, which have added to a car's sticker price. Also, contends Marvin Kosters, resident scholar at Washington's American Enterprise Institute, "the same reductions would have come anyway as a result of higher oil prices."

A preferred conservation strategy would be to rely on market incentives—invariably more effective than government mandates. Worthy of experimentation is the "feebate" idea proposed by the Rocky Mountain Institute. Under this concept, which likely will become law in California this year, new-car buyers would either pay a fee or get a rebate when they register their vehicles. If they buy a more fuel-efficient car than owned before, they'd get a rebate; if not, they'd pay a fee. The fees would pay for the rebates.

Besides conservation, an econvironergy policy would emphasize development of alternative fuel sources (alternative, that is, to oil). They're needed. Many supporters of stepped-up oil development admit that. Says one, Sen. Alan Simpson (R, Wyo.), who represents an important oil-producing state: "It's time to realize there are things in this world other than oil."

> *"For all its energy profligacy, the nation has made remarkable progress in energy efficiency."*

Alternatives include a host of "renewable" forms of energy—solar, wind power, geothermal, and biomass, for example. Among others: nuclear power; synthetic fuels such as oil from shale and gasified or liquefied coal; "clean" coal (with its sulfur and nitrogen, but unfortunately not carbon, pollutants reduced); and natural gas.

In the long term the nation should—and will—be fueled extensively by renewables. Several of the technologies, in fact, are carving market niches now. Controversy arises, however, over exactly how—and how aggressively—to speed their development.

Environmental groups call for increased federal direct grants, tax subsidies, and mandates. That sounds logical; clearly, the sooner the nation unhooks from oil, the better. But there's danger in doing it too quickly. Force-feeding technologies can result in wrong choices and would impose an unacceptable economic cost. Besides, observes Carleton Jones, Houston-based director of the Arthur D. Little consulting firm's energy practice, "every time the government steps in, it slows things up."

Indeed, in the past, most new energy sources have been developed with little government intervention. Points out Murray Weidenbaum, director of the Center for the Study of American Business at Washington University in St. Louis: "Changes in consumer demand from whale oil to kerosene to gasoline did not result from an act of Congress or a subsidy from Treasury. The major force for change was successive movements in the price of different forms of energy."

To be sure, however, federal R&D backing for renewables is advisable. Spending for renewables' research, which has been raised admirably by Mr. Bush, should be increased still more.

Yet, even such an advocate of renewable fuels as Mr. Wolf at the Alliance to Save Energy isn't asking for *new* money; he'd be content to see federal R&D funds shifted toward renewables from nuclear and clean coal, which now get heaviest emphasis. Some other analysts,

IW'S 3-POINT PROGRAM

Incentives for conservation, alternatives, and exploration.

Unlike education, drugs, urban poverty, or many other public-policy issues, the solution to the U. S.'s energy problem is clear-cut.

As President Bush's energy-strategy proposals move into Congress, IW suggests simultaneous actions in three major areas that legislators need to take to reverse the nation's growing dependency on foreign oil.

Here are highlights of IW's *econvironergy* program—linking energy needs with considerations of the economy and the environment:

1. Energy conservation. Corporate average fuel economy (CAFE) standards should be phased out in favor of a free-market incentive approach, perhaps along the lines of the "feebate" concept that has been proposed in California.

Utility regulations should be changed to offer incentives for efficiency gains by electricity users. A broad-based energy tax—preferable to a politically unsalable gasoline-tax hike—should be considered for use if incentive approaches fail.

2. Alternative forms of energy. Research should be stepped up, but not force fed, on solar energy and other "renewable" sources.

So should work on electric and hydrogen-powered cars. "Clean coal" development should continue, as well as work on synthetic fuels. Natural gas, hailed by environmentalists, should be encouraged as a transition fuel, especially in fleet vehicles; regulation of natural-gas pipelines should be eased.

Another transition technology—nuclear power—should be revived; Washington should speed the licensing process and conduct research on spent-fuel disposal and standardized plant design.

3. Domestic oil and gas exploraton. The most promising area of the Arctic National Wildlife Refuge coastal plain, a tiny part of the refuge, should be opened to drilling to learn if economically recoverable oil is present. So should selected tracts of the Outer Continental Shelf. Leasing revenues from each should be spent on renewable fuel research.

meanwhile, recommend shifting federal aid (most of it currently in the form of subsidies) away from ethanol fuel. They argue that although ethanol is a renewable form of transportation fuel, producing it results in a net energy loss. Opponents also say ethanol delivers less mileage than gasoline and worsens urban smog.

In supporting individual technologies, it's clear the government should follow these precepts:

• It should resist political pork barrel in its funding decisions.
• Rather than emphasizing subsidies, it should focus on performing basic research and providing seed money for demonstration projects.
• Rather than trying to pick winners and losers, it should back a diversified portfolio of technologies, assuring a level playing field for each. Then, if a particular technology appears to be breaking out of the pack, it should get an extra push.

The R&D effort should extend not only to renewables, but also to those based on fossil fuels such as clean coal (the U. S. has vast coal resources) and synfuels. The aim of such research, notes Michael Koleda, president of the Washington-based Council on Alternate Fuels, shouldn't be merely to increase domestic energy sources to "back out" oil imports, but also to lower the cost of imports. "If it looks as if we're serious about alternate fuels, OPEC oil producers will be quaking in their boots that they'll lose our market," he declares.

Since transportation accounts for such a huge percentage of U. S. oil consumption, an important area of research should also be storage batteries; if they could be made smaller, nonpolluting electric cars would have greater range and practicality. Even more promising: development of hydrogen-powered cars using hydrogen produced directly from water through solar or other renewable energy sources.

Unfortunately, even if the U. S. were to make an all-out push to bring them on, alternative forms of energy couldn't play significantly greater roles anytime soon. That's why natural gas—a fuel liked by both environmentalists and energy developers—looms so important. It's the perfect interim fuel, ample in supply and easier on the environment than oil or coal. But the Federal Energy Regulatory Commission needs to speed the certification of new pipelines to make the fuel more available.

Use of natural gas in the utility and industrial sectors—already spurred by last fall's clean-air legislation—should be expanded. Vast potential also lies in the conversion of fleet vehicles (13 million of them are on the road in the U. S.) to natural gas.

Among utilities, another transition technology that could prove of significant help is nuclear. Although Japan and European countries rely heavily on nuclear power, safety fears and high costs have given it a bad name in the U. S. No nuclear powerplant is even on order.

However, rising concern about global warming and acid rain resulting from the burning of fossil fuels is prompting at least some environmentalists (most still cringe at the thought) to look more favorably on nuclear power. The government can help revive the technology by: conducting research on safe disposal of spent fuel; shortening the time required for licensing nuclear plants; assisting in the development of a new generation of safer plants of standardized design; and by leading a public education campaign on the advantages of nuclear power.

Unfortunately, development of alternative fuels—even the transition fuels—will take years. What

can be done in the meantime?

"We need to buy time," asserts G. Henry M. Schuler, director of energy-security programs at the Center for Strategic & International Studies, Washington. And one of the best ways of doing that, he indicates, is by "prompt exploration of domestic supply sources," including the heretofore environmentally sacrosanct Arctic National Wildlife Refuge. Even in a time of price uncertainty such as now, such exploration would attract risk capital, Mr. Schuler believes.

Opening ANWR to drilling wouldn't be the environmental cataclysm often pictured. Although ANWR itself is about the size of South Carolina, notes API President Charles DiBona, the area affected—on the refuge's coastal plain—is only the size of Washington's Dulles Airport. "With ANWR," he says, "there's likely to be a reasonable supply of energy in the '90s, and we [the oil industry] can keep prices down until technologies are developed to solve the long-term problem. The alternative is to raise energy prices relative to other economies or impose regulatory restraints that will stifle growth."

It would be folly to suppress the role of the automobile in the U. S. But surely Americans can be enticed to drive less.

Certainly the oil industry should be held to the strictest environmental standards in drilling ANWR. Drilling *should* proceed, though; the Interior Dept. estimates a 46% chance that the refuge holds economically recoverable oil. The nation needs at least to know how much oil is truly there; a decision to produce it could be made later.

Similarly, moratoriums against drilling on the Outer Continental Shelf (OCS) should be lifted in selected promising areas.

To make these environmental sacrifices more palatable, Congress should require that income from ANWR and OCS leases be spent on research on renewable forms of energy. The money also could be directed to R&D on "enhanced oil-recovery" techniques to coax more production from existing oil fields.

The debate over ANWR and OCS likely will be the hottest environmental fight in the 102nd Congress. There's danger that legislators will become so ensnarled in this controversy that they won't get around to other, larger issues in the Triple E triad.

That would be a tragedy. The U. S. cannot blunder along without an econvironergy policy any longer.

THE NEGAWATT REVOLUTION

Using existing technology, says this expert, we can save three fourths of all electricity used today.

Amory B. Lovins

Amory B. Lovins, a physicist, directs research at Rocky Mountain Institute, a nonprofit resource policy center in Snowmass, Colorado.

Ron Perkins's colleagues at Compaq Computer Corporation were incredulous. Perkins, manager of facilities resource development, believed Compaq's next Houston office building could be designed to use a fourth less electricity per square foot than the previous one. "Why make it so hard on yourself?" they asked. "Our designs are already excellent. Savings of 5 or 10 percent would be pushing your luck." But a year later, when construction was completed, the building's use of electricity proved a third less than previous designs, thanks to the new electricity-saving technology that Perkins had harnessed. The payback period on the energy-saving equipment was only a few years. Perkins has since designed another building in which he expects to slash electric use by still another fifth.

The innovative equipment that Perkins uses in his buildings—chiefly advanced lighting fixtures, electronic ballasts, high-efficiency lamps, and lighting controls—are part of a flood of astonishingly cheap and powerful electricity-saving techniques. They have the potential to add many tens of billions of dollars a year to business's bottom line.

Some companies are already enjoying the rewards that come from saving electricity. For example, Southwire, the largest independent rod, wire, and cable business in the United States, found itself facing hard times in the early 1980s as the energy-intensive heavy industry was squeezed between market prices and manufacturing costs. The company responded by cutting its total energy use per pound of product by half in eight years. The energy reductions—about a 60 percent savings in gas and 40 percent in electricity—yielded virtually all of the company's profits during the tough years of 1980 to 1986, and may have saved 4,000 jobs at 10 plants in six states. Southwire continues to make further improvements, which still generally pay back investment in fewer than two years. Even in less energy-intensive industries, savings from energy efficiency can be dramatic. A couple of years ago, a large company had an energy manager at one of its plants who was achieving annual energy savings of $3.50 per square foot. "That's nice—a few million extra on our bottom line," said one of the company's executives. He then added, in the same breath: "But I can't really get excited about energy. It's only a few percent of our cost of doing business." Such thinking is stalling energy improvements throughout the corporate world. Installing energy-efficient equipment may not be sexy, but the savings are real. If this executive's company had achieved similar savings at all its facilities worldwide, its total net would have gone up by 56 percent.

Industry has already seen major savings from its fuel-conserving programs initiated during the Arab oil embargo of the early '70s. Yet companies spend more than twice as much on electricity as on oil. Unbeknownst to many in industry, in the past few years there have been tremendous advances in electric effi-

From *Across the Board*, September 1990, pp. 18-23. Reprinted by permission of the author and *Across the Board*, The Conference Board, 845 Third Avenue, New York, NY 10022.

ciency. Electricity-saving technology is evolving so quickly that most of the best options now on the market didn't exist last year. Today, you can save twice as much electricity as you could five years ago, at only a third the real cost. Practically every building, however modern, can be made much more efficient.

American companies have a $93 billion annual electric bill, with 25 to 45 percent of the total going toward lighting—about three fourths of it directly and a quarter to counteract the heat generated by the lights. In most commercial buildings, lighting consumes more than a third of the electricity used—upward of half when the cooling load is considered. Yet according to studies by Lawrence Berkeley Laboratory (the leading national lab on saving energy in buildings) and Rocky Mountain Institute, 80 to 90-plus percent of this lighting energy could be saved by fully converting to today's most efficient lighting equipment.

The vast range of efficient lighting hardware now available fits almost any need, providing unchanged lighting levels with less glare, more pleasant and accurate color, no flicker, and no hum. Upgrading a typical office fluorescent lighting system can be accomplished by installing computer-designed reflectors, which deliver virtually the same light from half as many lamps; new lamps that give off more light per watt and nicer color; sophisticated high-frequency electronic ballasts, which start and regulate the current of the lamp and can now power four lamps instead of two; and several kinds of controls. As a result, a company will need only half as many lamps and a quarter as many ballasts, which should save it 70 cents per square foot on maintenance costs—nerly half the total cost of the upgrade. Typical direct energy savings are about 70 to 90 percent, and including the 35 to 40 percent "bonus" for saved space-cooling, most paybacks are well under two years.

Even juicier savings come from converting incandescent lamps, such as the ubiquitous floodlamps in can fixtures, to compact fluorescent lamps. These lamps can cut lighting bills by 75 to 85 percent, and they last 4 to 13 times as long, thereby more than paying for themselves just by reducing maintenance costs on replacement bulbs and the labor needed to install them.

Other improvements can boost lighting savings by another third or more, including better maintenance, lighter-colored finishes and furnishings to distribute light better, top-silvered blinds and glass-topped partitions to bounce sunlight three times as far into buildings, polarizing lenses that make reading easier by almost eliminating glare, half-watt electroluminescent panels to replace 30- to 50-watt EXIT signs, and miniature tungsten-halogen spotlights for displays.

Together, these commercially available lighting innovations have the potential to save about a fourth of all the electricity in the country, at a net cost somewhat

less than zero. In fact, Rocky Mountain Institute estimates that because the amount saved on maintenance costs would be more than the cost of the electricity-saving devices, the average cost of replacement will be about *minus* 1.4 cents per kilowatt-hour. In the United States this would displace 120 Chernobyl-size power plants costing about $200 billion and eliminate more than $30 billion a year in utility operating costs. This may be the biggest goldmine in the whole economy.

Opportunities nearly as dramatic abound in every other electricity-consuming device. Together, they can cut U.S. electricity consumption by another half.

After lights, motors are probably the next fattest opportunity. Motors use at least two thirds of industrial electricity and some 53 to 60 percent of all the electricity in the country—more than $90 billion a year worth, or about 2 percent of our gross national product. In fact, making the electricity to run U.S. motors now uses more fuel than is consumed by all U.S. highway vehicles.

A typical big industrial motor consumes electricity costing some 10 to 20 times its own total capital cost per year. Over a motor's life, a 1 percentage point gain in efficiency typically adds at least $10 per horsepower to the bottom line. Direct efficiency gains averaging about three-and-a-half percentage points are currently possible, which for a motor-intensive company, such as a paper mill, can create enough savings to turn around a foundering firm.

Two measures that have gained wide acceptance are buying only high-efficiency new motors, which can now save twice the electricity that they could a decade ago, and using electronic speed controls. Immediately replacing a standard induction motor with a high-efficiency model has many advantages. In addition to cutting electricity costs, the replacement will last twice as long because it runs cooler and has better bearings, will need fewer capacitors to boost the motor's "power factor" (the fraction of electricity fed into the motor that actually turns it rather than heating it), and will work better with adjustable-speed drives.

Electronic speed controls have become popular because many machines, especially pumps and fans, need to vary their speed to match production needs. Before electronic adjustable-speed drives became widespread and affordable, output was usually varied by running the pump or fan at full speed while "throttling" its output with a partly closed valve or damper—like driving with one foot on the accelerator and the other on the brake. Today, electronic speed controls can eliminate this waste. When you need only half the flow from a pump, you can save almost seven eighths of the power because its energy needs vary as roughly the cube of its flow. In all, electronic adjustable-speed drives can save 14 to 27 percent of total U.S. motor energy, with paybacks of a year or two. Only a few percent of this opportunity has yet been grasped.

Currently, most engineers consider just these two measures, ignoring the other half of the total electricity-saving potential in motor systems. At Rocky Mountain Institute we have identified 33 kinds of further improvements that could be made to motor systems, comprising the choice, maintenance, sizing, and controls of motors and the systems that supply electricity and transmit torque from the motor to the driven machine. Implementing all 35 of the improvements can cut the motor systems' use of electricity in half, for a potential national savings equivalent to 80 to 190 giant power plants. (This figure doesn't even take into account the potential for another 50 percent savings on the remaining electricity bill from improving the machinery that the motors are driving.) Because you pay for only seven of the 35 improvements—the rest are cost-free by-products—the average payback on the doubled efficiency is only about 15 months.

Capturing the savings depends on simultaneously doing many things right. For example, to double a motor's lifetime, the motor's shaft must be kept precisely aligned with the shaft it's driving or the bearings will fail prematurely, and the bearings themselves must be lubricated by someone with clean hands to prevent dirt from getting into the grease and eating the bearings—both simple steps that frequently aren't taken in American industry.

Companies should switch from V-belts, which stretch, slip, and require so much tension to stay in place that they harm the bearings, to "synchronous" belts, which have teeth that engage sprocket lugs so the belt doesn't slip, and fiberglass or Kevlar bands inside so it doesn't stretch. Not only would such a conversion save about 5 to 15 percent of the transmitted energy, but it would cost about *minus* a dollar per kilowatt-hour because of immense maintenance savings.

Maintenance itself must also be improved. Poor maintenance ruins costly motors, wastes energy, and needlessly incurs downtime costs that can exceed $10,000 per hour. Measurements by General Electric Company suggest that in the United States between $1 billion and $2 billion worth of electricity is wasted each year by the damage done to the iron cores of motors as a result of poor repair practices used to remove old windings. An alternative technique using only gentle warmth to loosen old windings for removal causes no damage and is faster and cheaper, but is known to relatively few motor repairmen. Another step to improve the state of motor maintenance would be to make lubrication and other motor upkeep a white-lab-coat profession, with a "motor doctor" who makes house calls and administers precisely metered dosages of special medicine to motors.

When it comes to energy efficiency, details matter. Jim Clarkson, the mastermind behind Southwire's dramatic savings, found that before executives toured plants, motors were often given a coat of shiny new paint. Over the years, so many coats built up that the heat couldn't get out. Today, you can't repaint a Southwire motor without first stripping off the old paint. Clarkson also discovered that of the typical 6 percent power loss between the meter and the machinery, three fourths could be saved, with a payback of around two years, just by installing wire twice as fat. The wire in most big buildings, it seems, is chosen by low-bid electricians told to meet the local building code, which is meant only to prevent fires, not to save money.

Many energy-saving techniques require no investment. Clarkson found he could save Southwire 10 percent of its motor electricity bill by turning off idling motors. At some machines he installed a red light that went on when high peak-period utility charges approached, and told the operators that if they would take a coffee break when the light went on, the company's overall profits would be more than if they kept working.

Almost every other electricity-consuming device holds potential savings as well. Replacing a desktop computer with an equally sophisticated laptop model can save up to 95 percent in electricity—enough to pay for the difference in cost for the laptop—while improving safety, portability, ergonomics, and space use, and eliminating the need for a costly uninterruptible power supply. Simple improvements to such common office devices as laser printers and photocopiers can save most of their energy and help avoid multimillion-dollar investments to expand air-conditioning capacity to handle machine heat in older buildings. For both these devices, for example, "cold fusing"—setting the toner onto the paper with a cold compression roller rather than a hot drum—saves 90 percent of the electricity, eliminates fumes and warm-up time, and gives twice the life with half the maintenance. Further savings can be gained by installing controls that turn such machines off or into a standby mode when not in use.

Making windows more efficient also saves money. Most buildings use plain glass, so you're hot in summer and cold in winter. But new "superwindows" provide year-round comfort. Some let in 60 percent of the visible light, thereby displacing electric light and the heat it produces, while admitting only two percent of the sun's heat. Other windows, designed for cold climates, can insulate up to six times as well as double glazing, and can even gain more heat than they lose in the winter while facing in any direction, including north. At Rocky Mountain Institute's research center, we have no furnace in a climate that goes down to −47°F. Our cold-climate windows not only permit us to do without a furnace, cutting winter heating bills by $1,000 a month, but also have reduced our building's net capital cost. The reason: We saved more by eliminating the furnace and ductwork than it cost us to install the superwindows and superinsulation. Fully used, superwindows could save the United States four million barrels worth of oil and gas per day, at costs of a few dollars per barrel—far cheaper than drilling for more.

Increasingly popular superefficient appliances are another fountainhead of savings. There are refrigerators and freezers on the market that consume 10 to 20 percent of the usual amount of energy, commercial refrigeration systems that save more than 50 percent, and televisions and high-performance showerheads that save 75 percent. The collective results can be astounding. My 4,000-square-foot home's lights and appliances cost only $5 a month to run—a 90 percent savings over normal bills. Installing new technology has also resulted in a more than 99 percent savings in space and water heating and a 50 percent reduction in water usage. Best of all, the payback period for my home's improvements was only 10 months, and that was with 1983 technology.

Belt-tightening: Electric drives can be made 5 to 15 percent more efficient by switching from V-belts (top) to synchronous belts, such as the Poly Chain GT (bottom) made by the Gates Corporation.

What do all these opportunities add up to nationwide? A comprehensive study by Rocky Mountain Institute suggests that if the thousand or so best electricity-saving innovations now on the market were fully installed in U.S. buildings and equipment, they'd save about three fourths of all electricity now used, at an average payback of slightly more than one year, while providing unchanged or improved services.

Some of these innovations are now becoming popular. Sales of many kinds of electricity-saving devices are more than doubling every year. Advanced windows, for instance, have gone from 1 percent to more than 60 percent of the insulated-glass market in just a few years. More than 20 million compact fluorescent lamps are expected to be sold this year.

Yet progress in converting to electricity-saving technologies has so far been much slower than it should be. A major obstacle to efficiency is the indifference or outright opposition of about a third of the utility industry. Some utilities have exemplary (and highly profitable) programs to help their customers use electricity more efficiently, but others are still trying to sell more electricity, not less. This reflects a basic misunderstanding of their business. Customers don't want kilowatt-hours; they want services such as hot showers, cold beer, lit rooms, and spinning shafts, which can come more cheaply from using less electricity more efficiently. Good programs to save commercial and industrial electricity cost only about a half cent per kilowatt-hour, which is severalfold cheaper than just operating a coal or nuclear plant, and 10 to 20 times cheaper than building a new one.

Many utilities, conditioned by a century of rising sales and revenues, still forget that, like any other business, they can make money on margin instead of volume. This is true even for utilities with overcapacity. If it's cheaper to save electricity than to make it, then a utility should save it regardless of how much capacity it has, because capacity is a sunk cost, whereas marginal variable costs can still be saved. New regulations in California, New York, and Massachusetts are encouraging such choices by decoupling utilities' profits from their sales and letting them keep part of the savings as extra profit, thereby directly rewarding efficient behavior. A dozen more states are developing similar incentives for their utilities.

A second obstacle to efficiency is that many electricity-saving devices are purchased by people who won't be paying their running costs and thus have little incentive to consider efficiency when comparing prices. Furthermore, most customers don't know what the best efficiency buys are, where to get them, or how to shop for them. Business customers have trouble conveniently buying integrated packages of efficient equipment; only a handful of companies can do everything to your lighting systems and do it right, and nobody as yet offers such a service for completely overhauling your motor systems.

Perhaps the most critical obstacle to overcome is the "payback" gap" between consumers and utilities. If you invest your own money to save energy in your business or home, you'll probably want it back within a couple of years, implying a real discount rate upward of 60 percent a year. In contrast, if a utility has to build or expand a power plant to meet increased demand, it'll probably use a 20-year payback horizon, or about a 5 or 6 percent real annual discount rate. The utility's great technical and financial strengths, low information costs, diversified risk portfolio, and steady cash flow allow it to take a more relaxed view of investments than consumers can.

Although these respective discount rates are rational for each party, for the American economy their tenfold payback gap makes us invest too little in efficiency and too much in new power plants, misallocating some $60 billion a year.

Many utilities are seeking to equalize the disparity in discount rates between them and their customers by financing efficiency via concessionary loans, rebates, and even gifts. Southern California Edison Company, for instance, has given away more than 800,000 compact fluorescent lamps because it's cheaper than operating the company's existing power plants. Utilities are also beginning to explore leasing electricity-efficient lamps and motor systems to consumers. For example, a 20-cent-per-lamp-per-month charge on a consumer's electric bill lets him pay for the efficiency improvement over time, exactly as he now pays for power plants.

Rocky Mountain Institute has come up with an innovative way to foster such efficiency gains: creating negawatt markets. Negawatt markets would treat saved electricity as a commodity, just like copper, wheat, and pork bellies. Negawatts (saved watts) would be subject to competitive bidding, arbitrage, and secondary markets. Some entrepreneurial utilities even want to become "negawatt brokers" and create spot, future, and options markets in saved electricity. Such markets could be highly profitable: Arbitrageurs make money on spreads of a fraction of a percent, but the spread in discount rate between utilities and their customers is closer to 1,000 percent.

Perhaps the strongest incentive to create negawatt markets is their win-win solution to many environmental problems. Because it's now generally cheaper to save fuel than to burn it, global warming, acid rain, and urban smog can be reduced not at a cost but at a profit. A 1989 Swedish State Power Board study found that by using electricity twice as efficiently, Sweden could fulfill the electorate's mandate to phase out the nuclear half of the nation's power supply while simultaneously supporting 54 percent growth in real gross national product, reducing the utilities' carbon dioxide output by a third, and cutting the total cost of electrical services by nearly $1 billion per year. This finding is all the more encouraging because Sweden has a severe climate, a heavily industrialized economy, and perhaps the world's highest aggregate energy efficiency to start with.

In the United States, a conservative study by the American Council for an Energy-Efficient Economy found that reducing sulfur emissions from Midwestern power plants by 55 percent through scrubbers and fuel-switching would cost about $4 billion to $7 billion. Yet it also found that by using electric savings to pay

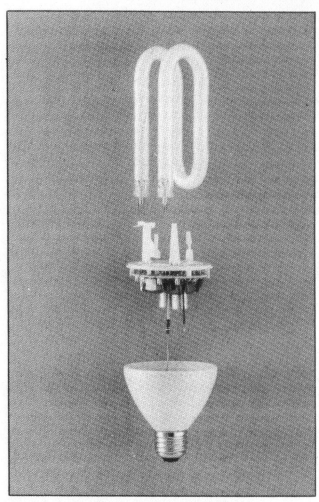

Bright Idea: Replacing a 75-watt incandescent bulb with an SL fluorescent bulb from Philips (above) cuts electricity use by 76 percent.

for the cleanup, that cost would change to a $4 billion to $7 billion profit.

Today, the best energy investments provide the most environmental protection. A study by Rocky Mountain Institute found that a dollar spent on nuclear power will displace less than a seventh as much coal-fired electricity as would spending the same dollar on efficient use of electricity. This means that each dollar spent on nuclear energy will result in the release of at least six units of extra carbon that would not have been released if it had been spent instead to improve electric efficiency. From this perspective, nuclear power makes global warming worse. Most of global warming, Rocky Mountain Institute analysts believe, can be abated by advanced energy-saving techniques at a net profit of about $200 billion per year.

Given the negawatt markets' profit opportunities, why are a market-oriented Administration and many in the business community opposing aggressive abatement of energy-related pollution such as acid rain and global warming? Probably because they think abatement will cost extra. Eminent economists running computer modeling studies have shown costs running into the trillions of dollars for reducing fossil-fuel combustion by, say, 25 percent by 2005. But these models base costs on economic theory, while ignoring the results of real-life efficiency programs. What those economists presumably have in mind is that because fossil-fuel use declined when energy prices quadrupled after the Arab oil embargo of 1973, a decline today in fossil-fuel use must be accompanied by similar price increases. As a result, their models only ask how high energy prices need go, based on historical elasticities, to reduce fossil-fuel use by a given amount. They forget that major efficiency gains are cost-effective at well below current fuel prices. It is a national tragedy that a few noted economists' ignorance of the empirical costs of energy efficiency has so widely spread the myth of costly environmental protection that it threatens to paralyze energy-efficiency and anti-pollution programs, thereby blocking major profit opportunities for the private sector.

Energy efficiency ultimately represents a trillion-dollar-a-year global market. American companies have at their disposal the technical innovations to lead the way. Not only should they upgrade their plants and office buildings, but they should encourage the formation of negawatt markets. And they should let the United States Government know that the best energy policy for the nation, for business, and for the environment is one that focuses on using electricity efficiently—for it's the only policy that makes economic sense.

HERE COMES THE SUN

The technology exists today to produce most of our energy from the sun, wind, and heat from the earth. Tapping these sources, though, will require a vigorous public commitment to push renewable energy into the mainstream.

Christopher Flavin and Nicholas Lenssen

Christopher Flavin is vice president for research and Nicholas Lenssen is a research associate at the Worldwatch Institute.

Imagine an energy system that requires no oil, is immune to political events in the Middle East, produces virtually no air pollution, generates no nuclear waste, and yet is just as economical and versatile as today's.

Sound like a utopian dream? Hardly. Scientific and engineering breakthroughs now make it practical to begin producing our electricity, heating our homes, and fueling our cars with renewable energy—the energy of the sun, the winds, falling water, and the heat within the earth itself.

The conventional wisdom among government leaders, energy experts, and the public at large is that we are stuck with dependence on fossil fuels—whatever the cost in future oil crises, air pollution, or disrupted world climate. But, with continuing advances in technology and improvements in efficiency that make it possible to run the economy on reduced amounts of power, a renewables-based economy is achievable within a few decades.

In California, the future has already begun to emerge. The state that always seems to be a decade ahead of everyone else is once again ushering in a new era. The current revolution is subtle, yet momentous. It's evident in the spinning white wind turbines on the hills east of San Francisco and the glinting mirrored solar-thermal troughs set in rows in the Mojave Desert.

Since the early 1980s, California has built no coal or nuclear power plants and has been harnessing renewable energy and improving energy efficiency with a vengeance. The state gets 42 percent of its electricity from renewable resources, largely from hydropower, but also 12 percent from geothermal, biomass, wind, and solar energy—virtually all of it developed in the past decade.

But, for all its success, California's energy revolution is a bit one-dimensional. Its electricity system has been altered, but its cars and homes are still powered largely with fossil fuels. The next step is to find a way to run the whole economy on renewable energy sources.

The missing link is hydrogen—a clean-burning fuel easily produced using renewable power and conveyed by pipeline to cities and industries thousands of miles away. Hydrogen shows great promise as the new "currency" of a solar economy. It can be used to heat homes, cook food, power factories, and run automobiles. Moreover, the technologies to produce, move, and use hydrogen are already here in prototype form.

The challenge of creating a clean, efficient, solar-powered economy is essentially that of reducing the cost of the various constituents of a solar-hydrogen system—from the manufacturing costs of wind turbines to the efficiency of new automobiles.

Here, the pace of progress will be heavily influenced by government policies. The change will come slowly if governments continue to shower favors on fossil-fuel based energy sources. To encourage the adoption of alternative energy sources, policymakers will need to reduce subsidies, raise taxes on fossil fuels, increase research funding on new energy technologies, and provide incentives to private industry for renewable energy development.

Power from the Sun

Renewable resources now provide just 8 percent of the energy used in the United States, but government scientists estimate that renewables could supply the equivalent of 50 to 70 percent of current U.S. energy use by the year 2030 if the government got behind the effort. This estimate is based on the abundance of renewable resources and the technological advances made in tapping them since the mid-1970s.

Such improvements have reduced the cost of renewable energy technologies by 65 to 90 percent since 1980, a trend that is projected to continue through the 1990s [see **Table 1**]. Increasingly, as governments begin to consider the full costs of fossil fuels (including air pollution and threats to national security), renewables look like a bargain.

Renewables' ability to go head-to-head with coal, oil, and nuclear power has credence even among some utility executives. Greg Rueger, senior vice president of California's Pacific Gas & Electric Company, the nation's largest utility, says "many renewable-generation options are technically feasible today, and with encouragement can prove to be fully cost-competitive...within 10 years."

Solar energy probably will be the foundation of a sustainable energy economy, because sunlight is the most abundant renewable energy resource. Also, solar energy can be harnessed in an almost infinite variety of ways—from simple solar cookers now used in parts of India to gleaming solar collectors on rooftops in Beverly Hills.

Using sunlight to generate electricity has been a dream of scientists and energy planners since the early 1950s, when the first practical photovoltaic cell was invented. This device converts the sun's rays directly into electric current via a complex photo-electric process. Photovoltaic technology has advanced for four decades now, making it possible to convert a larger share of sunlight into electricity—as much as 14 percent in the most advanced prototype systems. Manufacturing costs also have fallen, making this technology a competitive energy source for some limited applications.

Photovoltaic solar cells are now widely used, for example, to power electronic calculators, remote telecommunications equipment, and electric lights and water pumps in Third World villages. These and dozens of other uses created a $500-million market for photovoltaics in 1990, with sales projected to double every five years. The 50 megawatts' worth of cells produced in 1990, though, is only sufficient to power about 15,000 European or Japanese homes.

Table 1: Costs of Renewable Electricity, 1980-2030[1]

Technology	1980	1988	2000	2030
	(¢ per kilowatt-hour)			
Wind	32[2]	8	5	3
Geothermal	4	4	4	3
Photovoltaic	339	30	10	4
Solar Thermal				
trough with gas assistance	24[3]	8[4]	6[5]	—[6]
parabolic/central receiver	85[7]	16	8	5
Biomass[8]	5	5	—	—

[1]All costs are averaged over the expected life of the technology and are rounded; projected costs assume return to high government R&D levels. [2]1981. [3]1984. [4]1989. [5]1994. [6]Estimates for 2030 have not been determined, primarily due to uncertainty in natural gas prices. [7]1982. [8]Future changes in biomass costs depend on feedstock cost.
Source: Worldwatch Institute, based on Idaho National Engineering Laboratory et al., *The Potential of Renewable Energy*, and various sources.

During the past two decades, the cost of photovoltaic power has fallen from $30 a kilowatt-hour to just 30 cents. (This figure is composed almost entirely of manufacturing costs, since solar power requires no fuel.) This is still four to six times the cost of power generation from fossil fuels, so further reductions are needed for solar power to be competitive with grid electricity.

In April, the Texas Instruments Company of Dallas announced plans to produce a new solar cell that costs half as much to manufacture as existing models. Several American, Japanese, and European companies expect advances in other photovoltaic designs, cutting manufacturing costs, and improving efficiency levels.

Photovoltaics are already the most economical way of delivering power to homes far from utility lines. This technology will soon become an economical way of providing supplementary utility power in rural areas, where the distance from power plants tends to cause a voltage reduction that is otherwise costly to remedy.

New applications will spur further cost reductions, which is likely to lead to widespread use of solar cells. As they become more compact and versatile, photovoltaic panels could be used as roofing material on individual homes, bringing about the ultimate decentralization of power generation. Around the same time, perhaps a decade from now, large solar power plants could begin to appear in the world's deserts—providing centralized power in the same way as do today's coal and nuclear plants.

Another source of centralized electricity is solar-thermal power, a technology already proving its viability in California's Mojave Desert. Luz International of Los Angeles has developed a solar-thermal system using large mirrored troughs to reflect the sun's rays onto an oil-filled tube, which in turn superheats water to produce the steam that drives an electricity-generating turbine.

Since the mid-1980s, Luz has installed 350 megawatts of solar systems across three square miles of southern California desert— enough to electrify 170,000 homes. The Luz systems turn 22 percent of incoming sunlight into electricity, which is higher than for any commercial photovoltaic system so far. And because they are of modular design, they can be built on a variety of scales.

Solar-thermal electricity is now produced for about 8 cents per kilowatt-hour, close to the cost of that from fossil fuels in California, where extensive pollution controls are required. However, because it relies on mirrored concentrators, solar-thermal power is only practical where there is intense, direct sunlight—conditions found only in arid regions.

Photovoltaics, which are much more effective in hazy or partly cloudy conditions and can be installed even on a very small scale on residential rooftops, are likely to become the more common power source in the long run. Still, both solar technologies will play important roles.

Contrary to what their critics charge, solar energy systems won't require unusually large areas of land to power the economies of tomorrow. In fact, they need less space to produce a megawatt of electricity than does coal-fired power when the land devoted to mining is factored in [see Table 2]. One-quarter of U.S. electricity needs could be met by less than 6,000 square miles of solar "farms," according to the Electric Power Research Institute in Palo Alto, California, the research arm of the U.S. electric utility industry. That's about the area of Connecticut, or less than 8 percent of the land used by the U.S. military.

Table 2: Land Use of Selected Electricity-Generating Technologies, United States

Technology	Land Occupied
	(square meters per 1,000 megawatt-hours, for 30 years)
Coal[1]	3,642
Solar Thermal	3,561
Photovoltaics	3,237
Wind[2]	1,335
Geothermal	404

[1]Includes coal mining. [2]Land actually occupied by turbines and service roads.
Source: Worldwatch Institute, based on various sources.

More Renewable Options

Another form of renewable energy, wind power, can be captured by propeller-driven turbines mounted on towers in windy regions. Though wind power has a rich history in areas such as Holland and the American Great Plains, it has been taken seriously as a major energy source only since the late 1970s.

Technological advance in the design of wind turbines brought down the cost of wind electricity from more than $3.00 a kilowatt-hour in the early 1980s to the current average of just 80 cents. By the end of this decade, the cost of newer models is expected to be around 50 cents per kilowatt-hour, while the cost of coal-fired power will rise above 50 cents as a result of tightening pollution standards.

Most of the cost reductions for wind energy stem from experience gained in California, which has 15,000 wind machines producing about $200 million worth of electricity annually, enough to power all the homes in San Francisco. Denmark, the world's second-largest producer, received about 2 percent of its power from wind turbines in 1990—but still only about one-fifth of that produced in California.

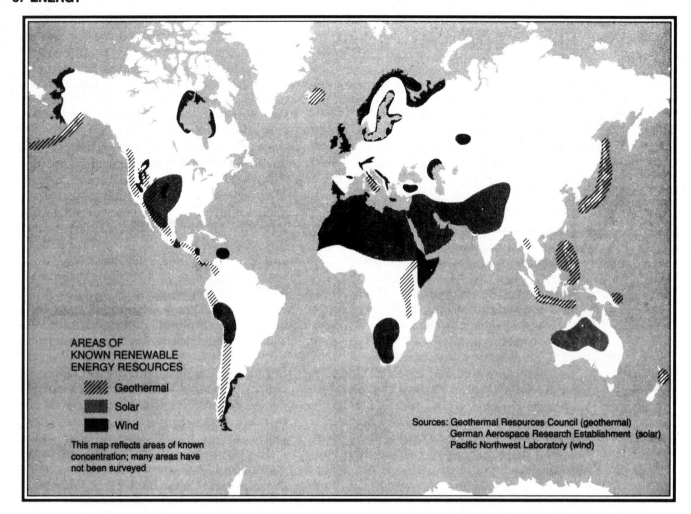

AREAS OF
KNOWN RENEWABLE
ENERGY RESOURCES

///// Geothermal

■ Solar

■ Wind

This map reflects areas of known
concentration; many areas have
not been surveyed

Sources: Geothermal Resources Council (geothermal)
German Aerospace Research Establishment (solar)
Pacific Northwest Laboratory (wind)

Wind power could provide many countries with one-fifth or more of their electricity. Some of the most promising areas for wind energy are in North Africa, the western plains of the United States, and the trade wind belt around the tropics—including the Caribbean, Central America, and southeast Asia. In Europe, the largest wind farms will likely be placed on offshore platforms in the turbulent North and Baltic seas.

U.S. government studies show that one-quarter of the country's power could be provided by wind farms installed on the windiest 1.5 percent of the continental United States. A windy ridge in Minnesota, located less than 400 miles from Chicago, could provide one-quarter of the power the city now uses.

Most of the best land for wind power in the United States is grazing land in the western high plains—costing no more than $40 an acre. If "planted" in wind turbines, an acre of this land could generate $12,000 worth of electricity annually while cattle still graze below. One reason it's not being developed is that the region already has more electrical generating capacity than it can use.

Of course, any energy source has its drawbacks, and wind power development, with its rows of towering turbines, will need to be limited in scenic areas, particularly on coastlines. Further efforts are also needed to reduce the incidental bird kills that plague some wind farms.

The energy captured from burning crop residues, wood, and other forms of "biomass" also could play a role in certain locales. It is already the main energy source in scores of countries. In the future, 50,000 megawatts of generating capacity, 75 percent of Africa's current total, could come from burning sugarcane residues alone. However, biomass energy development is likely to be constrained by environmental issues, including the heavy demands already being placed on much of the world's forests and croplands.

Another potential source of power is geothermal energy—the heat from the earth's core. Already, El Salvador gets 40 percent of its electricity from the earth's natural heat, Nicaragua 28 percent, and Kenya 11 percent. Most Pacific Rim countries, as well as those along East Africa's Great Rift Valley and around the Mediterranean Sea, sit atop geothermal "hotspots." Virtually the entire country of Japan lies over an enormous heat source that one day could meet much of the country's energy needs.

Geothermal energy is not without its environmental drawbacks, including the underground sulfur it tends to release, and development will have to be limited in ecologically sensitive areas. However, this still leaves a vast resource potential, particularly as engineers develop techniques to drill deeper and deeper into the earth's crust.

The Hydrogen Solution

If renewable energy is abundant and economical, then why isn't it being harvested on a larger scale? The answer stems in part from the difficulty of storing and moving energy from ephemeral, intermittent sources such as the sun and the wind. While oil can be moved from remote areas by tanker, and coal by barge, sunshine is hard to carry to far-off cities.

Electric power produced by renewable energy could be stored and transported to the user, but at some cost. Electric batteries are expensive, heavy, and must be recharged frequently. Power lines also are costly, generate potentially dangerous electromagnetic fields, and lose energy over long distances due to resistance in the lines. Nonetheless, extended transmission of electricity is already common: California, for example, relies on hydropower produced nearly 1,000 miles away in British Columbia.

Given the limits on moving electricity, it makes more sense to convert renewable power to a gaseous form that is cheap to transport and easy to store. Hydrogen is an almost completely clean-burning gas that can be used in place of petroleum, coal, or natural gas. It releases none of the carbon that leads to global warming. And it can be produced easily by running an electric current through water—a process known as electrolysis.

Hydrogen can be transported almost any distance with virtually no energy loss. Over distances greater than 400 miles, it costs about one-quarter as much as sending electricity through a wire. Gases are also less risky to move than any other form of energy—particularly compared with oil, which is frequently spilled in tanker accidents and during routine handling.

Hydrogen is much more readily stored than electricity—in a pressurized tank or in metal hydrides, metal powders that naturally absorb gaseous hydrogen and release it when heated. Years' worth of hydrogen could be stocked in depleted oil or gas wells in regions such as the U.S. Gulf Coast.

Moreover, hydrogen can provide the concentrated energy needed by factories and homes. It can be burned in lieu of natural gas to run restaurants, heat warehouses, and fuel a wide range of industrial processes. Around the home, new hydrogen-powered furnaces, stove burners, and water heaters can be developed that will be much more efficient than today's appliances.

The gas can also be used to produce electricity in small, modular generators that turn out heat and power for individual buildings. Such co-generating plants would produce far less pollution than today's power plants do in getting a similar amount of electricity to individual users. Hydrogen can be used to run automobiles, using either an internal combustion engine or, more efficiently, a fuel cell [see "Green Machines"].

Eventually, hydrogen fuel could be even more prevalent than oil is today. The gas could become cost-competitive as a transportation fuel within the next two decades. Solar or wind-derived electricity at 5 cents per kilowatt-hour—achievable by the late 1990s—could produce hydrogen that would sell at the pump for about the equivalent of a $3 gallon of gasoline. While this is more than Americans are now charged to fuel their cars, it is less than the price most Europeans pay.

The transition to hydrogen as a major energy source likely will be eased by the ability to mix hydrogen with natural gas up to a one-to-ten ratio with minimal alteration of the existing infrastructure of gas pipelines, furnaces, and burners. Thus, as natural gas reserves are gradually depleted and prices rise in the early part of the next century, hydrogen can gradually be worked into the mix.

The Global Solar Network

Solar energy—whether transmitted through electrical lines or used to produce hydrogen—could become the cornerstone of a new global energy economy. All of the world's major population centers are located within reach of sun- and wind-rich areas [see map]. The U.S. Southwest, for example,

Green Machines

In both industrial and developing countries, automobiles are now a major consumer of petroleum and a chief cause of the stifling air pollution that plagues many cities. Finding a way to power transportation systems on non-polluting renewable energy sources is essential to a sustainable energy future.

Electric automobiles are one way to achieve this end. Electric cars produce virtually no air pollution in cities and, of course, they require no oil. While electric cars have been around since before the advent of the gasoline engine, the size of their batteries and the need for frequent recharging have limited their appeal.

But during the past year, spurred by tightening air-quality standards, a half-dozen auto manufacturers have announced plans to develop electric cars and vans during the 1990s. Fiat already has its Elettra on the market, the first mass-produced electric passenger car sold by a major manufacturer.

General Motors announced last year that it will introduce an electric sports car, known as the Impact, in 1993. Aerodynamic and sporty, the Impact accelerates rapidly and has a top speed of 100 miles per hour. But its batteries need to be recharged every 120 miles—making the car impractical for anything but city driving. More convenient electric cars are likely to be produced later in the decade as better batteries come along.

The U.S. Congressional Office of Technology Assessment estimates that the United States has sufficient existing power plant capacity to operate more than 10 million of these vehicles. Even if electric cars were used for one-quarter of U.S. auto travel, total electricity use would rise only 7 percent and overall air pollution emissions would drop significantly.

The battery-powered automobile is not the ultimate in "green" cars, however. Automakers are also considering vehicles fueled with hydrogen, the cleanest burning fuel, which produces only water vapor and small amounts of nitrogen oxides. The German companies BMW and Mercedes-Benz have built and test driven prototype hydrogen-powered cars; French and Japanese automakers are launching projects of their own.

Only minor modifications are needed to make a gasoline-powered engine run on hydrogen, although storing hydrogen remains a problem. Tanks that contain compressed gas are bulky, and hydrides—a chemical storage system—are expensive, more suitable for large trucks or buses than passenger cars.

Fuel cells offer a solution. They chemically combine hydrogen or natural gas with oxygen to produce electricity, which then can be used to run an electric motor. This process is, in effect, the reverse of electrolysis, which likely would be used to produce the hydrogen in the first place.

Because fuel cells are more than twice as efficient as today's internal combustion engines, cars won't need to carry as much hydrogen, greatly easing storage problems and reducing fuel costs. Fuel cells are currently being developed in both public and private R&D programs; several prototypes are in use. Researchers are still at work on reducing the cost of fuel cells and developing a compact version that could be used in cars.

A new generation of natural gas-powered vehicles will make a first step toward hydrogen-powered fleets, since the technologies behind both are very similar. A half-million natural gas vehicles are now on the road worldwide, and the United States alone could have 4 million by 2005 due to clean air laws. Researchers also have found that hydrogen mixed with natural gas produces dramatically less pollution than natural gas alone. In recent tests, a one-to-seven mixture lowered hydrocarbon emissions by more than 50 percent and nitrogen oxide emissions by 75 percent.

Still, "green" cars will not solve all the problems created by over-reliance on individual automobiles. To effectively deal with congestion, traffic fatalities, unlivable cities, and other impacts of car-dominated transportation planning, nations will need to move toward public transit systems. Subways, light-rail systems, and high-speed inter-city trains are efficient modes of transport that already are electrically driven. These transit modes—powered by renewable energy—could be greatly expanded in the years ahead.

could supply much of the country either with electricity or hydrogen.

Although renewable energy sources are regionally concentrated, they are far less so than oil, where two-thirds of proven world reserves are found in the politically unstable Persian Gulf area.

Wherever renewable resources are abundant, hydrogen can be produced without pollution and shipped to distant markets: from the windy high plains of North America to the eastern seaboard; from the deserts of western China to the populous coastal plain; and from Australia's outback to its southern cities.

For Europe, solar-power plants could be built in southern Spain or in North Africa. From the latter, hydrogen would be transported along today's natural gas pipeline routes into Spain via the Strait of Gibraltar, or into Italy via Sicily. Within Europe, today's expanding pipelines and electrical networks would make it relatively easy to distribute the energy.

To the east, Kazakhstan and the other semi-arid Asian republics might supply much of the Soviet Union's energy. In India, the sun-drenched Thar Desert in the northwest is within 1,000 miles of more than a half billion people. Electricity for China's expanding economy could be generated in the country's vast central and northwestern desert regions.

While pipelines must be sited to avoid ecological damage or accidents, their overall environmental impacts are minimal, especially where natural gas pipelines already exist, as between Wyoming or Oklahoma and the industrial Midwest and Northeast. The pipelines themselves will need to be modified or rebuilt to accommodate any shift to hydrogen, since the gas has properties that corrode some metals.

Germany leads the effort to develop solar-hydrogen systems. It has demonstration electrolysis projects powered by photovoltaic cells already operating in Germany and Saudi Arabia. Germany spends some $25 million annually on hydrogen research projects, according to Carl-Jochen Winter, a scientist with the German Aerospace Research Establishment (by contrast, the United States devotes less than $3 million). Experience in transporting hydrogen comes from a 120-mile pipeline in Germany that transports hydrogen produced from fossil fuels for use in industry.

Getting There from Here
In the end, major energy transitions tend to be driven by fundamental forces, either the evolution of new technology or problems facing society, such as population growth, resource depletion, or climate change. These forces, it can now be argued, are pushing the world toward a solar-hydrogen economy.

But the pace of change will inevitably be determined in part by government policies, most of which are now biased to favor today's energy systems. Will political leaders cling obstinately to the status quo or will they begin encouraging the development of new energy systems? In California, for example, it was a series of state policy changes made in the late 1970s and early 1980s that cleared the way for a new era.

Many state and local governments now encourage use of renewables through regulations and incentives. National governments are also moving toward new policies—under strong pressure from voters. Some European countries have raised gasoline taxes, other governments are taxing cars that pollute, and still more are forcing change via regulation. These are policies that work best in consort.

Higher taxes on fossil fuels is one way to accelerate the energy transition. The carbon taxes levied so far in countries such as the Netherlands and Sweden are an important step, but they haven't been set high enough to cause major shifts in the choice of energy technologies.

To make a real difference in energy habits, a carbon tax would have to be large enough to replace at least a quarter of today's taxes. One possibility is to lower income taxes as carbon taxes are raised. The voting public might accept higher taxes on gasoline, coal, and other fuels in exchange for more take-home pay.

Another approach that can level the playing field is including environmental costs within the electric-utility planning process. If environmental costs were added to construction costs in considering what kind of power plants to build, new coal-fired capacity would become economically unattractive compared with renewable energy sources.

The state of Nevada, for example, decided in early 1991 to tack on a hefty environmental charge when utilities license new coal-fired power plants. Part of the charge is for the potential costs of climate change, the rest is attributed to pollution costs, such as for the extra medical care required when air pollution damages people's lungs. As a result, the coal-based power that now dominates the state is likely to shift in the future to energy-efficiency programs, natural gas, geothermal heat, and sunshine.

National governments also can speed the transition to a sustainable energy future by providing modest incentives for the building of renewable energy systems. New energy technologies have in the past been subsidized by governments—hydropower and nuclear power are obvious examples. In this case, the subsidy can be justified due to the avoided environmental damage that results from renewable energy development. It is worth remembering that California state tax credits for renewable energy development helped spark the renewables boom of the early 1980s.

The U.S. Congress is now considering a subsidy for the generation of renewable electricity. An incentive of just 2 cents per kilowatt-hour—equivalent to about 25 percent of the average retail price of power in the United States—would be sufficient, according to market analysts, to spark a boom in renewable energy development. The cost to the taxpayer would be about $1 billion over five years—a fraction of 1 percent of the nation's annual power bill.

A re-orientation of research and development programs is also called for. In 1989, the leading industrial-country governments spent just 7 percent of their $7.3-billion in energy research funds on renewable technologies. Most of the rest went to nuclear energy and fossil fuels. Drastically trimming breeder-reactor and nuclear-fusion programs would free billions of dollars to accelerate the commercial development of new technologies.

Most countries still have a long way to go in reforming their outdated energy policies, but there is a new sense of urgency about future energy sources as the public reacts to the threats posed by greenhouse gases building in the atmosphere.

There are, for example, 23 countries committed to limiting carbon emissions. More are likely to follow as a United Nations-sponsored treaty to protect the world's climate is readied for signing in Rio de Janeiro in 1992.

Some of the biggest obstacles blocking change in many countries are caused by the politicians who are captives of today's energy industries. The halls of the U.S. Congress, for example, are filled with lobbyists for powerful energy industries—ranging from oil to coal to nuclear power—and their policy agenda predominates. Ironically, while their political power remains intact, these industries have been automated and no longer provide many jobs. As more such positions are eliminated in the 1990s, the political position of these industries is likely to weaken.

In the end, the key to overcoming political barriers is to demonstrate that a solar economy would have major advantages over today's fossil fuel-based systems. California again provides a good example. It already has greatly reduced its fuel bills and begun to clear its skies as a result of the energy policy changes begun more than a decade ago.

As California's political leaders seem to understand, a solar future is just too attractive to be ignored. Indeed, a solar economy would be healthier and less vulnerable to oil price gyrations of the sort that have shaken the world in recent decades. And a solar future is the only practical energy future that would be environmentally sustainable—eliminating the greenhouse gases now threatening the planet's health.

Energy Crops for Biofuels

Research now under way aims to make cultivated trees and grasses an important source of fuel for transportation and electricity generation.

Janet H. Cushman, Lynn L. Wright, and Kate Shaw

Janet H. Cushman, Lynn L. Wright, and Kate Shaw work in the Biofuels Feedstock Development Program, Environmental Sciences Division, Oak Ridge National Laboratory, Oak Ridge, TN.

The "amber waves of grain" in "America the Beautiful" may be replaced by "rows of poplar trees" or "silver waves of grass," if energy crop researchers meet their goals. Rows of trees may replace rows of corn, and fields of soybeans or cotton may be interspersed with fields of switchgrass. Growing crops for energy may become as important as growing crops for food.

In laboratories and fields across the United States, common trees and grasses are beginning to receive the same kind of intensive study as that directed toward solar collectors and fuel cells—and for much the same reasons. Scientists have recognized that trees and grasses can be biologically "engineered" to become more efficient collectors and storers of solar energy. They have also recognized a wonderful potential versatility of plant matter ("biomass"[*]) as an energy source—it can either be burned

directly to release heat or be converted to a variety of readily useable fuels, including methane, ethanol, and hydrogen. Biomass conversion technologies include both those of direct combustion to produce heat and those for producing liquid or gaseous fuels. [See "Biomass Conversion Technologies."]

Burning biomass to produce heat is an ancient and still widespread practice. At present, however, only small-scale applications that have access to very low-cost biomass burn it to produce electricity.

For conversion of biomass to a liquid fuel, especially ethanol, existing technologies use as feedstock only select components of a plant, such as the starch from corn kernels or the sugar from sugarcane. The stalks of both corn and sugarcane, made primarily of cellulose, have been waste materials.

However, a new approach to better using the full potential of biomass energy is rapidly developing. This approach is expected to offer a near-term, relatively low-cost, and renewable alternative to coal or oil on a significant

scale. The idea has been generated both by new discoveries about the growth potential of trees and grasses and by technological developments that promise enhanced efficiencies in converting biomass to useable energy.

Of special interest is the research to develop cost-effective processes for converting cellulosic material, including wood and grass as well as corn and sugarcane stalks, to ethanol. The mastery of this technology would make possible the production of liquid fuels for transportation from a wide range of fast-growing plants. Other research is developing combustion processes with enhanced efficiencies.

A new urgency for developing biofuels has resulted from the recent concerns about the possible role of increased levels of atmospheric carbon dioxide in causing global warming. In 1987 almost 30 percent of the carbon dioxide released in the atmosphere in the United States came from transportation fuels. Coal-fired electricity-generating facilities contributed another 36 percent. If biofuels were to replace a significant portion of the fossil fuels being used, the rate of carbon dioxide buildup would be substantially reduced. Indeed, carbon diox-

[*] Whereas *biomass* in its general definition includes both plant and animal matter, in the field of biomass energy the term refers only to plant matter.

■ Experimental plots of cultivated hybrid poplars in the midst of Wisconsin farmland foreshadow a time when farmers will raise crops both for food and for fuel.

ide would still be released by the burning biomass, but that carbon dioxide would have been removed from the atmosphere through photosynthesis during the short lifetime of the plant being used as feedstock. In contrast, the carbon dioxide released by burning fossil fuels was removed from the atmosphere millions of years ago. Burning biomass thus yields no net change in the level of atmospheric carbon dioxide, except possibly for carbon dioxide released by the machines that tend energy crops or carbon dioxide released in the production of pesticides and fertilizers used to raise the energy crops.

The DOE initiative

In 1978 the U.S. Department of Energy's (DOE) Biofuels Feedstock Development Program, managed by the Oak Ridge National Laboratory (ORNL) in Oak Ridge, Tennessee, started ex-

ploring the variety of plants that could be used as biomass feedstocks and the types of land most suitable to cultivate such crops. Today the program coordinates an extensive network of research on crop development, maintains a unique data base on energy crop production, and performs resource and economic analysis regarding energy crops ranging from trees to grasses. This work is also coordinated closely with the cellulosic conversion technologies being developed by DOE's Solar Energy Research Institute in Golden, Colorado.

The oil crises of the 1970s drove home the fact that wood might provide transportation fuel. In 1977 the DOE called for research proposals "to improve the productivity and increase the cost efficiency of growing and harvesting woody plants for fuels and petrochemical substitutes." This initiative was able to build on the pioneering experiments of foresters and university researchers in the late 1960s and early 1970s to grow trees as crops.

Twenty-seven proposals, most from universities, were selected for the new Fuels and Chemicals from Woody Biomass Program. Before long, investigators across the country were scrutinizing 140 different species, trying to learn how to establish energy crop plantations—which trees

■ Four-year-old sycamore trees fall before the "scythe" of a tree harvester in research trials conducted by North Carolina State University and Scott Paper Company.

grew best, where, and under what conditions. In the Pacific Northwest, Reinhard Stettler, a geneticist at the University of Washington, and Paul Heilman, a soil scientist at Washington State University, joined forces on poplar research. Their research has also gained the support of the U.S. paper industry as a few farsighted companies have already initiated programs to begin to meet their pulp and energy needs from sustained-

■ **The U.S. Biofuels Feedstock Development Program today includes research on the energy crop potential of grasses (at HECP sites) and trees (at SRWCP sites). After widespread screening trials, research at many less-promising tree-growth sites has been discontinued (inactive SRWCP sites).**

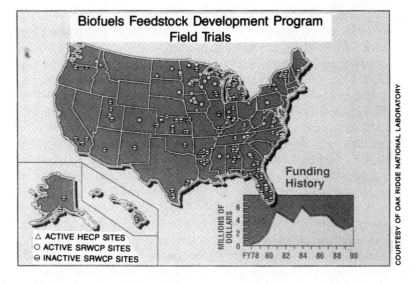

Biofuels Feedstock Development Program Field Trials

△ ACTIVE HECP SITES
○ ACTIVE SRWCP SITES
⊖ INACTIVE SRWCP SITES

Funding History

COURTESY OF OAK RIDGE NATIONAL LABORATORY

yield tree farms using Stettler and Heilman's hybrid fast-growing tree varieties.

By 1984 the first phase of DOE biofuels research had pinpointed 25 tree species with high growth potential. Two years later, in response to the need to focus on fewer trees, investigators selected poplar, eucalyptus, black locust, sycamore, sweetgum, and silver maple as models for woody crops. Along the way researchers learned that successful "tree farming" requires genetically superior clones or seedlings, careful site selection, inten-

sive site preparation before planting, and a variety of weed control techniques. Densities of between 800 and 1,600 trees per acre were settled on as being most cost-effective for harvest cycles of 5 to 10 years. The concept name was purposely changed to "short-rotation woody crops" to signify that the biomass being produced should be thought of as a standard agricultural product.

Herbaceous energy crops

Also, in 1984 the DOE decided to expand its energy-crop research efforts to include nonwoody herbaceous plants. The department, in collaboration with ORNL, sought proposals for herbaceous energy crops that might comple-

ment and supplement wood energy crops, and by 1985 five herbaceous energy crop screening studies were set up in the Southeast, the Midwest, and the Lake States. Mindful of the debate over the use of cropland, DOE concentrated on species suitable for marginal cropland. Thereby an increase in energy crop production would not significantly reduce food production.

The researchers had originally plunged into the herbaceous energy crops work assuming that the effort would focus on the selection of a few species for

a standard production system. But they soon discovered that herbaceous plants, with their diverse growth forms, offer an array of interchangeable production options. Appropriate crops could be developed for any kind of land that becomes available for energy feedstock production.

Land quality is a variable from place to place and climatic conditions change from year to year. In order to optimize land use in any given region and year, from both an economic and environmental perspective, it is desirable to have a variety of energy crop choices. Cellulosic herbaceous energy crops greatly expand the choices. Like short-rotation woody crops, which are perennial and therefore require that specific areas be dedicated to their production, many herbaceous energy crops are also perennial. These plants, because they can effectively minimize soil erosion, are also good choices for land prone to erosion. Further, the equipment and farming practices for these crops are similar to those used for forage crops.

As a complement to perennial herbaceous energy crops, annual herbaceous energy crops, like most food crops, are desirable because they allow flexibility in changing crops annually to adapt to market demands or climatic variation. Again, the equipment and farming practices needed for annual energy crops are similar to those used for many food crops. Sorghum is the most promising of the annual cellulosic energy crops tested to date. However, like all row crops, sorghum should be grown only on relatively flat cropland to reduce soil erosion. Leguminous plants are also good cellulosic crops that help optimize land use, and, being nitrogen-fixing plants, they replenish the soil with nitrogen. In order to select a full complement of choices, several species are being grown on sites in four areas of the eastern United States with different soil qualities, fertilization rates, and weed-control measures. The results of the first phase of those trials reveal that herbaceous crops hold tremendous potential as biofuels.

Other characteristics of herbaceous crops that need to be considered when developing management systems is when and how they should be harvested and how they should be handled after harvesting. Crops like sorghum, cane varieties bred for bio-

mass rather than for sugar, and other tall tropical grasses grow best during summer and should be harvested in late summer or early fall. These crops also have thick stems which means they will not dry easily and thus must be stored green if storage is required. If storage costs are to be avoided, other varieties of cellulosic feedstocks should be considered. For instance, thin-stemmed grasses can be harvested and stored like hay in large round bales that may be left for several months in the field.

While switchgrass, the most promising of the thin-stemmed grasses tested, grows best in summer, many other thin-stemmed grasses grow best in spring and fall. By including several crops in the feedstock supply mix for a conversion facility, including short-rotation woody crops, the supply of fresh material can be extended.

Although farmers have experience with several herbaceous crops, their successful production for energy purposes will require a different mind-set. Farm-

Biomass Conversion Technologies

Biomass conversion technologies are both old and new, traditional and innovative. Biomass, the world's first fuel for heating and cooking, meets almost 90 percent of the energy requirements of many developing countries even today. Worldwide, biomass provides 14% of the total energy needs, and in the United States, according to the National Wood Energy Association, it provides 5 percent of all energy consumed.

The nation's approximately 200 biomass energy facilities predominantly use residues from forestry and agriculture to produce steam or electricity through conventional combustion processes. Although the existing technology may be cost-effective when biomass is cheap or free, it is often only 20 to 25 percent efficient in converting biomass to energy. New, electricity-generating combustion systems, which may have biomass to energy conversion efficiencies of 35 to 45 percent, are under de-

velopment. These systems are better suited for cellulosic biomass, including wood, grasses, and the stalks of corn and sugarcane.

In the whole-tree-burner concept, entire trees, dried with waste heat, would be batch-cut into four- to eight-feet-long segments and fed into a large furnace for electricity production. Steam-driven turbines, similar to ones that propel jet engines, could produce electricity efficiently and economically using gasified biomass feedstocks. These new biomass electric developments could be demonstrated and proved cost-effective within the next 5 to 10 years.

Liquid fuels from cellulosic biomass may not be cost-effective for another 10 to 20 years, depending on oil prices. Methanol production from wood is already possible using a high-temperature gasification process, but it hasn't been attempted on a large scale. A nearly commercial-size biomass gasifier demonstration project, test-

ing wood and herbaceous crops, will begin construction in 1992.

The technology for commercially fermenting cane sugar and corn starch to make ethanol has long been practiced. However, processes for converting cellulosic biomass, only a part of which is sugar, to ethanol are still under development. The sugar portions (cellulose and hemicellulose) must first be separated and broken down into simple sugars before traditional fermentation processes will work. The most promising methods for separating and breaking down the components appear to be enzyme-catalyzed processes in which the enzymes are produced by yeasts or bacteria. These enzyme-catalyzed processes present tremendous opportunity for technology improvement through genetic engineering, which could make ethanol's market price competitive with that of existing fuels.

—*J.H.C., L.L.W., K.S.*

■ "Energy cane," bred not for sugar but for biomass, is a prime energy crop candidate in regions of the southern United States where it can be grown. Harvest here is by students from Auburn University who will carefully collect all biomass as a step in evaluating biomass productivity.

ers will have to grow the crops in ways that will optimize cellulose production rather than the nutrient content or seed production. This means, for instance, that thin-stemmed grasses will need to remain longer in the field before being harvested as opposed to grasses for standard forage crops. Because herbaceous energy crops can be interwoven into standard farming operations, the farmers' interest could come swiftly if energy markets were developed.

There are also other noncellulosic energy crops that farmers may choose to grow. For instance, rapeseed, whose oil (canola) is normally sold for human consumption, can also be used to replace diesel fuel. It is particularly valuable as a crop in the Southeast since some varieties can be grown over the winter on the same fields that support a summer food crop.

Future research

Energy crop researchers continue to look for genetic factors that can improve total biomass productivity. Scientists are trying to determine if the solar collection efficiency of a stand of trees can be enhanced by altering the size and orientation of the tree leaves, or by changing the overall shape of individual trees. Evaluations of the structure and growth rate of roots are answering questions about the ability of different trees to optimally utilize soil nutrients. Investigations into how some trees survive better under drought conditions involve looking at the chemistry of the molecules that the trees manufacture. This research is providing information about tree biology that has implications far beyond energy issues. Similar research is also being done on herbaceous plants.

A major emphasis of genetic research on poplars will be to develop a large number of clones

■ A thick-stemmed, annual grass (corn) on the right and a thin-stemmed, perennial grass (switchgrass) on the left illustrate the wide variety of grasses that can be grown as energy crops. While fuel can presently be produced from the starch in corn kernels, recent research indicates that corn stalks may also be a valuable energy source.

Strides in Technology

Between 1980 and 1990, short-rotation woody crop investigators made some remarkable strides in developing new research techniques. Several short-rotation woody crop species were among the first trees to be propagated from tissue cultures in order to replicate the exact genetic makeup of superior trees.

Tissue culture, called micropropagation or microcloning, is important to the development of energy crops because it is a precursor of new methods of genetic improvement. In taking a tissue culture, a researcher extracts a piece of plant tissue—a half-inch stem, perhaps—and stimulates it to grow a new complete plant. All model short-rotation woody crops have been successfully tissue–cultured. And in a first for tree research, genetic screening field trials were established with tissue–cultured silver maple plantlets.

Hybrid poplar, often referred to as "the white rat tree," has been in the forefront of the emerging field of biotech-nology. Even researchers who prefer to work on conifers have used hybrid poplars as a "model" species for developing new techniques.

Hybrid poplars have been regenerated from single cells by using culture methods developed for regenerating plants from tissues. The ability to regenerate plants from single cells has allowed scientists to test individual cells for possible genetic differences. For example, clumps of cells of poplar clones are being subjected to such treatments as exposure to toxic compounds and those surviving are regenerated into complete plants. After further testing of the young plants to determine which of them has truly achieved improved tolerance to the compounds, tissue culture techniques can be used to clone several thousand copies of the tolerant plants.

Genetic transformation, which involves the transfer of DNA, or genes, from one species to another, was also first achieved in trees with a hybrid poplar clone. Currently, at least three genetically transformed poplar genotypes are being grown in the lab or field to test whether the transferred genes remain active throughout the plant's growth.

The first complete gene map for a tree is expected to be completed in 1992 and again, it will be a hybrid poplar. This work is being cofunded by private industry, the Department of Energy, and the state of Washington. One of the products of the work will be the development of a method for DNA fingerprinting. "We want to take a leaf's fingerprints and be able to tell what clone it's from," says Chuck Wierman of Boise Cascade, a paper and wood products company cofunding the research. This is a very practical goal since it is easy to mix up clones in nursery operations, and planting the right clone can make the difference between a profit and a loss.

—*J.H.C., L.L.W., K.S.*

that have resistance to pests and diseases. This will be done both through traditional breeding procedures and by experimenting with the insertion of specific genes for insect and pathogen resistance.

A major thrust of future energy crop research will be to ensure that crop management systems will be environmentally be-nign and sustainable. By testing species mixes, researchers hope to discover which planting patterns reduce erosion and which combinations create favorable wildlife habitats. Short-rotation woody crops or perennial herbaceous energy crops may offer greater biological diversity in both plant and animal species than in most cropping systems, particularly if appropriately landscaped.

It is clear that if biofuels are to meet a significant portion of the nation's energy demand, large amounts of land will be needed to cultivate biofuel feedstocks. Preliminary analyses of current and future projections of overall land use trends indicate that sufficient land may be available in the United States to produce enough feedstock to supply about 20 percent of the nation's energy needs. These analyses need to be expanded and improved by performing test case studies for a number of specific locations around the United States.

Of course, the United States is not alone in its biofuel research efforts. Several European countries are examining biofuel production and conversion systems. Sweden has already made a national commitment to increase the use of biofuels to produce electricity instead of building nuclear facilities. Programs aimed at working directly with farmers are already under way in Sweden.

Developments in biotechnology, though presently unrelated to energy crops, may have a major impact on the use of energy crops of the future. As we know, plants can produce a wide range of chemicals in small amounts. If genes responsible for producing specific chemicals can be isolated and inserted into the fast-growing energy crops, then trees and

COURTESY OF OAK RIDGE NATIONAL LABORATORY

The lush growth of grasses and trees converting solar energy through photosynthesis to a form of chemical energy that lends itself to conversion to fuels holds great promise for the world's energy future.
■ *Above:* Switchgrass, a perennial that is harvested annually like hay.
■ *Right:* Hybrid poplars in their second year of growth.

COURTESY OF OAK RIDGE NATIONAL LABORATORY

grasses may become not only sources of biofuels, but also chemical manufacturing plants of the future. The possibilities offered by biological engineering open whole new realms of research and practical application.

Pollution: The Hazards of Growth

Of all the massive technological changes that have combined to create our modern industrial society, perhaps none has been as significant for the environment as the "chemical revolution." Although the chemical industry reminds us that life itself is not possible without chemistry, the fact remains that the largest single threat to environmental stability is the proliferation of chemical compounds for a nearly infinite variety of purposes, including the universal use of organic chemicals (fossil fuels) as the prime source of the world's energy systems. The problem is not just that literally thousands of new chemical compounds are being "discovered" or created each year, but that the long-term effects of these compounds in an environmental system are often not known until an environmental disaster involving humans or other living organisms occurs. The problem is exacerbated by the time lag that exists between the recognition of potentially harmful chemical contamination and the clean-up activities that are ultimately required. We were warned of the ecological consequences of certain pesticides nearly twenty years before DDT and other offending chemicals were banned. In addition, the damages that could result from PCB poisoning were known well in advance of the Love Canal tragedy in New York. The expanding function of chlorinated fluorocarbons (CFCs) in the deterioration of the atmospheric ozone layer has been suspected for more than a decade and recently confirmed. And certainly the role played by fossil fuel combustion as a contributing factor in the enhanced greenhouse effect has been discussed by scientists for many years.

A critical part of the process of dealing with chemical pollutants is the identification of toxic and hazardous materials, a problem that is intensified by the myriad of ways in which a vast number of chemicals can enter environmental systems. Pollution from chemical compounds takes one of two forms: direct or accidental. Direct impacts result from the intentional applications or disposals of potentially hazardous materials or chemically-treated substances from industrial, commercial, agricultural, or domestic sources. Even worse problems are posed by the accidental entry into the environment of hazardous chemicals. This latter danger is particularly great for communities housing industrial plants manufacturing or using hazardous chemicals, since few local communities have extensive environmental regulations to protect themselves against accident. Government legisla-

tion and controls are an important part of the correction of the damages produced by toxic and hazardous materials such as DDT or PCBs or CFCs in the environment, or in the limitations on fossil fuel burning. Unfortunately, as evidenced by most of the articles in this unit, we are losing the battle against these harmful chemicals regardless of legislation, and chemical pollution of the environment is probably getting worse rather than better. This is expressed in the title of the lead article in this section, "It's Enough to Make You Sick." In this article, Susan Stranahan, a member of the editorial board of *The Philadelphia Inquirer* and a frequent writer on health topics, takes a public health approach to the pollution problem. Using as her baseline the first Earth Day (1970), Stranahan notes that 20 years of research into the ways in which pollution affects our health has brought us a great deal of knowledge but precious few solutions. The role of chemicals in public health is also the focus of the second selection in this unit, although the focus of "The Greening of Industry" is the role played by toxic waste disposal in environmental pollution. In his ground-breaking analysis of industrial toxic waste, Ken Geiser, director of the Toxics Use Reduction Institute, makes a powerful case for "sustainable industry." While we often think of the term "sustainable" in terms of energy use, Geiser makes the point that it can equally apply to chemical compounds. In "From Ash to Cash: the International Trade in Toxic Waste," environmental issues writer Ron Chepesiuk describes a little-known component of the toxic waste problem: the exportation of toxic waste from developed countries where laws prohibit their disposal or storage to countries that either have no such laws or are desperate enough for money to act as receptacles of the wastes of the industrial world. Chepesiuk suggests that, while the toxic waste has been profitable for many Third World countries, many of these countries are becoming "increasingly sensitive about their role as wastebaskets" for the United States and other industrial countries.

While the previous articles in the unit deal with pollution at the global or regional scale, "Will the Circle Be Unbroken?" narrows the focus of the pollution problem to the settlement scale. Authors David Weir and Constance Matthiessen examine the rural villages of Central America and describe the particular dangers of agricultural chemicals utilized in the banana plantations of Costa Rica and other Central American countries. The authors note that the continuing manufacture and use of agricultural chemicals, particularly pesticides, is one of the greatest environmental dangers facing Third World farm laborers. Much of the problem can be attributed to the lack of governmental control and intervention on the part of governments of developing regions. A prime example of what can happen as a result of the failure of governments to be proactive in their control of chemical use is offered in the last two articles in the unit. In Eastern Europe, where industry has proceeded unchecked along the classic lines of Marxist environmental philosophy, the air, water, and soil are suffering from pollution that is alarming both in itself and as a predictor of what could happen in other industrial regions. In "Eastern Europe: Restoring a Damaged Environment," Richard A. Liroff of The Conservation Foundation describes the pollution of Eastern Europe—perhaps the most polluted major region of the world—and lays the blame squarely at the feet of the governments who have failed to heed the warnings. Liroff notes that, although the savaged environment presents a real challenge for the development of stable economic systems, there are options open to both Eastern European governments and those who would aid their "greening." If countries such as the United States can provide structured environmental assistance as part of their foreign aid package to Eastern Europe, there may be some hope for recovery of the stressed systems of the region. Problems similar to those of Eastern Europe exist in the Commonwealth of Independent States (the old Soviet Union). In "Environmental Devastation in the Soviet Union," journalist James Ridgeway surveys the environmental situation in Russia and her sister republics of the old Union. He presents a grim picture of a ravaged environment. As in Eastern Europe, however, an increasing environmental movement and growing environmental awareness on the part of Russian citizens (coupled with an increase in Russian nationalism) suggests that the mechanisms for changes in attitudes toward the environment are there. Ridgeway even suggests that much of the current liberal political movement in Russia has its roots in Russian environmentalist movements of the mid-1960s. It would be ironic if the destruction of the governmental system responsible for such massive environmental destruction were indeed contributed to by environmental philosophies.

The pollution problem would appear to be one that is impossible to solve. Yet solutions exist: massive cleanup campaigns to remove existing harmful chemicals from the environment and to severely restrict their future use; strict regulation of the production, distribution, use, and disposal of potentially hazardous chemicals; the development of sound biological techniques to replace existing uses of chemicals for such purposes as pest control; the adoption of energy and material resource conservation policies; and more conservative and protective agricultural and construction practices. We now possess the knowledge and the tools to ensure that environmental cleanup is carried through. It will not be an easy task, and it will be terribly expensive. It will also demand a new way of thinking about humankind's role in the environmental systems upon which all life forms depend. If we do not complete the task, the support capacity of the environment may be damaged or diminished beyond our capacity to repair it. The consequences would be fatal for all who inhabit the surface of this planet.

Looking Ahead: Challenge Questions

What are some of the key things that researchers over the last 20 years have learned about the relationships between pollution and public health? Where do enormous gaps still exist in our knowledge of that relationship?

What is meant by the term "sustainable industry," and how does it relate to the problem of industrial toxic waste? What are some of the ways in which industries can become "greener" by changing the ways they deal with toxic materials?

Why do environmental laws contribute to increasing toxic waste problems in Third World countries? How can such countries balance their needs for hard currency with their desire for clean environments?

How can current practices of the use of agricultural chemicals imperil both agricultural laborers and entire environmental systems? Can making the American public aware of the misuse of these chemicals reduce the reliance upon them in developing nations?

What are some of the most serious pollution problems in Eastern Europe, and what steps have been taken to alleviate these problems? Are there parallels between the situation in Eastern Europe and in the eastern portions of the North American continent?

What are the similarities between the environmental situation in Eastern Europe and that in the old Soviet Union? What role has the environmental movement played in the internal politics of Russia in the recent past?

It's Enough To Make You Sick

*After two decades of
research, we know much more
about how pollution affects our
health—and also what
to do about it*

Susan Q. Stranahan

*Susan Q. Stranahan, who is on the editorial
board of* The Philadelphia Inquirer, *writes
often about health issues.*

April 1970. All across the nation, many
Americans were discovering that pollution
in its various forms was not merely an
annoyance but a serious threat to their
health, and they were mobilizing to do
something about it.

That month, officials in the United
States and Canada banned fishing on Lake
Saint Clair and imposed a partial fishing
ban on Lake Erie. For nearly three decades after World War II, industries in
both countries had dumped millions of
pounds of mercury and other poisons into
the Great Lakes watershed, even though
mercury's toxic properties were already
well known. Indeed, a century earlier,
Lewis Carroll made the Mad Hatter in

Alice's Adventures in Wonderland a victim
of mercury poisoning. Yet no one sought a
solution until large numbers of fish caught
in Lake Saint Clair showed mercury levels
14 times the amount deemed safe for human consumption.

Twenty years ago in April, federal officials also announced a ban on the home
use of the herbicide 2,4,5-T and halted its
use as a defoliant in Vietnam. The action
came in response to reports of miscar-
riages, birth defects, respiratory difficulties and neurological disorders among
people living near areas where aerial
spraying of the chemical was commonplace.

A few days after that announcement,
Edward Simmons was manning a broom
near the South Street Seaport Museum in
Manhattan, helping to spruce up the city
for something called Earth Day. "Last
year I read the papers and figured I didn't

have very long to live," Simmons, then 13,
told a reporter at the time. "But now I'm
beginning to think maybe I won't die so
soon, because more people are seeing
what's wrong—and something's got to
happen."

Something did happen. The next two
decades produced a tidal wave of new
laws, scientific research and technological
advances, public awareness and activ-
ism—all geared toward making the
United States a cleaner and healthier place
to live. But the same period has also
brought discoveries of environmental
problems no one dreamed of in April
1970, dilemmas reflected in the host of
recent additions to the average American's
vocabulary: acid rain, Superfund, carcin-
ogens, mutagens, radon, dioxin, global
warming; in place names that need no further identification: Times Beach, Love
Canal, Bhopal; and in a list of new prod-

ucts expressed as ominous-sounding initials: PCBs, DDT, EDB and TCE.

To the exuberant celebrants of 1970, cleaning up the nation's pollution was achievable. Today, most environmental pragmatists agree that the less-complex problems have been solved, while those that remain will be extraordinarily difficult to bring under control.

Without doubt, the United States has made great strides in protecting its population in the past 20 years. "You only have to go to central Europe or Latin America to see what America would be like if we hadn't imposed any [environmental] controls," says Ellen K. Silbergeld, a specialist in public health and biomedical research at the Environmental Defense Fund. But with each step, it becomes more obvious that we have only begun to understand the true extent of the toll pollution is taking on our lives. Most Americans today come in daily contact with a variety of chemicals suspected of posing health risks. Still, scientists cannot say with certainty at what levels exposure to most of them causes cancer, neurological disorders or birth defects—or whether they cause them at all.

"We're sort of in the cave-painting era in terms of what we know [about environmental health risks]," says Marvin Schneiderman, former associate director of the National Cancer Institute. Of the 48,000 or more chemicals registered with the Environmental Protection Agency, almost no health research has been done on 38,000 of them. About 1,000 have been tested to measure acute risks, while only half that many have been studied for their cancer-causing or reproductive effects, or for their ability to cause genetic change.

At the same time, however, scientists have become more sophisticated detectives, capable of predicting which types of chemicals are most likely to threaten our health. "We also have gotten better at estimating the potency at which they cause problems," explains Ila Cote, who is associate director of EPA's Health Effects Research Laboratory in North Carolina.

Twenty years ago, many health standards were based on isolated, large doses of a single pollutant. A big dose of a chemical solvent, for example, could sicken or asphyxiate a worker. Two decades of research have taught scientists that chronic exposures even at extremely low levels can produce irreversible health problems. For example, prolonged exposure even to tiny quantities of ozone, a major component of smog, can cause serious lung damage. One study showed a few hours of exposure to a level of ozone smaller than the amount in the air on a bad day in Los Angeles can

Poisoned Drinking Water: One Town's Solution

THE NIGHTMARE started in 1983, when the Hickey family of Long Pine Road in Whately, Massachusetts, noticed mysterious rashes erupting on their children's skin. Having ruled out allergies and skin diseases, the Hickeys checked their water supply. Sure enough, their well water was contaminated, laced with high levels of Temik and ethylene dibromide—pesticides which were used liberally on nearby tobacco and potato fields.

For Whately (population 1,500), the bad dream was just beginning. Like many other communities across the nation, its citizens suddenly found themselves facing an environmental calamity. When the state of Massachusetts checked private wells supplying as many as 175 homes in the low-lying agricultural area of East Whately, it found virtually all were contaminated. So for the next three years, 175 families lined up at an elementary school in West Whately to fill water containers from the school's outdoor tap. (The water supply at the school, located on a hilltop overlooking the Connecticut River Valley, was unaffected by lowland contamination.)

"One poor woman with a station wagon came every two days," Frank Marchand, chairman of the Whately Water Resource Protection Committee, told the Associated Press. "She filled up so many milk jugs they stretched from the front seat all the way to

the back window." The health of the entire community was at stake, but nothing short of a miracle or a mountain of cash could bring its residents safe water. In a dramatic show of unity, they managed both.

Townspeople knew there was a deep and plentiful aquifer, untapped and uncontaminated, lying beneath 50 feet of clay. But if each of the 175 families dug its own well to tap that aquifer, in the words of town Water Superintendent William Smith, it would amount to "175 puncture wounds in that clay layer"—wounds that could allow pesticides to leak down and contaminate the deeper aquifer.

At an emotional town meeting in 1986, residents passed a bond issue to pay for digging a single well to supply all of East Whately's families. That, along with grants from two state agencies and a one-time hookup fee of $3,500 to $5,000 for each family, gave the town the $4 million it needed for the new system, which began providing uncontaminated water in 1987.

Today, the long lines at the elementary school have vanished, and the use of pesticides has dropped dramatically. Says Paul Fleuriel, chairman of the Whatley Water Commission, the new water system is "as close to a permanent solution as we have in this modern day." It's also an object lesson in the potential rewards of concerted community action.—Bill Lawren

result in chest pain, coughing and nausea.

Researchers have learned that some segments of the population—notably the young and the elderly—are especially susceptible to pollutants, and, therefore, that exposure standards must be calculated accordingly. Nitrates from fertilizers that have infiltrated groundwater pose a greater health risk to infants than to adults, for instance. And current air pollutant standards for sulfur dioxide are set to protect asthmatics, who are particularly sensitive to that chemical.

Over the past 20 years, scientists also have honed their ability to predict through laboratory experiments the effects of some chemicals on human health. In many cases, regulators no longer have to wait for sickness or death to strike before lim-

iting exposure. The chemical 1,3-butadiene, widely used in synthetic rubber and plastic pipes, was considered safe until studies showed it produced cancer in laboratory animals. (It is now thought likely to be carcinogenic also in humans.)

Despite such progress, enormous gaps remain in the body of scientific knowledge. Research on the combined effect of exposures to several pollutants is almost nonexistent. What's more, the bulk of health-based environmental standards treat single pollutants in individual media: a standard for lead in air, for example, does not take into account that a person also is exposed to lead in drinking water and food. Nor do scientists understand the synergistic action of chemicals—the ways in which they combine or break down to form different, possibly far more potent, substances inside the body.

Is America a riskier place to live today than it was in 1970? Some vital signs say no. Within the past 20 years, for instance, U.S. life expectancy has increased by 4.4 years for males and 3.6 years for females. But if risk is measured by the quantity of toxic chemicals streaming into the environment, the hazards are greater than ever. "The levels of chemical usage and production have increased dramatically in the past 20 years," says Jerry Poje, staff toxicologist for the National Wildlife Federation, "and they are projected to increase even more in years to come."

More than a ton of toxic waste is produced each year for every man, woman and child in the country, says the EPA. Of that amount, 135 billion pounds are discharged into the nation's waterways, while billions more are spewed into the air. In 1987, for instance, nearly 3 billion pounds of airborne pollutants were released. That figure accounts for only about half of the nation's top 30,000 industrial polluters.

At least 2,000 Americans die each year from cancer attributed to polluted air, according to government figures. Recently scientists have identified new sources of airborne pollution, such as radon, a naturally occurring radioactive gas. Radon levels in some areas are so high that the gas has been ranked as the second-leading cause of lung cancer after cigarette smoking. Fumes from household products and building materials also pose a serious threat, and in some buildings—including the EPA's headquarters in Washington, D.C.—workers have taken ill from breathing the indoor air. Meanwhile, ozone pollution reached unhealthful levels in many cities in recent summers, and experts predict the problem will only grow worse.

As Americans have witnessed one environmental calamity after another during the past two decades, legislators have responded with a flood of new laws and regulations. For example, widespread reports of illegal dumping of hazardous wastes triggered passage of the Toxic Substances Control Act in 1976. Similarly, the discovery of toxic chemicals seeping from the ground in the Love Canal neighborhood in Niagara Falls in 1977 prompted the 1980 Superfund law, which was created to clean up the worst toxic dump sites.

With legislation has come some success. The ban on DDT in 1972 has begun lowering levels of that pesticide in humans and wildlife populations. Laws requiring

unleaded gasoline and catalytic converters in new cars starting in 1975 greatly improved air quality. And the decision in 1977 to prohibit manufacture and sale of PCBs has reduced that threat. But, as Ellen Silbergeld points out, these victories came about only on the strength of government-imposed bans. "Will it take that kind of intervention to make a difference with other chemicals?" she says.

Critics argue that the government fails to protect the American public from many risks because it allows large numbers of new products to enter the environment without first examining the health consequences. David Doniger, an attorney with the Natural Resources Defense Council in Washington, D.C., contends the government should test new chemicals just as it tests new drugs. "It's perfectly okay [legally] to fill the environment and expose people to these chemicals without knowing they're dangerous," he says. "It's as though ignorance is equated with safety."

Statutes often pay lip service to a more precautionary approach, and in some cases courts have forced the federal government to adopt controls. But the high cost of testing (borne by taxpayers, not producers), along with powerful resistance from the chemical industry, have effectively blocked implementation of a screening policy. Nowhere is that more apparent than in the area of pesticides.

Around 2.6 million pounds of pesticides are applied to fields each year in this country, to say nothing of the millions of pounds of other chemicals applied to crops, lawns and gardens. Pesticides, such as insecticides and herbicides, are the only toxics purposefully introduced into the environment, yet research on their effects is extremely limited. In 1987, the National Academy of Sciences estimated that pesticide contamination in food could be responsible for up to 20,000 cancer cases annually. Ironically, last summer the Academy found that farmers could be just as productive using fewer chemicals.

Cleaning Up A Toxic Playground

WHEN THE STUDENTS in Barbara Lewis' class at Jackson Elementary School in Salt Lake City learned four years ago there was a hazardous waste site three blocks from their school, they were shocked. After all, some of the youngsters used that abandoned, barrel-strewn lot as a playground.

What if the chemicals at the site contaminated the groundwater? What if someone set fire to one of the thousands of corroding barrels, causing it to explode? It quickly became clear that the site posed a health threat not only to the students, but to the entire community—and nothing was being done about it. That's when the sixth-graders in Lewis' special problem-solving class decided to take matters into their own hands. As Lewis later wrote in a magazine article, "I had no idea I was unleashing a tiger."

With coaching from their teacher, the youngsters began polling people living near the site and, through surveys of their classmates, discovered that 32 of them had played in the poorly fenced lot at one time or another. They also asked the Utah Department of Health when the site was slated for cleanup. No time soon, they were told: there were no funds, and the site had yet to be nominated for cleanup through the federal Superfund program.

But the Jackson students, who came from

Salt Lake City's poorest neighborhood, weren't about to take no for an answer. By letter and by phone they pestered the health department, the mayor, the Environmental Protection Agency and the firm storing the barrels on the lot. As momentum picked up, Lewis' fourth- and fifth-grade students got in on the act. Within a year, the company had moved the barrels, which contained a variety of toxic industrial chemicals and pesticides.

But there was still no money for cleaning up the soil, contaminated to a depth of 17 feet. So the students held a white-elephant sale, raising $468 toward cleaning up the lot. A letter-writing campaign to corporations brought in another $2,200.

When they tried to turn the money over to the state, they hit another snag: there was no legal mechanism for designating private funds to clean up a specific site. Undaunted, the children drafted their own bill creating a state Superfund law, then personally lobbied the state legislature. The bill passed both houses without a dissenting vote.

The students of Jackson Elementary taught Salt Lake City—and the nation—a valuable lesson about the war against hazardous wastes. As 12-year-old pollution-buster Kory Hansen summed it up, "Kids can make a difference."—Nancy Shute

"Essentially what we've done is allowed an unregulated, uncontrolled and, in my opinion, unethical experiment to be performed on the American public and our ecosystem," says Jerry Poje. In California's San Joaquin Valley, the world's richest, most intensively cultivated farm belt, the "experiment" may have backfired.

Seven percent of all pesticides used in the United States are applied in the vicinity of McFarland (population 6,200), an area resident Connie Rosales calls a "toxic fishbowl." The community's well water is tainted with DBCP, a now-banned pesticide known to cause cancer. For reasons still not understood, 16 children in McFarland have developed the disease. Since 1975, nine of them have died, most between 1982 and 1985—a rate eight times higher than normal for a town McFarland's size. Says Rosales, "Our kids have been the canaries in the coal mine."

The people of the San Joaquin Valley

aren't the only ones paying for the nation's legacy of chemical use. Pat Brown and her family were among the 1,000 who fled their homes in Niagara Falls a decade ago when chemicals such as benzene, dioxin and trichlorethylene were discovered seeping from a long-abandoned canal that had been used as a dump for almost 22,000 tons of chemical waste. Love Canal, named for its builder William Love, has come to symbolize the costs of industrial negligence. But it also illustrates the chasm between what science can offer in the way of assurances on the health effects of pollution and what it cannot.

State and federal governments conducted studies in the Love Canal neighborhood, but so far none has proven any link between exposure to the chemicals and cancer or other ailments. Unsatisfied with the official findings, Pat Brown conducted her own, unofficial study. She was shocked at what she learned.

Working with the Ecumenical Task

Force of Niagara Falls, which helps communities cope with environmental crises, Brown surveyed ten households. She found every home has at least one, and as many as three, cancer victims. Children from those homes are now giving birth to children with health problems and birth defects. "But," Brown adds scornfully, "we have to call this 'housewife data.'"

In September 1988, New York's top health official announced that about 250 Love Canal homes could be reinhabited. Six months later, scientists discovered toluene and PCBs in the soil of an area thought to be free of contamination. The resettlement plan was put on hold. "We are careful not to say 'safe'" when referring to the area, explained James Melius, director of New York's Division of Occupational Health and Environmental Epidemiology. "One thing we are learning about chemicals is that if we think something is safe today, we may learn a few years later that we were wrong."

The Greening of Industry

Making the Transition to a Sustainable Economy

Instead of controlling toxic wastes, industry must shift to safer materials and cleaner technologies.

Ken Geiser

KEN GEISER participated in drafting the Massachusetts Toxics Use Reduction Act and serves as the director of the Toxics Use Reduction Institute at the University of Lowell. In the past year, he has promoted new chemical management policies across the country and in several international forums. He received his master's and doctoral degrees from MIT's Department of Urban Studies and Planning.

Once belching smoke and foamy rivers visibly symbolized industrial pollution. Today, as concern moves to the global accumulation of invisible and persistent toxic compounds, the ecological impacts of chemical contaminants are inspiring a reassessment of the basic elements of industrial production. Whether the issue is hazardous waste, acid rain, the depletion of the ozone layer, the greenhouse effect, or the scarcity of fresh air and water, the root of concern is the same: the inappropriate use of materials and technologies.

The result is an international consensus around the need to reconsider conventional approaches to environmental regulation. In this country, state governments are taking the initiative with "toxics use reduction" laws that focus on cutting toxic material inputs rather than controlling chemical wastes. The U.S. Environmental Protection Agency (EPA) has endorsed "pollution prevention" strategies over conventional pollution control. And in Europe, the same philosophy underlies the "precautionary" approach and the notion of "clean technology."

By whatever name, these policies suggest a new paradigm in environmental policy, one that entails transforming the materials and technologies of production. The goal of this paradigm is to devise a system of sustainable industrial practices that can be implemented without posing undue environmental risks now or in the coming decades.

The concept of sustainability first emerged regarding agricultural policy. Sustainable farming aims to lessen the need for chemical pesticides and fertilizers in favor of practices that work with natural ecological cycles to improve the soil and increase the pest resistance of plants. While the shape of sustainable industry is still emerging, several features will be critical:

☐ Technologies appropriate to the desired ends.
☐ Safe and environmentally compatible materials.
☐ Products that meet basic social needs and some individual wants.
☐ Low- and no-waste production processes.
☐ Safe and skill-enhancing working conditions.
☐ Energy efficiency.
☐ Resource conservation to meet the needs of future generations.

The transition to sustainable industry will not be easy. Consider the hurdles to lessening society's dependence on chlorine-based hydrocarbons. Each year, the world processes 24 million tons of ethylene dichloride, the chlorinated hydrocarbon produced in the largest volume. Of that, 87 percent is needed for making vinyl chloride, essentially all of which is polymerized into polyvinyl chloride. PVC has replaced wood, cotton, copper, paper, and other materials in products ranging from toys, house siding, and sewer pipes to textiles and food packages.

But PVC manufacturing creates huge volumes of hazardous wastes. What's more, vinyl chloride, a known carcinogen, is directly linked to liver angiosarcoma among workers in plants that make PVC. And lethal smoke emitted when PVC building products burn raises the risks of fatalities from house fires. It was because of such concerns that Sweden, Denmark, and Switzerland took steps in 1990 to ban PVC in some products and applications.

The challenges to moving rapidly to sustainable energy use are similarly daunting. The 1987 Toronto Accord, signed by over 50 nations, calls for cutting

Reprinted with permission from *Technology Review*, August/September 1991, pp. 64-72. Copyright © 1991.

*The move toward sustainable industry stems
from the shortcomings of 25 years of pollution control regulations.*

carbon-dioxide emissions 20 percent by the year 2005. But carbon-based fuels account for 85 percent of the global energy supply. The world took nearly 50 years to move from wood to coal as its primary energy source and another 50 to move to oil.

The goal of replacing common substances like PVC or oil in our daily lives suggests the scope of the revolution taking place in the thinking about chemical hazards. The new paradigm for environmental protection will extend into decisions about the materials society consumes, the technologies for manufacturing goods, and the responsibilities of government and industry to protect the biosphere. In essence, sustainable industry means converting the material basis of society.

Shifting Assumptions

The move toward sustainable industry stems from the shortcomings and failures of 25 years of pollution control regulations. Conventional environmental protection policy seeks to safeguard the public by setting conditions and limits on the release of contaminants. Except for some exposure prohibitions in occupational health and safety laws, governments have focused on wastes, emissions, and air and water quality while rarely intervening in the decisions of private firms about their selection of materials or technologies. In other words, the regulations observe the traditional boundaries between public and private property, applying only to substances released into the public air and water.

Conventional regulations, such as those developed under the Clean Air and Clean Water acts, are based on the assumption that the environment has an unlimited capacity to assimilate small amounts of contaminants with negligible risk. By establishing emission standards, this approach proposes that the environment dilutes or transforms chemicals so thoroughly that they become mere traces and do little damage, and that contaminants, once diluted or altered by the environment, can't reconcentrate or re-form. The assumption is summarized in the adage "The solution to pollution is dilution."

However, many synthetic materials, especially those based on chlorine or heavy metals, resist degradation. DDT, PCBs, mercury, and cadmium compounds stay toxic for decades. Even when these substances are released within permitted levels, ecological forces slowly move them into the quiet eddies of the environment. The substances collect in the upper atmosphere, the depths of aquifers, and the sediment of lakes and oceans.

These concentrations are all the more worrisome in food chains. Some persistent chemicals easily bond to

fats in living organisms. If the organisms are elements of primary food chains, the chemicals can concentrate further as more complex organisms consume them. The most recent National Human Adipose Tissue Survey, conducted in the early 1980s, found over 100 such contaminants in tissue samples taken from Americans regardless of their occupation or where they live.

Additional proof that assimilation-based regulations have failed arises from studying environmental quality data. In 1988, Barry Commoner evaluated 15 years of pollution control regulations in the United States. Reviewing annual EPA air samples and water quality data from nearly 400 sampling stations, he found that environmental quality had improved only slightly and in some cases had worsened. The rate of air quality improvement, which averaged 1.52 percent per year before 1980, dropped to 1.16 percent per year in the 1980s. Despite nearly $100 billion spent to clean up water, quality had deteriorated or remained constant at over four-fifths of the test sites.

Commoner did offer one hopeful finding. Pollution levels had fallen for DDT and PCBs in wildlife, for mercury in the Great Lakes, and for strontium 90 in the food chain, and lead in the air had dropped by as much as 70 percent. All these cases had one thing in common: production or use of the substance was phased out or stopped. Rather than relying on assimilation, Commoner says, "the best way to stop toxic chemicals from entering the environment is not to produce them."

At the same time that such findings have discredited the idea that assimilation is enough, other features of the regulatory approach have also gone through a reassessment. Industry uses well over 65,000 chemicals, with nearly 1,000 new substances added each year. Adequate toxicological data are available on fewer than 1 percent of all industrial chemicals. Federal agencies face the nearly impossible task of setting "acceptable" limits on releases of each substance.

The need to go beyond assimilation and chemical-specific regulations challenges the distinctions both between public and private property and between wastes and materials. Such distinctions are giving way to a view that integrates industrial and environmental needs into a single vision of sustainable industrial development.

The details of how industry needs to change its processes to realize this vision have yet to be worked out. But the mechanisms for this movement are beginning to appear as state governments and private firms promote pollution prevention and toxics use reduction.

States Take the Lead

The paradigm of sustainable industry is evolving on

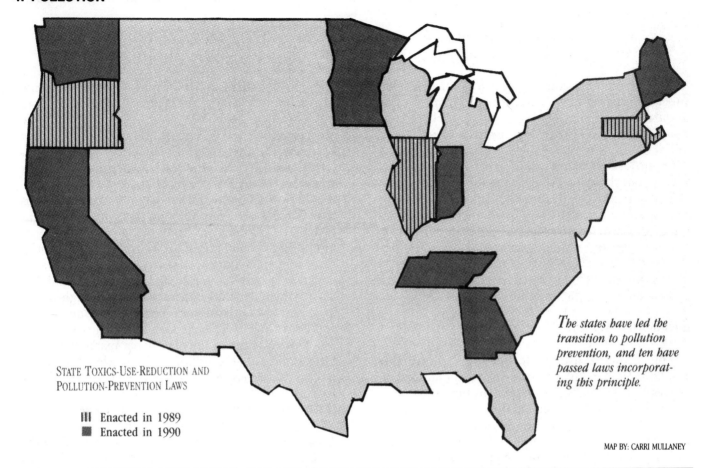

STATE TOXICS-USE-REDUCTION AND
POLLUTION-PREVENTION LAWS

||| Enacted in 1989
▓ Enacted in 1990

*The states have led the
transition to pollution
prevention, and ten have
passed laws incorporat-
ing this principle.*

MAP BY: CARRI MULLANEY

several fronts. In the United States, one place it started
was in federal policy to address hazardous chemical
wastes. The 1976 Resource Conservation and Recovery Act placed the highest priority for managing waste
on reducing its generation.

However, public and private policymakers failed to
turn in earnest to options that cut the use of toxic
materials until the mid-1980s, when operating and liability costs of waste disposal facilities soared and
resistance grew to siting new ones. In *Serious Reduction of Hazardous Waste,* an influential 1986 report
from the congressional Office of Technology Assessment, Joel Hirschhorn advocated shifting waste
management policy from pollution control to pollution prevention. The following year, the EPA established
a special Office of Pollution Prevention to guide agency programs toward this goal. In its first three years,
the office prepared a policy statement on pollution
prevention, organized several conferences, and awarded
about $10 million to state technical-assistance pro-
grams. But only in the past year has the prevention approach begun to take firm hold within the federal
government.

In the absence of a strong commitment in Washington, state governments have led the transition to pollution prevention. By the mid-1980s, North Carolina,
Minnesota, New Jersey, and other states had begun
offering technical assistance to firms attempting to

reduce their waste streams through preventive actions.
By the end of the decade, several states had enacted comprehensive laws along these lines. The new laws bypass
debates over acceptable levels of toxicity and the risks
of specific exposure levels or releases. They rest on a
simple argument: the use of every toxic chemical should
be reduced or eliminated.

In 1989, Massachusetts and Oregon took the first
step, enacting model laws aimed at toxics use reduction. In all, ten states have now passed laws incorporating some features of these programs (*see the map*), and
at least four more are considering them. Six state laws
roughly follow the Massachusetts definition of toxics use
reduction: "in-plant changes in production, processes,
or raw materials that reduce, avoid, or eliminate the use
of toxic or hazardous substances or generation of hazardous by-products per unit of production."

Such laws neither require risk assessments nor establish thresholds for chemical exposure. Instead, they set
up reporting and planning responsibilities for firms that
handle specified toxic substances. Most states base their
toxic chemical list on the 300 or so substances cited
as extremely hazardous under the 1986 federal Emergency Planning and Community Right to Know Act.
Many of the laws set a statewide goal for stemming the
use of toxic chemicals or the amount of hazardous
wastes generated, but they leave it up to firms to de-

Pollution prevention foreshadows the idea of sustainable industry by focusing on materials policy.

sign their own "facility plans" for how to meet the goals. In other words, the plans give firms the responsibility to reduce the use of toxics.

Consider the Massachusetts law, which aims at halving the state's volume of hazardous waste by 1997. Every firm that manufactures, processes, or uses any of the listed chemicals in amounts greater than established levels must prepare a plan for using less of them. The goals a firm selects and the schedule it pursues are determined by those preparing the plans. The act suggests, but does not prescribe, the technical options a firm should consider (*see the box on page 126*). Although the plans are confidential, company managers must release summaries of them along with the goals they have adopted. After 1994, when the plans are due, each firm must report on its yearly progress in meeting its goals.

In the first year of the program, over 600 Massachusetts plants have reported that they are required to prepare plans. Digital Equipment, Monsanto, AT&T, and a few other companies have volunteered to try out the planning process before the required date. The state has set up an Office of Technical Assistance to assist small and medium-sized firms, and has started an intensive combined media inspection program in one region. A Toxics Use Reduction Institute has opened at the University of Lowell to develop professional training programs and study new technologies and safer material substitutes. To aid those who will prepare facility plans, the institute is preparing a training curriculum that should be available this fall.

In contrast to fierce debates over most environmental legislation, the Massachusetts law was not forced upon an unwilling business community. Instead, it was drafted through intense negotiation between industry leaders and environmentalists. The business coalition—the Associated Industries of Massachusetts and small-business trade associations—wanted a narrow bill that linked waste reduction to the siting of a hazardous-waste treatment facility. The environmental coalition—the Massachusetts Public Interest Research Group and various academics—sought a bill with an array of policy options from requiring firms to report chemical use to giving the state authority to demand the phase-out of certain chemicals. For over four months, the negotiators debated and drafted each line of the bill. The discussions united both sides in support of a proposal that the state legislature passed unanimously. The collaboration has continued through the process of writing regulations and developing the procedures that firms will engage in to consider changing their production technologies and their selection of materials.

Other state laws vary in emphasis, goals, and specificity. The Maine law commits the state to reduce both toxic chemical use and toxic releases. Indiana encourages but doesn't require facility plans. Massachusetts, Washington, Tennessee, Oregon, and Maine firms must publicly report chemical use annually, while those in Illinois and Indiana can report privately. Most states provide small businesses with technical help in preparing plans.

These laws stretch the boundaries of pollution prevention in several significant ways. First, they focus on chemicals in use rather than on wastes. Second, firms must set goals and make plans, not just comply with an emission limit. This emphasizes creativity and innovative technology. Third, the laws encourage continuous improvement, not simply reaching a regulatory threshold.

Prevention and Precaution

These state laws to reduce the use of toxics are often spoken of as pollution prevention, yet most seek to lessen the need even for chemicals that don't affect waste streams or whose risks aren't definitely established. Thus, toxics use reduction foreshadows the idea of sustainable industry by focusing on materials policy as much as on environmental or waste policy. A similar policy shift is occurring abroad as well.

In 1987, the environmental ministers of northern Europe, meeting to discuss the deterioration of the North Sea, endorsed a far-reaching "precautionary principle." They backed an approach that would "require action to control inputs of [toxic] substances even before the causal link has been established by absolutely clear scientific evidence." The North Sea ministers asked for action "when there is reason to assume that certain damage or harmful effects on the living resources of the sea are likely to be caused by such substances."

Like toxics use reduction, the precautionary principle contrasts with policies that simply seek to keep risks to a supposedly manageable level. "In practice, [precaution] implies that emissions to the environment are to be avoided wherever possible," explains Konrad von Moltke, a specialist on European environmental policy. Von Moltke notes that there is no mention of assimilation or thresholds. "Even when no environmental effects are discernable, the avoidance of emissions is preferable to allowing them to occur."

Precaution shifts the burden of proving safety from those who would protect the environment to those who would release chemicals into it. And like toxics use reduction, it bases decisions about releases on available options, not on the environment's assimilative ca-

Techniques for Reducing Toxics Use

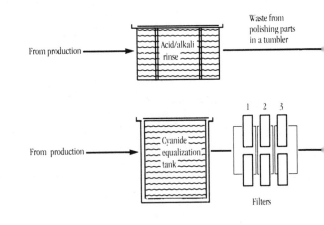

FOLLOWING the lead of the Office of Technology Assessment's 1986 report, *Serious Reduction of Hazardous Waste*, 10 states have passed laws that encourage firms to prevent pollution and to reduce their dependence on toxic materials. Companies can comply through various technological means, including:

Substituting material inputs: Firms can substitute less toxic materials for more toxic ones either in a final product or as intermediaries in production. Substituting water-based cleaning agents for chlorinated solvents is one example. Many major newspapers have substituted soy-based ink for petrochemical-based ink in color printing, thereby shortening cleanup time, cutting hazardous waste, and improving working conditions.

Reformulating processes: Redesigning production equipment or processes can reduce the need for toxics. For example, simple mechanical processes can replace some that depend on toxic chemicals. The Air Force blasts plane bodies with reusable plastic pellets to eliminate the need for hazardous paint-removing solvents before repainting. Automating production processes such as scheduling, temperature control, or metering chemical additives can reduce toxics use by improving yields and lowering the proportion of products that don't meet specifications.

Redesigning products: Better product designs can reduce toxic constituents or the need for chemicals in processing. For example, "natural color" paper and low-chlorine white paper, both of which are marketed as environmentally friendly, reduce the need for chlorine in pulp and paper processing. In Switzerland, the Product Life Cycle Institute has developed criteria for extending product durability, repairability, and reusability, all of which conserve energy and resources and reduce toxic chemical use and waste.

Improving operations and maintenance: Simple improvements in housekeeping, storage, and handling or in maintaining equipment can help. Better monitoring can catch leaks and emissions from equipment. For example, careful attention to pressure levels in spray surface coating can prevent overspraying.

Recycling in-process substances: Production methods that recycle, clean, or reuse chemicals lessen the amount of feedstock required. "Closed loop" processes, such as those being implemented in electroplating, continually clean and reuse plating baths. However, sending chemicals to commercial recyclers does not constitute toxics use reduction because it doesn't necessarily reduce toxics use.

pacity. The new adage is "When in doubt, don't throw it out."

Clean Technology

Prevention, precaution, and toxics use reduction direct industrial practice away from high-risk procedures. Still, there remains a need for transitional steps that would move current industries into more sustainable production systems. Toward fulfilling this need, several European nations have taken up the idea of clean technology, a phrase that became common in Germany, Denmark, and the Netherlands in the mid-1980s.

In contrast to "add-on" or "clean-up" technology, clean technology consists of low- or no-waste production equipment that conserves energy and materials. The Commission of the European Communities, recognizing the multiple components of the concept in 1985, defined clean technology as "any technical measure taken . . . to reduce, or even eliminate at [the] source, the production of any nuisance, pollution, or waste and to help save raw materials, natural resources, and energy."

The governments of France, Germany, Denmark, Norway, Finland, and the Netherlands have all taken steps to promote clean technology. In some countries, these measures include disseminating information and providing economic incentives, such as research subsidies and tax savings. Other countries have gone further and created official bodies to promote clean technology, invest in research, and adjust regulations so industry can more easily adopt cleaner modes of production.

France has established a Clean Technology Commission to coordinate promotion and development efforts. In 1984, the commission provided about $3.4 million to 60 pilot projects ranging from recycling solvents and reclaiming used oil to redesigning electroplating tanks and plating operations. A survey of over 600 French clean-technology projects conducted by the Organization for Economic Cooperation and Development

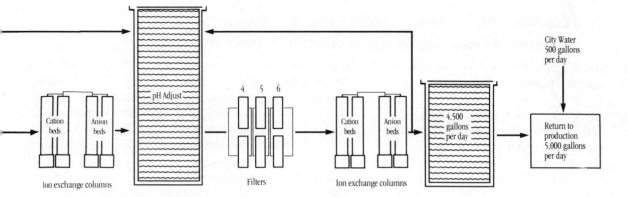

Wastewater Purification Subsystem

In the purification subsystem, hydrogen peroxide destroys cyanide. Progressively smaller particulate filters remove solids like dirt and oils, and carbon filters remove organic compounds. Ion exchange resins remove salts and metals. Sodium hydroxide is used to adjust the pH in spent acids, and muriatic acid does the same for bases. The water that returns to production is 40 times cleaner than city water and thus improves the quality of Robbins's plating process. Since the subsystem contains parallel filters and resins, manufacturing doesn't stop for preventive maintenance.

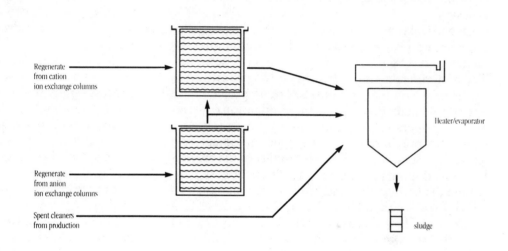

Metal Recovery Subsystem

Ion exchange resins from the wastewater subsystem produce "regenerate" laden with metals. In the second subsystem, this regenerate goes to separate cation and anion storage tanks. The two streams later combine in an evaporator where water is boiled off, leaving more metal and a small amount of salt sludge—7 gallons per year, all of which is recovered, compared to 4,000 gallons per year in 1986, all of which was landfilled. The sludge, resins, and some filters go to recyclers for further reclamation of base and precious metals.

Annual Savings in 1989 over 1986

ITEM	PERCENT REDUCED	SAVINGS
Water use	47.7	$18,000
Wastewater treatment chemical use	81.8	$ 8,000
Hazardous waste disposal	89.0	$14,000
Laboratory analysis costs	86.9	$26,000
Regulatory fees	—	$ 5,000
ANNUAL SAVINGS		$71,000

Payback Period: 1.69 years

The transition to clean production offers an opportunity to unite human health and industrial productivity.

(OECD) found that 67 percent saved on raw materials, 65 percent conserved water, and 8 percent cut energy use.

In the Netherlands, the Department of Clean Technologies in the Ministry of Housing, Physical Planning, and the Environment promotes the idea in existing industries and encourages the development of clean-technology processes and products for export. As part of their national environmental policy, the Dutch have established a program that requires a firm to assess the environmental impact of its products throughout their useful life and during their disposal. The Dutch government is also funding research into substitutes for CFCs, chlorinated pesticides, cadmium-bearing fertilizers, and asbestos.

Of particular note is a regional clean-technology program set up near Rotterdam and Amsterdam by the Netherlands Organization for Technology Assessment. The program focuses on 10 firms and 35 priority areas for clean-technology options. So far, 45 projects have been implemented, including the elimination of several rinsing operations, a transition to cyanide-free electroplating, purchasing new packaging equipment, and replacing resin sampling systems. Twenty of the projects have saved money; the others didn't affect cost.

One of the best-documented experiments in clean technology began in Landskrona, Sweden. Researchers from a technical institute at the University of Lund identified seven small to medium-sized companies for an intensive clean-technology program. After thoroughly analyzing current operations, the Lund team recommended over 100 different technology changes. For example, one lamp plant totally eliminated trichloroethylene by changing to biodegradable cutting oils, simple detergents for cleaning parts, and alkaline rinses.

Denmark's Cleaner Technology Support Program started in 1987 with a three-year budget of about $14 million and last year received an increase to $62 million for another three years. The program mainly funds pilot projects, many initiated by the national Technologisk Institute, in the metal, wood products, and food industries. It also supports projects that seek replacements for CFCs and mercury and other heavy metals.

At the regional level, the European Community has established an information clearinghouse to advise firms on adopting clean technology and to promote the adoption of low-waste equipment. For several years, the EC has also awarded prizes for clean technologies and products, like "no clean" metal parts manufacturing and cadmium-free batteries.

Toward Sustainable Industry

Despite such government advocacy, clean technology is not sweeping Europe, nor is it becoming common practice in the United States. A 1987 OECD survey found that clean-technology projects accounted for no more than 20 percent of all industry investments made for environmental protection. The situation has only gradually improved since then.

For a host of reasons, industry has preferred conventional add-on (end-of-pipe) technology to installing entirely new production equipment. Paralleling the findings of the U.S. Office of Technology Assessment, the OECD found that add-on equipment tends to cost less up front, even if process changes save money over the long haul. Add-ons are also readily available, and distributors prefer selling such devices, which often come with service contracts. Add-on equipment is often widely accepted and well proven, so officials accept its purchase as a sign of strong intentions to comply with regulations. Finally, add-ons don't affect production much and can fail without disrupting it, but faulty process equipment can slow or shut down an entire production line.

While many of the advantages of add-ons reflect the conservatism of managers, some are due to limited experience with clean technologies. For sustainable industry to take hold, then, manufacturing will need information on how to implement clean technologies as they become available.

In other words, sustainable industry requires both new technology and new management concepts. Too often advocates of clean technology focus on developing and diffusing equipment while neglecting a broader, holistic approach to materials, products, processes, and technologies. "Industries should cooperate with scientists, citizens, and governments to close the life cycle of products as far as possible," advises Frank van den Akker, director of the Dutch Department for Clean Technologies. While cost-effective clean technologies can sometimes accomplish this, "in some places, completely new production technologies and a new structure and way of living have to be worked out."

Sustainable industry encompasses the entire social, economic, and technological system by which we produce goods. This expansive context is also called "clean production," a term coined at a 1989 meeting of the United Nations Industry and Environment Program on low-waste and no-waste technology. The systems-wide perspective of clean production and sustainable industry merges prevention policies, the precautionary principle, and clean technology. Eschewing a singular focus on wastes, materials, or technologies, it unites these elements into an integrated view that regards economic and environmental goals as equal determinants of a healthy society.

From this point, it is a short jump to considering the entire industrial system as an environmental and health issue. The design of new production processes would take into account both occupational and community health. The consumption of materials, water, and energy would be evaluated in determining production efficiencies.

In short, policies to promote sustainable industry would consider the risks of materials throughout their full life cycle—from synthesis or extraction through processing, distribution, and application to final disposal. The use of existing materials would be carefully tailored to fit into natural ecological systems. The design and selection of new materials would be consciously directed toward enhancing the quality of the environment and public health.

The materials revolution of the twentieth century brought forth a cornucopia of products, yet in the rush to supply markets there has been less-than-adequate accounting for the risks either to people or to nature. The transition to clean production offers an opportunity to unite human health and industrial productivity. The challenge for sustainable industry is to develop the knowledge, techniques, and materials to guarantee that tomorrow's enterprises are as safe and clean as they are productive.

FROM
ASH TO CASH
The
International Trade
in Toxic Waste

RON CHEPESIUK

RON CHEPESIUK *writes about environmental issues from Rock Hill, South Carolina.*

Call it a toxic memorial, a monument to loose laws and fast money. On a rural Haitian beach stand rows upon rows of barrels filled with 3,000 tons of municipal incinerator ash from Philadelphia, dumped one night by a barge called the Khian Sea. The boat had entered port with a permit to unload fertilizer; a crewman even stuffed his mouth with a handful of the flaky black cargo to prove its harmless nature. A hundred workers began unloading the ash, building a heap only yards from the incoming waves. They unloaded 3,000 tons out of more than 13,000 on the barge before the Haitian government intervened, ordering the ash back onto the barge. Under the cover of darkness, the Khian Sea disappeared.

Six Haitian governments have come and gone since that night in 1988, but the ash remains, although some has washed out with the tides or blown off with the winds. A stranger from Long Island came by and shoveled the ash into barrels, intending to ship them to upstate New York, but somehow the barrels never left, nor are all of

> "It costs from $250 to $300 per ton to dispose of waste under the current U.S. environmental regulations; some developing countries will accept waste for as little as $40 a ton."

From *E Magazine*, Vol. 11, No. 4, July/August 1991, pp. 31-37, 63. Reprinted with permission from *E*, the Environmental Magazine, P. O. Box 6667, Syracuse, NY 13217. (800) 825-0061.

them even covered. Fertilizer? Hardly. The ash contains some of the most toxic chemicals known to humankind—dioxin and furans. A Greenpeace investigation found it may also contain such toxic heavy metals as lead and cadmium, and may be spiced with mercury and arsenic. Everyone from Greenpeace to the local Haitians to the City of Philadelphia has cried foul, but still the ash sits. The "fertilizer" importer, Cultivators of the West, turns out to have been owned by two brothers of the cruel Haitian Colonel Jean Claude-Paul, an associate of former dictator "Baby Doc" Duvalier. The Khian Sea reappeared in Philadelphia, then crossed the Atlantic to West Africa, hooked around into the Mediterranean and down the Suez Canal into the Indian Ocean, finding a welcoming port nowhere. Finally, it pulled into Singapore with a new name, The Pelicano, and an empty hull.

"For all intents and purposes, the incinerator ash disappeared into the Indian Ocean," says Jane Bloom, senior staff attorney of the Natural Resources Defense Council (NRDC). Philadelphia lays the responsibility with the middleman, Joseph Paolino and Sons, who lays it on the barge owner, Amalgamated Shipping, who says they sold the boat to Romo Shipping while the ash was still aboard. The tangled case has gone to court, but who knows how much longer the barrels will sit by the shore. Some Haitian environmentalists have sent envelopes filled with ash to the City of Philadelphia and the Environmental Protection Agency marked "Return to Sender," but the real solution will take more than a postage stamp.

Over the past two decades the international trade in toxic waste has grown alarmingly as the industrialized nations run out of room at home. Greenpeace reports that from 1968 to 1988 more than 3.6 million tons of waste — solvents, acetone, cobalt, cadmium, medical and pharmaceutical waste, and perhaps some low level radioactive waste—got shipped to less developed nations. "This figure should be interpreted as a minimal total and the tip of the iceberg," says Jim Vallette, Waste Trade Coordinator for the group. "Greenpeace suspects the figure is much higher."

The voyage of the Khian Sea was one of the most notorious episodes of this trade, but sadly the world is now dotted with similar stories. In Koko, Nigeria, 3,800 tons of highly poisonous waste, including poten-

"**The bottom line is, why should we allow something to happen in someone else's country that we wouldn't allow to occur here in the U.S.?**"

tially lethal polychlorinated byphenyls (PCBs), were found in drums at an open site in town, the gift of a local businessman who forged his cargo papers and bribed the Koko port officials. Several hundred barrels recently washed up on the Turkish shore where some curious locals opened the lids only to suffer nausea and skin rashes. A few barrels even exploded. In fact, these barrels of trouble have rolled down the economic slope to any number of poor countries: black South Africa, the former East Germany, China, Romania, Poland (where some barrels were sold as varnish), Thailand and others. Greenpeace publishes an inventory as thick as a phone book with these schemes, which can stretch the imagination. One company tried to dump millions of used tires off the Marshall Islands, billing them as a new form of underwater reef for the fish. Another firm almost sweet-talked Guatemala into building 100 kilometers of roads with incinerator ash.

Toxic wastes shipments aren't widespread, but environmentalists — as well as U.S. government and developing country officials—fear they could become common unless stronger measures are taken. The charges of "toxic terrorism" and "garbage imperialism" have already been made, as less developed nations fear, in the words of one African official, that they'll change from "the industrialized world's backyard to its outhouse."

"It's insulting and humiliating to the African countries, coming after several decades of exploitation in the colonial era," said one Nigerian official at a recent United Nations meeting in New York City. Nigeria's military government grew so incensed at discovering that toxic and radioactive waste had been dumped at one of its ports in July 1988, that a government spokesman warned that those found guilty of importing waste might face the firing squad.

Over 83 countries have banned waste imports. Such organizations as CARICOM,

a 13-nation association of English speaking Caribbean countries, and the Economic Community of West African States (ECWAS) have taken strong stands against the trade. In January 1991 all of Africa, except for South Africa and Morocco, adopted a sweeping ban under the Organizations of African Unity (OAU) which may be the toughest in the world. Ghana, Togo, Liberia and the Ivory Coast have enacted their own heavy penalties and fines, jail terms and clean-up costs for people convicted of importing toxic waste. Ivory Coast laws call for prison terms of up to 20 years and fines of up to $7.6 million. Ghana has created a "Toxic Task Force" to check chemical imports at customs.

Back in 1988 one international toxic waste shipment led to a major diplomatic row when the Guinean government jailed a Norwegian counsel, fined him $600 and placed him on a six-month suspended sentence after a Norwegian ship dumped 15,000 tons of incinerator ash from Philadelphia. The Norwegian firm, Bulkhandling Inc., arranged the deal with a local company ostensibly importing the waste as "raw material for bricks." The Guinean government, however, aborted the deal and ordered an investigation after Greenpeace alerted them that the boat carried toxic waste, not construction material.

Some industrialized nations do recognize the problem, especially those in Western Europe. In Brussels, the European Parliament has unanimously called for a ban on large-scale exports of hazardous waste to poor countries, while the European Economic Community (EEC) has joined 66 nations from Africa, the Pacific and the Caribbean in curbing the international shipment of toxic wastes.

"You can't build a dump on a heap of sand, and you can't build an ally on a heap of toxic waste," says United States Senator Robert Kasten (R-WI). He fears that exported toxic waste may create an overseas Love Canal-like catastrophe. If a U.S. company is at fault, the incident could create the same international backlash as Union Carbide's deadly chemical leak in Bhopal, India. "The bottom line is, why should we allow something to happen in someone else's country that we wouldn't allow to occur here in the U.S.?" Kasten says.

Even without a catastrophe the export of toxic waste may come back to haunt the West. For instance, toxic materials could be mixed with the fertilizers used on such crops as spinach and lettuce that wind up on our dinner tables.

But the economic drive behind the international trade in toxic waste can be overwhelming. And, ironically, the growing clout of environmentalists in the U.S. and Western Europe is partly to blame. Strict laws now regulating toxic waste disposal at home have pushed the price up considerably. Wendy Greider of the Environmental Protection Agency's (EPA) International Office says that it costs $250 to $300 a ton to dispose of these wastes at home, while some developing countries may accept them for as little as $40 a ton.

The international toxic waste trade "became a trend in the late 80s as opposition to landfills and incinerators gained momentum in the U.S. and Europe," says Greenpeace's Vallette. And the countries with the toughest laws also produce most of the waste. The West generates about 90 percent of the estimated 325 to 375 million tons of hazardous waste produced worldwide every year. West Germany alone produces 80 to 90 million tons and can only dispose of a third of it at home. And each North American generates one ton of toxic waste each year.

> "Most likely, waste exports to developing countries will be dumped into the environment to re-create the problems already found in industrial countries."

This waste disposal crisis has spawned a slew of new companies that have turned to financially strapped nations in Africa, Latin America, the Caribbean and the Pacific as repositories for Western refuse. Fortunes are being made by enterprising and unscrupulous businesses that know how to take advantage of the gray zones in international law. Michael Yokovitz of the Organization for Economic Cooperation and Development says that international trade in toxic wastes is a "get rich quick scheme," adding, "It's easy. Anyone can do it—they don't even need to know about wastes." "They are not technical people," says the EPA's Greider. "The president of the company may be a lawyer or someone who has worked for a shipping company for many years with no expertise in biology or toxicology."

Toxic waste brokers often operate from post office box numbers in such places as Gibraltar, Switzerland, Liechtenstein and the Isle of Man. They register as a private company and buy the wastes with no questions asked. After they find a receptive dump site in a poor country, they rent a ship, hire a crew and set sail.

Enticed by the prospect of millions of dollars for their economy, developing countries often accept shipments of toxic waste with little understanding of the health and environmental dangers involved. "Most developing countries have neither the technical capability nor the regulatory

Mexico's Maquiladoras—Free Trade, or Foul Play?

— Ron Chepesiuk

Take a drive along the Mexican side of the U.S. border, past the "maquiladoras" assembly plants that dot the dry landscape, and you may come upon rows and rows of trees with plastic bags in place of leaves. The barren branches have no life left of their own—only the refuse they collect from the wind. Stop at a maquiladoras, and you may find the workers' shacks built right up against the factory's chain link fence, homes built of two by fours, shipping pallets or tar paper held on by bottle cap rivets. The back seat of an old Chevy may have been transformed into the dining table. A hairless dog might run by festering with red sores. Poverty isn't pretty anywhere, but at the maquiladoras there's an added heartbreak: the poisoning of the landscape because of the lax handling of toxic wastes at the factories. Marco Kaltofen, who has made such a drive for the National Toxics Campaign Fund, says, "To make yourself a surface water sample from Mexico, drain a couple of cups from the bottom of your hot water tank where it's got lots of iron, add four or five tablespoons of used motor oil, spray in every aerosol can from your house, and add a little Drano for taste."

The maquiladoras present an awful conundrum. Mexico views them as the price to be paid for economic development. Since 1982, plunging oil prices have drastically lowered Mexico's standard of living, forcing the country to devalue the peso and accrue $100 billion in foreign debt. The maquiladoras are now the country's second largest source of foreign currency behind petroleum. More than 1,900 plants now line the border from Tijuana on the Pacific to Matamoros on the Gulf of Mexico, importing raw materials from the U.S. and shipping back the finished product free of any import duties. The owners range from such giants as RCA, IBM, Chrysler and General Electric, to small companies, manufacturing everything from furniture to sun glasses to carburetors. They start workers, mainly women, at $27 for a 48-hour week, but they've still attracted half a million impoverished people from the rural provinces of Mexico.

"It's a trade-off," Mexican environmental engineer Rene Franco told *The Progressive* magazine in 1988. "I don't like to see my country becoming a wastebasket for the United States...and some companies use us that way. But on the other hand, I think we always have to be aware of the fact that, for all the jobs created, we will be paying some kind of environmental price."

Environmental activists in the United States, though, have begun to ask why, if we import goods from the maquiladoras, we can't export some of our environmental standards. Many of the workers regularly handle toxic chemicals without proper safety clothing or adequate training, they say, and many of the factories are filling desert ravines, garbage dumps, even market stalls with toxic chemicals and other wastes. The toxic materials include: carcinogenic trichlorethylene, a common de-greaser used by the electronics industry; poisonous copper cyanide; paint strippers and thinners; and PCB contaminated wastes. These wastes may be spewed out into the dismal squatter camps that have no running water or sewage facilities.

In 1987, the two countries signed an agreement requiring U.S. companies and their subsidiaries to ship their waste to the States. They must also notify the Environmental Protection Agency in writing before doing so. Yet, Kathleen Shimmer, chief of the office's health and

infrastructure to ensure the safe handling of toxic waste," NRDC's Bloom explains. "Most likely, waste exports to developing countries will be dumped into the environment to re-create the problems already found in industrial countries."

But the potential income from some proposed deals can be awfully tempting. In Papua New Guinea, for example, the government of the province of Oro negotiated a deal with Global Telesis Corporation, a firm from California, to build a $38 million detoxification plant to handle 600,000 metric tons of toxic waste a month from the West Coast. The agreement, had it been accepted, would have generated an income approximately six times the annual budget of the province. But the deal fell apart under pressure from the national government, and over concerns that Global Telesis would not be able to raise the necessary funding.

"Given the dire conditions existing in Third World countries, it's understandable how some might be led into wishfully

> "Most developing countries have neither the technical capability nor the regulatory infrastructure to ensure the safe handling of toxic waste."

thinking that the waste they are accepting poses no dangers for their people or their environment," says Dr. Noel Brown of the United Nations Environment Programme (UNEP). He worries about the impact of large, unregulated shipments of toxic waste overseas, but he cautions against sweeping generalizations. "Some countries have reported incidents," he says. "But we have to look at the situation on a country-by-country basis."

Environmentalists and officials from developing countries say Western countries must tighten up their laws on toxic waste exports. In March 1989, 105 countries, under the auspices of UNEP, met for three days in Basel, Switzerland to write the Basel Convention on the Control of Transboundary Movements of Hazardous Waste and their Disposal. "It's a first step," says Brown. "The Basel Pact provides the world with the first clear set of principles that would control shipments of toxic waste across national borders. It protects Third World countries while at the same time setting well established obligations for both the exporting and importing countries in dealing with the problem."

The Basel Convention hasn't been without controversy. Many countries and experts on the thorny problem of international law say that the pact is cloaked in vague language and riddled with loopholes. A lot of important points still have to be addressed or the treaty will be rendered ineffective. These include the definition of international waters, the

emergency planning, said in May 1989 that only ten such notifications were made in the last two years.

"The agreement has not been enforced," says Dick Kemp, director of the Naco, Arizona-based Border Ecology Project, a private nonprofit advocacy group. "Besides, it's filled with loopholes. For example, under the terms of the agreement, U.S. corporations are allowed to 'donate' their wastes to Mexican charities, which then can sell them to unlicensed recycling plants in the country. But these plants have little understanding of the wastes they are acquiring."

"There is a tremendous amount of waste that is unaccounted for," Kemp adds. "No one really knows how much toxic waste is being generated."

Nor does the waste know national borders. Last October, Santa Cruz County, Arizona, had to declare a state of emergency after millions of gallons of raw sewage flowed into its system from Mexico, the latest incident in a problem that goes back two decades. In the little town of Nogalas, Arizona, downstream from some maquiladoras, the hepatitis rate has shot up to 20 percent over the national average.

Some U.S. lawmakers take these prob-

lems so seriously that they have warned President Bush that they will oppose his proposed free trade agreement with Mexico unless action is taken on the hazardous waste problems. A number of labor and environmental groups in both countries have demanded that any free trade agreement links foreign industrial and occupational health protection. They recently formed the Coalition for Justice in the Maquiladoras which has developed a Maquiladoras Standards of Conduct to be presented to businesses in the region. "We want to send a message—moral behavior knows no borders. What would be wrong in the United States is wrong in Mexico, too," says Susan Mika of the Coalition. "It is profoundly unfair that these wealthy and successful corporations should look the other way while their workers live in crushing poverty and while raw sewage runs in the streets of their colonies. Surely these companies have some responsibility to improve these appalling conditions." The standards of conduct call upon companies to adhere to existing environmental and occupational safety laws and to help replace the squatters camps with decent housing.

The problems of the maquiladoras can

be duplicated in any poor country hoping to become a manufacturing haven for the West, from the Philippines to Haiti to the Dominican Republic. Environmentalists like Dick Kemp want to see a two-pronged attack on the problem. "We need to find mechanisms that are going to work not just in Mexico, but in other developing countries where industrial nations are locating assembly plants," he explains. "One way to do that is to regulate unilaterally in the industrialized countries. The other way is for the industrialized countries to help build up the power of the local people who are in a position to help control the environmental problems that are occurring."

But the place to act is Mexico, and the time is now. President Carlos Salinas de Gortari of Mexico, fearing that the environmental crisis could block the proposed free trade agreement, shut down the country's largest oil refinery, which had poured smoke in its Mexico City neighborhood, and he toured the U.S. to reassure lawmakers his country is ready to crack down. If we don't clean up the maquiladoras, warns John O'Connor of the National Toxics Campaign Fund, we could be "turning the border into a two thousand mile-long Love Canal."

specific meaning of environmental safety, and the assignment of legal liability. "Once the pact is ratified, hopefully the ratifying countries will meet to work out the details," says Brown.

Here in the U.S. the Basel Convention has gotten a mixed reception. Senator Kasten believes, "The pact doesn't go as far as it should, but it's a step in the right direction." But Greenpeace's Vallette calls it a "glorification of the status quo" and a "justification for what should be an outlawed business." He adds, "The international agreement has no means of enforcement, relying instead on the weight of public opinion to curb the dumping."

The U.S. has an ambiguous system for handling the export of toxic waste. Under current law, the EPA must notify and receive the consent of the destination country before a shipment of hazardous waste can leave a U.S. port. The federal agency, however, has no authority to prevent a shipment from taking place, even if it doubts the country is capable of handling it. "Destination countries often don't regulate as tightly," says Greider of the EPA. "Once it gets there, we don't know what happens to it." The only wastes

prohibited from export are those with PCB levels above 50 parts per million.

There are no restrictions on those waste exports not technically classified as hazardous, such as Philadelphia's incinerator ash, household refuse, some waste oils and uranium mining tailing wastes, even though many of these can be quite toxic. But Vallette is pessimistic about improving the situation. "I don't see a lot of energy in Congress right now for dealing with the issue," he says.

Environmental groups such as Greenpeace believe that regulating the trade gives it legitimacy and respectability, so a complete ban is necessary. Several international organizations, such as the Organizations of African Unity, the 101-nation Non-Aligned Movement (NAM) and the International Organization of Consumer Unions, agree. The World Bank says it will not finance projects in any country receiving hazardous toxic waste from another.

Some experts, however, call the total ban approach shortsighted. "Waste exports are the wave of the future," says Bonnie Ram, a waste consultant formerly with the Federation of American Scientists. "While

much hazardous waste may not be appropriate for shipment to other countries, there are just some things we don't use that other countries may be able to use."

Noel Brown agrees, adding, "What is toxic in one country may not be toxic in another. So toxicity is not necessarily inherent in the product or in the handling. Some developing countries import toxic wastes from which they recover valuable products." But Brown could not provide any examples.

Despite their sharp disagreement over a total ban, both sides agree that the West needs to come up with more imaginative approaches to the waste disposal problem. UNEP and Greenpeace say that the key is reducing the waste at the source. "This will help to avoid the movement of wastes wherever possible," Brown explains. "Incentives must be provided and new technologies invented to encourage recovery and reuse."

Vallette adds, "The only real solution to the toxic waste problem is to reduce waste at its source...to stop it from ever being produced. The world needs to begin thinking about long-term solutions to the waste crisis."

Will the Circle Be Unbroken?

David Weir and Constance Matthiessen

Constance Matthiessen and David Weir are staff writers at the Center for Investigative Reporting in San Francisco. Weir is also author of The Bhopal Syndrome *(Sierra Club Books, 1987), coauthor (with Dan Noyes) of* Raising Hell *(Addison-Wesley, 1983), and coauthor (with Mark Schapiro) of* Circle of Poison *(Institute for Food and Development, 1981). Special thanks to Jorge Simán-Zablah. Lori Ann Thrupp, Luisa Castillo, and CIR interns Stephen Levine and Christopher Rivers also contributed. Research for this story was supported by the Fund for Investigative Journalism in Washington, D.C., and by the* Mother Jones *Investigative Fund.*

Editor's Note: Ten years ago, reporters from *Mother Jones* magazine and the Center for Investigative Reporting collaborated on a series of stories, "The Corporate Crime of the Century," which exposed corporate "dumping," the export of hazardous products from the United States to the Third World. The crime continues, with especially painful consequences in faraway villages and fields largely obscured from public view.

Last fall we sent two reporters and a translator into a remote plantation in Costa Rica to untangle the story of *los afectados* (the affected ones)—a group of more than one thousands workers who have paid a high price for this kind of dumping. What follows is the workers' story, the details of the corporate decision to expose them to a hazardous pesticide, and news about the legal effort to bring them a measure of justice.

> *"Look at the mess we've got ourselves into,"* Colonel Aureliano Buendía said at that time, *"just because we invited a gringo to eat some bananas."*
> —Gabriel García Márquez,
> *One Hundred Years of Solitude*

Río Frío is a town that never quite happened. Across mountains that separate Costa Rica's stormy Atlantic plain from the capital city of San José, it is set in the middle of a vast plantation, a clutch of flimsy buildings dwarfed by miles of banana plants shuddering in a muggy wind. We've put a call out over the radio, asking the men who call themselves *los afectados*—the affected ones—to meet us here. We're hoping that at least a few will turn up to discuss the turn their lives have taken. To our surprise, about 70 dark, wiry men with open shirts and heavy mustaches, many with machetes strapped to their belts, are already gathered in the community center, and more continue to arrive.

The sky blackens as morning stretches into afternoon and the rains return, hammering loudly on the open-air building's tin roof. Most of the men smoke and trade jokes, but one sits slightly apart. Pedro Carrillo Dover, clear-eyed and articulate, leans forward in his chair as he explains that he came to Río Frío 19 years ago hoping to break away from a dead-end past. The way out seemed to be the banana plantation run by Standard Fruit (owned, in turn, by Castle & Cooke, Inc., one of the world's largest food-production conglomerates). Río Frío was supposed to be just a way station, a place to make some money before getting on with things. When Pedro escaped from the town ten years ago, he never intended to come back.

Returning today makes him feel as if he'd never left. He's uneasy and ashamed, he says, looking around at the other men. "Do people on the outside know what has happened?" he asks quietly. "Do they know about us?"

"How could they?" breaks in another man who has been standing nearby. "This problem doesn't have a name. It's so big. You have to be inside it to understand what it is like."

MORE THAN A THOUSAND MILES AWAY, JACK DEMENT settles into a high-backed executive's chair in Castle & Cooke's boxy Dole skyscraper, located in one of the many shopping malls that smother the landscape around Boca Raton, Florida. Dement chain-smokes Salems, speaking with a gravelly voice and grumpy amiability about his job, which keeps him on the road "60 percent of the time," perpetually jet-lagged, inspecting company property from the Philippines to Honduras.

Dement's leather jacket, stained Polo shirt, and gruff manner set him apart from the gray-suited company men around him. He's been with the firm for 25 years, and he is a field man, not a paper shuffler. Among other responsibilities, Dement oversees which chemicals are used at the company's banana plantations; that's how his life converges with Pedro Carrillo's, even though the two men have never met.

The banana business has always been a tough one, mainly

Pesticide Dumps— And a Boomerang

he problem of "dumping"—the practice of shipping banned, hazardous chemicals and products to other countries—has persisted over the past ten years. In addition to the resulting illnesses and deaths elsewhere, there is often a boomerang effect for U.S. consumers, especially in the case of pesticides, since much of the produce we eat is grown in other countries. Here's a reporter's notebook:

Total volume, in pounds, of bananas imported by the United States from 1983 to 1985:...17,620,058,245

Total number of imported bananas tested for pesticide residues during the same period:...... 160

Percentage of pesticides used on bananas that can be detected:..........................less than 50

Percentage of pesticides exported from the United States that are restricted or illegal to use inside the country:................................. 25

Estimate of the number poisoned by pesticides annually:.................. approximately 1 million

Estimate of number killed by pesticide poisoning annually:........................... 20,000

Percentage of all pesticides that are used in the Third World: 20

Percentage of pesticide accidents that occur in the Third World: ... 50 percent of poisonings, 90 percent of deaths

Odds of being poisoned by pesticides in Central America as opposed to in the United States: 1,800:1

Percentage of produce imported by the United States and other industrialized countries that is pesticide-contaminated: .... 50 or more

Percentage of all U.S. food imports tested for pesticide residues: 1

Percentage of food imported by the United States that the government believes to be contaminated: 5.3

Number of known pesticides that regulators can test for residues: .... less than half

Percentage of food identified by the FDA as pesticide-tainted, but which still has been allowed to reach consumers: 60

Percentage of pesticides wasted during aerial application, causing only poisonings and pollution:..... 99

—*Craig Karmin*

because the commercial life of the fruit—from harvest to breakfast table—is short and unforgiving. One of Castle & Cooke's biggest pest problems at its banana plantations is the nematode, a microscopic worm that feeds on the roots of plants. By the late 1960s, the nematode population had risen dramatically. Luckily for banana growers, two of the world's largest chemical companies—Dow Chemical and Shell Oil—had come up with a chemical, dibromochloropropane (DBCP), to combat nematodes.

(In the late 1950s, separate scientific studies sponsored by the two manufacturers revealed that DBCP damaged the testicles and reduced the sperm count of laboratory animals. But over the subsequent 20 years neither company included this information on its product labels. Government officials aware of the results approved the DBCP labels despite these findings.)

"What made [DBCP] so fantastic was that it killed the nematodes without hurting the crop," Dement says. By 1971, Castle & Cooke was sending regular shipments of DBCP to its banana plantations in Costa Rica. Although he didn't know it at the time, Dement had helped make a decision that was to change permanently the lives of Pedro Carrillo and his fellow banana workers in a way that none of them could have imagined.

Banana plants are giant herbs that grow up to 25 feet high. They mature quickly and must be tended carefully along the way. After several months, a large purple bud forms at the center of the leaf cluster, then slowly opens to expose a double row of flowers that develop into "hands" of 10 to 20 "fingers" each. Over the following four months, these tiny bananas gradually fatten as they grow ripe.

Work on the banana plantation builds to harvest time, when the bananas are picked, sorted, fumigated, and boxed for the trip to foreign markets. At Río Frío and at Standard Fruit's other big plantation in the remote Valle de la Estrella (Valley of the Star), Pedro Carrillo and his coworkers say they were instructed to mix DBCP (which they called "Nemagon," the Shell trade name) with water and pour it into canisters called *pichingas* for transportation to the field. Then they filled their injectors, which resembled mammoth hypodermic needles, and injected DBCP into the soil around the base of the banana plants.

Workers told us they were frequently doused with the chemical when it spilled from the *pichingas* onto their backs and arms. "It felt hot at first, and stinging," one worker says. "Then it refreshed you, like ice on your skin. But it smelled awful, and it made us sleepy."

The men grew to respect the new chemical, even to imbue it with mythic powers. "All the frogs and the toads were gone in the valley after we used Nemagon," another worker from the Valle de la Estrella recalls. "We poured it on the snakes and the ants, and it made them crazy. The fish would die in the rivers."

PEDRO CARRILLO BEGAN WORKING WITH DBCP IN 1976 (the pay and the hours were better). By then he'd spent six years at Río Frío, but his aspirations were far from the banana fields: three compañeros shared his ambition, and in 1973, they had begun attending school every night after work. Walking six miles to school and back, the young men, still wearing work clothes stained with the milky latex of banana plants, talked about their dreams of escaping. "We realized that to triumph in life, one had to educate oneself," says Bisai

Fernández Delgado, one of the four. "And one couldn't have vices."

Pedro Carrillo met Iris Marin Jiménez at night school when they were in seventh grade. Iris had come to the plantation in 1972 to help her sister and brother-in-law, who owned a small food store on one of the *fincas*. He sat beside her in class one night, and they soon began to spend all their time together.

Iris is small, with a proud, pretty face, her eyes deep, dark, and watchful. "I always wanted to be a nurse; that was my dream," she says. But her work at the family store kept Iris away from her studies. After two years, she left school; by then she and Pedro were in love and she was pregnant.

Pedro continued his exhausting schedule of work and school until he got his diploma in December 1977. He moved with Iris and Felicia, their little girl, to take a job as a loan officer at the Bank of Costa Rica. "It has always been my goal to be better tomorrow than I am today," Pedro says. "So we moved away from there with a lot of dreams."

IN THE SUMMER OF 1977, WHILE PEDRO CARRILLO AND HIS coworkers were applying DBCP in Costa Rica, the wives of chemical manufacturing workers at an Occidental Chemical Company plant in a small town in California discovered— during conversations in the bleachers while their husbands played baseball—that they all shared a problem: none of them was able to get pregnant. After prodding from their wives, the men, who worked with DBCP, approached union leaders, who in turn contacted a doctor about the problem.

Soon, newspaper stories reported that DBCP had made dozens of chemical workers sterile at plants across the country, and was carcinogenic as well. Dow and Shell immediately suspended all production, and the Environmental Protection Agency (EPA) initiated its lengthy and cumbersome regulatory process to cancel the chemical's registration.

The news stories about DBCP made officials at Castle & Cooke nervous, since by then the chemical had become, as Jack Dement wrote, "the material of choice in all of our banana divisions." Even a decade later, Dement cannot help growing animated when he talks about DBCP. "I've never seen another chemical that increased our banana yields by 30 percent," he says, snapping his fingers, "just like that!"

In an internal memo dated August 16, 1977, Jack Dement worried that a ban on DBCP would drastically reduce banana yields, pointed out that alternative chemicals were only 65 percent as effective, and emphasized that "there is no evidence that people who apply the chemical, as opposed to those who manufacture it, have been rendered sterile or have been harmed in other ways." Dement's memo concluded that Castle & Cooke would continue using DBCP until it was banned in the company's areas of operation.

Dement's memo was only the first of a series of efforts inside Castle & Cooke to keep using DBCP in Costa Rica throughout the rest of 1977 and most of 1978. After the sterility link became public in the summer of 1977, Dow officials were reluctant to continue providing their stocks of the chemical to Castle & Cooke. But the giant food conglomerate pressured Dow, warning that a refusal to supply more DBCP amounted to a "breach of contract." Dow finally relented and sold its remaining inventory to Castle & Cooke.

The decision to continue using DBCP in Costa Rica for 15 months after the sterility link made news in the United States came from the "highest level of the company,"

admits retired executive vice president Leonard "Ted" Marks, Jr., over lunch at his country club near Palo Alto, California. A slight, balding man with impeccable manners, Marks remains a discreet and effective spokesman for the company. "You see, DBCP was so important to us. It wasn't a cover-up, believe me, but hope springs eternal. When the DBCP disclosures began, it was like we were on a freeway, going 65 miles per hour, and suddenly there was a sign, 'Detour now.' Well, we didn't do that. We thought, what's going on? What should we do?"

Marks, Dement, and other officials at Castle & Cooke now express regret over the problems caused by the chemical. In their defense, they say they didn't know about the 1950s animal studies linking the chemical to sterility. Other corporate executives involved in the DBCP saga continue to dispute evidence that it causes sterility. Clyde McBeth, one of those who helped develop DBCP for Shell, says he handled it without safety equipment and suffered no health problems. "Anyway," says McBeth, who has never been to Costa Rica or met any of *los afectados*, "from what I hear, they could use a little birth control down there."

WHEN MARIO ZUMBADO SALAS FIRST TRAVELED FROM RÍO Frío in 1977 to see physician Carlos E. Domínguez Vargas in San José, nobody in Costa Rica had yet heard that the chemical might cause sterility. But as more banana workers came to him because they couldn't have children, Domínguez asked Zumbado and several others to bring him labels from the chemicals they were applying for Standard Fruit. Then he sent the labels to his brother, who worked for a U.S. chemical company, and asked if he knew anything about the products.

By this time, Domínguez had determined that the men had seriously reduced sperm counts; many, including Zumbado, were completely sterile. When his brother reported back about the DBCP cases in California, Domínguez had the answer to the mystery at Río Frío.

Ten years later, Domínguez shows us a file that commemorates his role in the DBCP case. He recounts how, bolstered by the evidence from his brother, he approached the Costa Rican government to denounce DBCP. "I had the names of patients, I had a list of products used, and I had scientific documentation," Domínguez, who has a penchant for American westerns, recalls with glee. "So John Wayne goes in and starts firing!"

In late November, 1978—a year after workers in California discovered they were sterile—Domínguez met with Costa Rican officials to present his evidence. By December, Standard Fruit, under pressure from the government, agreed to stop using DBCP in Costa Rica. In Río Frío and the Valle de la Estrella, however, the continued use of DBCP (for over a year, even after the California disclosures) may have had tragic consequences. According to the Costa Rican government's chief consulting urologist, Carlos Calvosa Alegretti, as little as one hundred hours of exposure to DBCP can cause sterility, and he says over one thousand workers in Costa Rica alone may have become permanently sterile as a result of exposure to it. (San José attorney Marlene Chavez, who represents some of the workers, believes the total is closer to three thousand.)

AFTER THEY LEFT RÍO FRÍO, LIFE SEEMED TO COME TOGETHER for Pedro and Iris Carrillo. They had saved enough to buy a small house outside the little city of Guapiles, near a park lined with large, ancient trees. Pedro still dreamed of going on to university, but for the moment he was satisfied with his job at the bank.

But Felicia was almost six, and although Pedro and Iris tried, no more children came. By 1982, the problem had become a major issue in their relationship. Pedro wouldn't talk much about it, but Iris noticed his declining appetite for sex. He seemed to be constantly tired, and he complained of vague, persistent pains. Finally, in early 1983 he decided to see a doctor. He was surprised when the doctor who examined him seemed familiar with his complaint, and questioned him closely about the chemicals he had worked with at the plantation.

Carrillo was shattered when he learned the results of his semen analysis. The doctor informed him that he was completely sterile and that too much time had elapsed since his exposure to DBCP to hold out any hope of recovery. "I felt that a dream had died, and I didn't know who killed it," Pedro says six years later.

After that diagnosis, Pedro grew moody and uncommunicative. Iris felt him slipping away from her. He began to spend more and more time away from home. At times, he seemed unable to accept his sterility and he would angrily insist that he wanted a son—"someone who could understand me as a man"—as if he blamed Iris for what had happened to him. Iris, an intensely private person, had no one to turn to for help. Her world revolved around her family, and now that seemed to be slowly falling apart. Increasingly isolated and desperate, she came up with a plan that frightened her at first. But she decided she had to do something before it was too late.

BY LATE 1979, DBCP WAS BANNED IN THE CONTINENTAL United States. It was not illegal to continue sending the banned chemical overseas, however (nor, remarkably, is it today). In November 1979, a team of journalists from the Center for Investigative Reporting and *Mother Jones* reported that Castle & Cooke was doing just that—shipping DBCP to its banana plantations in Central America.

Early the following year, allegations about this "dumping" led to a congressional hearing and a demurral from executive Ted Marks: "It is Castle & Cooke's corporate policy that we will not use nor purchase [a] product treated with any pesticide which is not specifically registered for that use by the U.S. Environmental Protection Agency."

Though categorical in his public statements, Marks then assigned a midlevel official to investigate the company's overseas pesticide use in greater depth—and the findings of the report contradicted his assurances. The internal report, dated November 13, 1980, contained disturbing news that the company continued to use all kinds of hazardous pesticides, including DBCP, at various places around the world—with accidents and controversy the inevitable result.

When we showed Marks the report recently, he said he only vaguely recalled it, but thought that company officials had tried to solve the problems it uncovered. According to another former official who was at the meeting when the report was presented, however, all copies of the report were gathered up and discarded—a charge Marks disputes. (The source smuggled one copy out of the meeting and gave it to us.)

Perhaps the most disturbing portions of the report were indications that as late as 1980, three years after the sterility link had surfaced in California and two years after the first cases showed up in Costa Rica, Castle & Cooke was still using DBCP in an unidentified "Third World country." Other corporate and government documents indicate that once the chemical could no longer be used in Costa Rica, which has one of Central America's most democratic governments, the banana company simply shipped its remaining stocks, plus shipments still on the way, out of the country. The records document that the DBCP was sent to Honduras, a military state where no one was likely to raise a public protest. The circle of poison was not yet broken; according to several sources in Honduras—where Castle & Cooke continued to use DBCP at least through 1979—many workers there now complain of sterility as well.

THOUGH PEDRO AND IRIS DIDN'T KNOW ABOUT IT, MANY OF the households in Río Frío and Valle de la Estrella were splintering in the aftermath of exposure to DBCP. As is the case with many of the workers we interviewed, Mario Zumbado Salas's wife left him when she learned he was sterile. "I'm not ashamed to say that I cried for a month," he told us when we talked with him in the barber shop he now owns in Río Frío. His new wife is only 19, and he worries that she, too, will leave him. "When she sees an advertisement with a kid in it, she always says, 'Oh, what a cute kid.' That mortifies you. You look away, you pretend you don't hear."

Many of the sterile workers in Río Frío and Valle de la Estrella also complain of genital pain and impotence. Although doctors in Costa Rica and the United States say these symptoms are psychosomatic, impotence seems to be pervasive. "With a woman, I am no good anymore," several men told us. One woman says of her relationship with her husband, "We sleep together now like two brothers." Depression, alcoholism, and even suicide have also increased among the banana workers in the wake of the DBCP tragedy.

José Sosa came to Río Frío when he was very young, and at 29, he is completely sterile. His wife, María, who has a round, fresh face, looks as if she is about to cry. "My house is very sad," she says. Her father, also a banana worker, became sterile after she was born. "One is born to reproduce oneself, then die," says Sosa. "But here, one is born only to die."

ONE DAY IN EARLY 1985, IRIS TOLD PEDRO THAT SHE WAS pregnant. Since he had only a vague understanding of the invisible mechanism of sterility, he reacted with a mixture of shock and elation: maybe the doctor had been wrong. But as much as he hoped this was true, the worst doubts arose in his mind. Iris seemed remote and evasive all through the pregnancy, increasing his suspicions.

When the baby, a boy, was born, Pedro could not resist examining him closely. By the time Juan was a few months old the truth was inescapable: the boy's coloring was much lighter than his sister's, and he looked nothing like his father.

It was simply too much for Pedro. He disappeared for several days, and when he returned he fought with Iris, shouting and interrogating her obsessively. Iris withdrew from him, refused to answer his questions, and devoted all of her time and attention to the children. Her depressed, angry husband then became what he calls "another kind of being."

Over the next two years, all semblance of intimacy between Iris and Pedro vanished. He was transferred to one of the bank's offices on the other side of the country, and came home only on weekends—when he came home at all.

IT IS A SWELTERING AFTERNOON IN DALLAS, BUT FROM TEN floors up in attorney Charles Siegel's air-conditioned office, the streets below look hushed and gentle. With his rumpled shirt, irreverent manner, and resonant Texas drawl, Siegel seems out of place in this hermetic office with its high-backed leather chairs and lush carpets. Likewise, the state of Texas seems an unlikely refuge for the rights of the Costa Rican workers. But for Siegel, who has been crisscrossing the United States over the past five years trying to find a court that will accept the case of *los afectados,* his home state of Texas now represents the last hope.

Five years ago, Siegel's firm, Baron & Budd, which has one of the leading toxics litigation practices in the country, sued Dow and Shell on behalf of the sterile banana workers (in 1987, Castle & Cooke was added as a defendant in the complaint).

Siegel felt strongly about the sterility case from the beginning. "These workers have clearly suffered, and they've really been screwed. It's another example of dumping in the Third World. I think that going to court may be the most effective way, if not the only way, to get these companies to stop. It's been shown again and again: it's not legislation, but awards to victims, that have shaped corporate conduct."

Courts in Florida and California refused to hear the case. The Texas Supreme Court will soon decide whether the case can proceed to trial there. The court has already heard from an array of big guns. A barrage of amicus curiae briefs filed in support of Dow and Shell by Fortune 500 companies—including Exxon, Texaco, and Tenneco—raises the specter of a damaged business climate and unemployment if the case gets a hearing in Texas. Those corporate briefs insist further that the state will become the "dumping ground" for claims by foreigners, and the "courthouse for the world" if the workers' case is allowed to go to trial.

Siegel fired back, arguing that the banana workers "were employed by an American company on American-owned land and grew Dole bananas solely for export to American tables. The substance that rendered them sterile was researched, formulated, tested, manufactured, labeled, and shipped by an American company in America to another American company. Yet now, defendants claim that the one part of this equation that should not be American is the legal consequences."

DURING THE PAST DECADE, BY ALL ACCOUNTS, STANDARD Fruit's pesticide safety programs have improved substantially. But poisonings still occur at Rí Frío, and one of the chemicals responsible may ultimately prove as hazardous in its own way as DBCP. Early last year 45 workers were made ill by Temik, the best known of the six pesticides that replaced DBCP in the battle against nematodes.

Temik, which is among the most toxic pesticides used anywhere today, is a groundwater pollutant currently under investigation by U.S. scientists because of evidence that it may suppress the human immune system. (The EPA's pesticide division recently recommended barring the use of Temik on imported bananas, asserting that up to 1,500 infants and children in this country may be at risk each day from Temik residues on bananas.)

The parallel between DBCP and Temik, both heavily applied all over the globe despite scientific uncertainty about their dangers, indicates how little the world's highly technological food-production system has changed over the past 20 years. That system remains so deeply dependent on chemicals that pesticides continue to be considered innocent until proven guilty—often with tragic consequences.

Yet little progress has been made in the development of alternative methods for controlling pests. Jack Dement of Castle & Cooke insists that "nothing would make us happier than nonchemical control"—but places the blame on consumer demand for creating the marketplace pressure that ensures the continued use of potent chemicals. Puffing his cigarette and gesturing toward glossy blow-ups of flawless fruit, Dement points out that although DBCP was used to protect the plant itself, most pesticides are used only to prevent cosmetic damage. According to Dement, market surveys show that consumers won't buy fruit if it looks scarred or slightly damaged. "The number of bananas that are thrown away borders on a sin, just because the fruit isn't big enough, or has a few spots."

Cathleen McInerney Barnes of EPA counters: "It's easy to blame the consumer, but when is the consumer ever asked whether he or she would prefer perfect fruit, or fewer poisonings in the Third World? I'm not sure they would choose to pay that price."

PEDRO AND FELICIA WAIT FOR US AT THE BUS STATION IN Guapiles. At 12, Felicia is sweet-faced and shy; encouraged by her father, she tries out some English phrases she's learned in school. Pedro directs us to the outskirts of town, then down a muddy road to their house. It is a warm, humid evening, and the door to the house is open.

Juan is almost three years old now. Small and quick, with mischievous eyes, he runs to meet Pedro as we walk in. Iris invites us to sit down and brings a tray of sodas and beers. Together in the small, cozy living room, it would be a happy family scene, except for the coolness between Pedro and Iris. When she takes the children into the other room, he lights a cigarette and says that things are very bad.

"Felicia used to be a good student, but this situation has ruined her grades," he says. "She hears us fighting and she goes off and cries. She doesn't respect me as she used to. And the boy, I love him, but I can't forget. He is a rebel, violent and angry. I see him as a boy with a trauma."

Iris comes back into the room and the two eye each other warily. He immediately finds an excuse to leave, announcing that he is going to take the children outside. "It makes me very sad when he asks questions about the boy," Iris begins as soon as Pedro has gone. As she talks, she twists the knob on the arm of the chair and doesn't lift her eyes. "He keeps asking over and over, and I don't know what to say. Maybe what I did was a mistake, but I did it to bring us together."

Iris has never spoken to anyone about this before, and everything pours out at once. "He wanted a boy so much. I saw the child as a 'table of salvation' for us. I thought Pedro was going to be so happy. But it only made things worse."

IN *ONE HUNDRED YEARS OF SOLITUDE,* A U.S. COMPANY moves into an imaginary town in Latin America and takes over vast tracts of land for banana production. Over time, local workers grow discontent with bad working conditions and poor pay, and demand reforms. When the company ignores them, they take their case to court. But their demands are ultimately thrown out when company lawyers prove that "the banana company did not have, never had had, and never would have any workers in its service because they were all hired on a temporary and occasional basis . . . and by a decision of the court it was established and set down in solemn decrees that the workers did not exist."

The invisible worker—a character that appears in the works of Gabriel García Márquez, Pablo Neruda, and other American writers, both North and South—is a leitmotif in the traditional "banana republic" relationship between the United States and the region. That relationship is based on unseen, unknown, and uncomplaining campesinos in the South growing crops for consumers in the North.

Now that relationship is being challenged by *los afectados*. Many of the sterile workers and their families have a deep desire for North Americans to know their story. "I would like to say to people in the U.S. who eat bananas that they themselves are a part of our homes, of our lives—of our most intimate lives," says Marita Pérez, the wife of a sterile worker at Río Frío. "Nemagon has affected everything. It is not that they are eating the forbidden fruit, but in a way they *are*. That fruit—and the chemicals used on it—have caused many problems here."

If the banana workers fail in the Texas courts—and fail to attract international attention as well—it's back to business as usual for the global food-supply system that today provides U.S. consumers with a year-round diet of unprecedented diversity. And those field workers harmed along the way will be effectively erased from history, consigned once again to the state described by García Márquez—permanent anonymity.

FOR PEDRO AND IRIS, SOME DAYS ARE BETTER THAN OTHERS. On a recent Sunday, the family is spending a rare afternoon together; after lunch they take a walk in the park near their house. Pedro carries a soccer ball as Juan rides his tricycle along the path ahead of his parents.

Pedro and Iris are careful with each other, and they seem to have reached a sort of understanding. They discuss their plan: to move to San José at the end of Felicia's school year. "It will be a new start," Pedro tells us. "I think the negative feelings I have been living are my weakness."

Unlike most of *los afectados,* Pedro lifts his eyes when he speaks, and as he talks about the future it is possible to believe that he will be one of the few who succeed in leaving Río Frío —and all that occurred there—behind. "What happened was only an accident, but it changed our lives. Inside, I think I have the strength and determination to survive."

Just then Juan, who makes zooming noises as he rides too fast on his tricycle, yells out as he hits a bump in the pavement. Pedro gets to the boy first and helps him to his feet. Juan isn't hurt, but he's scared, and he cries as Pedro brushes him off and comforts him. As the boy's crying subsides, Pedro tosses him the soccer ball. Juan immediately forgets his tears and runs after it, then gives it a wobbly kick in the direction of his father. Pedro kicks it back gently, his pace matching Juan's as they pass the ball back and forth over a small patch of grass.

It would occur to no one watching, as Pedro bends over the excited boy, showing him how to kick the ball, that a chemical with strange powers had damaged this man and his family; nor would an onlooker be likely to notice the difference in coloring between the man and the child. As Pedro stands watching Juan race across the grass after the ball, no one would think they were anything but father and son.

For more information about pesticide use nationally and internationally—and alternatives to them:

Pesticide Action Network
North America Regional Center
PO Box 610
San Francisco, CA 94101
(415) 541-9140

Greenpeace
1436 U Street, NW
Washington, DC 20009
(202) 462-1177

Northwest Coalition for
Alternatives to Pesticide
PO Box 1393
Eugene, OR 97440
(503) 344-5044

National Coalition Against
the Misuse of Pesticides
530 7th St, SE
Washington, DC 20003
(202) 542-5450

Eastern Europe: Restoring a Damaged Environment

Richard A. Liroff

(Liroff directs the Eastern Europe Environment Program at World Wildlife Fund and The Conservation Foundation. Previously affiliated, these groups were formally merged in 1990 to form a private, non-profit conservation organization involved in research and environmental protection.)

Whole sectors of industry are producing things in which no one is interested, while things we need are in short supply . . . . Our outdated economy is squandering energy . . . . We have laid waste to our soil and the rivers and the forests our forefathers bequeathed us, and we have the worst environment in all of Europe today.

—Vaclev Havel, President of Czechoslovakia

President Havel's assessment of his nation's economic and environmental ills applies broadly across Eastern Europe. The Iron Curtain has been lifted to reveal truly appalling environmental conditions. Eastern Europe has been savaged by economic-development policies indifferent to the carrying capacity of its ecosystems and to the health and well-being of its citizens. The East's central planners have demonstrated they can be as environmentally callous and cavalier as the worst private-sector managers in market-oriented economies.

The United States, together with Western Europe, is supporting the economic and political transition of Eastern Europe. The United States should offer a balanced, integrated program of environmental and economic assistance that fosters full restoration of a healthy environment in Eastern Europe. Such a program would help reduce the region's contribution to global warming and encourage use of both American technologies and innovative approaches to pollution prevention.

The Environmental Challenge

The German Democratic Republic (GDR), Czechoslovakia, and Poland are among the world's largest emitters of sulfur dioxide (SO_2). Moreover, in Europe, as elsewhere in the world, air pollution does not respect political boundaries. The Eastern European states export from 59 percent to 74 percent of their SO_2 emissions. According to monitoring data, however, of the total amount of SO_2 deposited in these nations, 36 to 59 percent originates outside their borders.

The forests of western Czechoslovakia, southwestern Poland, and the southern GDR have been devastated. Budapest, Prague, Krakow, and other major cities routinely have air-pollution readings well above existing health standards.

Drinking-water supplies throughout Eastern Europe are heavily contaminated. Vast reaches of the Vistula River in Poland, which drains much of the country, are classified as unfit for use even by industry. The Baltic and Black Sea coasts are badly degraded by domestic sewage, agricultural run-off, and heavy metals and organic pollutants from industry. Water quality problems are both domestic and transboundary; domestic progress in combating pollution has been slow, and multilateral cooperation negligible.

Reprinted from *EPA Journal*, Vol. 16, No. 4, July/August 1990, pp. 50-55. *EPA Journal*, United States Environmental Protection Agency, Office of Communications and Public Affairs.

4. POLLUTION

The soil, too, is polluted. Industrial discharges have contaminated soils and domestic food supplies. In the Upper Silesia region of Poland, for example, lead, zinc, cadmium, and mercury levels in samples of garden produce are 30 to 70 percent higher than World Health Organization norms.

Eastern Europe's mines and industries yield prodigious amounts of solid and hazardous waste. Waste generators in Hungary reportedly dispose of over 500,000 tons of hazardous waste annually in illegal landfills. In addition, substantial amounts of hazardous waste have been shipped east from Western nations. The GDR has reported importing one million tons of waste annually, but Greenpeace contends that the amount of imports has been disguised and is really five times greater. Few safeguards have been developed to assure appropriate management of these wastes.

The devastation of the environment is revealed through effects on human health and welfare. In especially contaminated areas, statistics and anecdotal evidence show dramatically elevated rates of respiratory disease, reproductive and developmental problems, and shortened life spans. These areas include Poland's officially designated ecological disaster areas, the coal-burning and industrial areas of Czechoslovakia and the GDR, and industrial centers in Romania and Bulgaria. Upper Silesia, one of Poland's ecological disaster areas, has circulatory and respiratory disease rates that are, respectively, 15 and 47 percent higher than national norms.

Human populations have been removed from some contaminated areas, and in other areas people have been offered economic incentives to remain. For example, in Bitterfeld, GDR, labeled by *Der Spiegel* as "the dirtiest city in Europe," wages are comparatively high to attract workers to the area. Also residents are given extra money to buy vegetables, to compensate for the loss of contaminated home-grown produce.

Economic and Political Roots of the Problem

Eastern Europe's heavy industries are

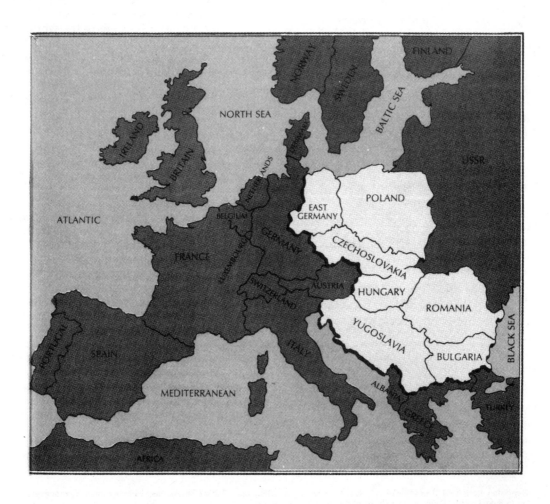

inefficient, requiring larger inputs of energy and raw materials than counterpart industries in Western Europe. On average, Eastern European economies use about twice as much energy and water per unit of Gross Domestic Product as do West European economies. Little has been spent on pollution control, and there is virtually no domestic pollution-control industry. Countries have been unwilling to spend scarce hard currency on Western pollution-control technology.

Eastern European societies have not been "societies of law" in the Western sense. Even when strong environmental laws have been enacted in these nations, they have had little impact, because decision-makers have not been held accountable for enforcing them. Watchdog nongovernmental organizations (NGOs) have rarely been tolerated.

The environment has had very low priority in planning. Heavy subsidies for inefficient industries and the lack of real pricing of goods have undercut any meaningful role for fines, penalties, and other economic tools in environmental protection. For example, Poland levies fines on polluters, but they constitute a very small percentage of clean-up costs. Until a few years ago, fines were negotiated away in talks between the environment ministry and other ministries.

Discharge fees in other Eastern European nations have had similarly small impact. State environmental functionaries generally have had little success policing state industrial enterprises. The fox has been guarding the chicken coop.

Guidelines for U.S. Programs

The principal U.S. programs to assist Eastern Europe in its economic, political, and environmental transition are authorized by Congress in the Support for Eastern European Democracy (SEED) Act of 1989. The U.S. Agency for International Development, EPA, the Department of Energy, the Peace Corps, and the Overseas Private Investment Corporation, among others, are authorized to launch multimillion dollar assistance programs in Poland and Hungary.

SEED's environmental programs include a regional environmental center in Budapest, Hungary, a report to be prepared by EPA on environmental problems and priorities in Hungary and Poland, and clean-coal retrofitting, air-monitoring, drinking-water, and wastewater projects for Krakow, Poland.

Budapest Conference Center

The Budapest Center, originating with President Bush, was established by the governments of Hungary, the United States, and Austria as well as those of the European Community. Housed in a 200-year-old silk mill in the old section of Buda, the Center's central mission is to help citizens, non-government organizations, the private sector, and government agencies in Central and Eastern Europe address problems threatening sustained economic growth in the region. Initially, it will focus on health, energy efficiency, and pollution prevention.

A key start-up project is an air-pollution monitoring network EPA is running in Krakow, Poland. EPA officials will also help upgrade that city's drinking water and wastewater treatment facilities.

In addition, EPA scientists, particularly in the Office of Research and Development, are collaborating with their East European counterparts under science-and-technology cooperation agreements with Poland, Yugoslavia, and Hungary.

4. POLLUTION

Reflecting the rapid pace of change in Eastern Europe, Congress is considering new legislation, SEED II, that will expand programs to other Eastern European nations and authorize U.S. participation in a new multilateral development bank for Eastern Europe.

The United States should observe the following guidelines as it expands its program in Eastern Europe:

● *Emphasize energy efficiency.* Many of Eastern Europe's environmental problems stem from mining and burning poor-quality brown coal. Slowing or reducing energy demand can reduce stress on the environment in a significant, cost-effective way. As

Eastern European governments restructure their economies and eliminate subsidies for energy, they will encourage energy efficiency. If the United States and other Western nations transfer efficient technologies, they will hasten this process.

● *Stress pollution prevention.* The United States should encourage Eastern European governments to promote process changes in manufacturing rather than "end-of-the-pipe" approaches to pollution control. The United States is only just now focusing on process changes in domestic environmental policy, after having emphasized end-of-the-pipe solutions for more than

---

Environmental Conditions in Eastern Europe

Poland

Mining and burning of coal lie at the root of many of Poland's environmental problems. It is the world's fourth largest producer of coal and seventh largest emitter of SO_2. Coal supplies 78 percent of the country's domestic energy needs. Coal-mining operations discharge 7,000 tons per day of salts into the headwaters of Poland's two major rivers, the Oder and the Vistula. Mines and industries produce large amounts of solid and hazardous waste, and severe pollution of land and water by heavy metals is reported.

Between the late 1960s and early 1980s, Poland's water quality deteriorated dramatically. The proportion of rivers classified as suitable for municipal water supply dropped from 33 percent to 6 percent, while the proportion so polluted as to be unfit even for industrial use rose from 23 percent to 38 percent. About 60 percent of the Vistula is unsuitable for industrial use.

The government has designated 27 "areas of ecological hazard," encompassing 11 percent of the country's area and just over a third of its population. Five of these are "areas of ecological disaster." The five include Gdansk on the Baltic Coast, the copper-mining and -refining region of Legnica-Glogow in west central Poland, and the contiguous, industrially impacted areas of Upper Silesia, Krakow, and Rybnik in southwestern Poland.

Bulgaria

Bulgaria, less industrialized and less dependent on coal for its energy, does not have as pervasive an air-pollution problem as other East European countries. But it does have "hot spots" of industrial pollution. Health statistics have only

recently been released. Bulgarians living near industrial complexes have markedly higher instances of numerous diseases and, in some cases, body tissue levels of heavy metals two to four times standards set by the World Health Organization.

The widespread harvesting of trees, the heavy contamination of air, water, and soil from industrial pollutants and agricultural chemicals, and other harmful practices have affected plants and animals as well. Forty percent of the country's bird species, 25 percent of its mammals, reptiles, and amphibians, and 20 percent of its plant species have been designated by the Bulgarian Academy of Sciences as endangered or rare.

Bulgaria's coastal resort trade is threatened by continuing decline in the quality of the Black Sea. The Danube River, which forms part of the boundary between Bulgaria and Romania, drains the agricultural, industrial, and municipal waste of eight highly industrialized countries into the sea.

Under the stress of these discharges, and as a consequence of reduced inflows of fresh water from rivers that have been dammed for energy and irrigation, the depth of the oxygen-rich upper fresh-water layer of the Black Sea has diminished. The U.S.S.R, Romania, and Turkey, all of whom also have coasts on the Black Sea, share in the problem.

Czechoslovakia

Czechoslovakia, dependent on brown coal for 60 percent of its domestic energy, is the sixth largest emitter of SO_2 in the world. Northern Bohemia bears the brunt of the impacts of coalmining and burning. Government studies leaked several years ago indicated that life expectancy in northern Bohemia is several years lower than the average for the balance of the

country, and rates of infant mortality, childhood illness, and respiratory illness are markedly higher. Those willing to work in the area for 10 or more years receive cash bonuses; skeptics among the citizenry label the funds "burial money." As in the German Democratic Republic, large swaths of forest are devastated by air pollution.

Czechoslovakia, like other East European nations, is reassessing the role of nuclear power in meeting its energy demands. The Chernobyl accident raised public concern throughout Eastern Europe about the safety of nuclear power, but at the time the effect of public opinion on government policies was limited. In January 1990, the government announced it was suspending plans to construct two nuclear reactors in Temelin, in southern Bohemia near the Austrian border. However, two others are scheduled to go on line there in 1992.

Prague, Czechoslovakia's capital, suffers from severe air-pollution problems, especially in winter. The pollution stems from auto emissions, household burning of coal, and factories. Prague's factories generate 11 percent of Czechoslovakia's industrial output. Prague's city planners cannot account for about 80 percent of the estimated 40,000 tons of hazardous waste produced in the city each year.

German Democratic Republic

The German Democratic Republic (GDR) is the richest nation in Eastern Europe, as measured in terms of Gross National Product per capita. But this conventional measure of economic well-being fails to capture fully the toll the GDR's industrial machine is taking on human health and the environment.

The GDR depends on brown coal for 70 percent of its domestic energy demand. It is one of the largest emitters of SO_2 in the

---

20 years. Because Eastern Europe is starting fresh with environmental policies, it can be a fertile testing ground for pollution-prevention approaches.

● *Promote Cost-Effective Technologies.* The U.S. government should also encourage Eastern European governments to develop cost-effective solutions to pollution problems at existing sources. Competing vendors of pollution-control technology have already descended upon Eastern Europe, hoping to sell the new governments technologies that can remove nearly all the pollutants from waste streams—but often only at a very high cost in money

or energy per unit of pollutant removed.

For existing sources, it might be better to promote broad use of alternative technologies that, pound for pound of pollutants, would cost less to operate. More cleanup of existing sources might thus be achieved at less cost. All new sources should be required to meet the most stringent Western standards.

● *Build Self-Reliance by Strengthening Education and Institutions.* Eastern Europe boasts a well-educated population and a rich tradition of scientific innovation. Unfortunately, in recent years, it has been cut off from the latest developments in environmental

---

world. Recently released data on air-pollution levels—previously kept secret—reveal that in such centers of heavy industry as Leipzig, Halle, and Bitterfeld, average annual levels of SO_2 in the air are five times the U.S. standard, and average annual levels of particulate are 13 times the U.S. standard.

The impact of pollutants on human health is readily visible. In the Pirna area near Dresden, children have unusually high rates of neurological and motor-development problems. Near the coal-processing facility of Espenhain, 50 percent of the children have respiratory ailments, and 33 percent suffer from eczema.

The GDR is the most industrialized nation in Eastern Europe but, by one estimate, as much as 60 to 70 percent of its chemical industry could be forced to shut down if it were subject to West German environmental standards. Much of the industry might be uneconomical to operate anyway. One chemical plant near Bitterfeld discharges 44 pounds of mercury into the Saale River each day, 10 times as much as the yearly discharges of mercury by the major BASF chemical facility in West Germany.

The GDR has been a major dumping ground for West German domestic and industrial wastes. West Germany is now being reunited with its wastes and will need to address this legacy.

Hungary

Oil and gas satisfy 60 percent of Hungary's energy demands, yet air pollution is a serious problem. It is especially pronounced in Budapest; automobile emissions are a major culprit.

Water pollution is a priority concern. Most of Hungary's water enters the nation in degraded condition from its neighbors. (This includes the Danube, which enters

Hungary after passing through West Germany, Austria, and Czechoslovakia.) Ground-water contamination from overuse of agricultural chemicals in Hungary poses a risk to public health. Water in hundreds of villages and towns is unfit to drink. Lake Balaton has been the focus of a major clean-up effort; much progress has been made in eliminating the phosphorus that contributed to the lake's decline.

The proposed Nagymoros Dam on the Danube River has been the most prominent Hungarian environmental issue in recent years. The dam, whose construction is nearly complete, is the lower portion of a larger hydroelectric project that will affect about 200 kilometers of the Danube. An upper dam is being built at Gabcikovo in Czechoslovakia. After having ignored several years of public outcry and scientists' forecasts that the project would disrupt the ground-water system supplying Budapest's drinking water, the Hungarian government recently agreed to abandon the dam. The final outcome remains to be seen.

Romania

Relatively little is known about environmental conditions in Romania because of the Ceausescu regime's veil of secrecy over environmental data.

Romania is much less dependent on coal for its energy than other East European nations. It relies on imports from the Soviet Union and domestic oil and gas for 64 percent of its energy needs.

The Danube Delta is Romania's most noteworthy ecological feature. One of the largest reedbeds in the world, and home to more than 160 breeding species of birds, the delta is a major stopping point for birds migrating between Europe, the Mediterranean, the Middle East, and Africa. In the late 1960s and early 1970s,

hundreds of thousands to over a million ducks were counted in the region during the winters. The Delta has been damaged by draining and diking, a product of Ceausescu's promotion of irrigated agriculture and construction of a shipping channel.

Pollution is greatest in five heavily industrialized regions. There is serious contamination of air, water, and soil by heavy metals. Characterizing the impact of two factories in the town of Copsa Mica, a *New York Times* reporter has written, "For about 15 miles around, every growing thing in this once-gentle valley looks as if it has been dipped in ink."

Yugoslavia

Yugoslavia's environmental problems have not drawn as much attention in the West as those of other East European nations.

As they enter Yugoslavia, the Danube and other rivers are substantially degraded, a result of upstream industrial, municipal, and agricultural discharges. Yugoslavia's largest internal river, the Sava, flows through the greatest industrial concentration in the country. The Sava is categorized as suitable only for irrigation and industrial uses or as requiring special treatment prior to any use.

Northern Yugoslavia is more industrialized than the south, thereby producing greater pollution, but concern about industrial discharges is found throughout the country. Yugoslavia's forests are subject to lower levels of SO_2 deposition than forests in the German Democratic Republic, Czechoslovakia, and Poland. Consequently, they have not experienced the substantial damage found in these other nations, but there is concern that levels of SO_2 may increase in the future.

science and technology. The United States should invest in the development and upgrading of environmental curricula, the training of officials at all levels of government, and the education of a new cadre of industrial managers.

Both ministries and enterprises will need assistance in taking an integrated approach to management that incorporates economics, technology, and administration. A landmark World Bank-funded program for Poland is a strategic example here. The Bank is funding the staffing and training of a new planning and investment unit in the Environment Ministry, the creation of a new regional regime for managing watersheds, and the development of environmental auditing for state-owned industries.

As part of its institutional investment, the United States should promote innovative economic approaches to environmental management. Domestically, it has relied heavily on "command-and-control" approaches to reduce discharges. These have been supplemented on a small scale with such economic incentives as the trading of air-emission privileges. Eastern European officials have expressed considerable interest in economic incentives, recognizing that these may yield environmental gains more efficiently than traditional regulatory approaches. They should be encouraged to combine such incentives with integrated approaches to reducing discharges to air, land, and water.

The investment in people should include the NGO community. Environmental awareness and organizing have grown dramatically in Eastern Europe, and nascent environmental organizations have been key players in the region's democratization. But environmental organizations suffer from lack of information and resources. In many cases, they lack the most basic supplies and equipment needed to carry out their activities. They need technical information, and they could benefit from training in organizing programs and reaching out to the public more effectively.

The United States should stress the urgent need for informing and involving the public as Eastern European countries revise their environmental programs. In Eastern Europe, until recently, environmental data have been treated as state secrets, and environmental dissent has been suppressed. The United States should push hard for freedom of information and involvement.

● *Encourage Environmentally Sound Investments.* The United States and its Western European allies should press an environmental agenda on the new, $12-billion European Bank for Reconstruction and Development. The United States holds a 10-percent share, the European Community nations just over 50 percent. The bank, which will be investing in both the public and private sector in Eastern Europe, is the first multilateral development bank whose charter requires promotion of "environmentally sound" development. But that mandate must be acted upon. Priority should be given to clean, efficient technologies and to ensuring that new investments are scrutinized for their environmental soundness.

The European Community nations are pushing for an environmental code of ethics for Western investors in Eastern Europe, and the United States should join in this effort. Occidental Petroleum, a U.S. corporation, provides a promising example in this regard: As a matter of corporate policy, the new facilities built in foreign lands must meet U.S. or local environmental standards, whichever are more strict.

The United States, together with other Western nations, should encourage development of a pollution-control industry in the East to serve both Eastern and Western markets. Strong U.S. support for strict enforcement of laws by new environmental administrators in Eastern Europe will further foster home-grown pollution-control industries.

The Imperative for Sustainable Development

Development strategies in Eastern Europe have failed in terms of both economics and the environment. Much of Eastern Europe is a wasteland. Tens and perhaps hundreds of billions of dollars will be required to restore and protect the environment.

One frequently hears the question, "Will the East Europeans be willing to pay for cleanup?" This question assumes a tradeoff between economic well-being and a sound environment. Trade-offs and hard choices undoubtedly will be necessary, but they should not be overstated. An enormous economic price already is being paid for environmental degradation. Poland's present pollution damage will cost the country an estimated 10 to 20 percent of its Gross Domestic Product.

Currently, many of the worst polluters may be the most inefficient operations. And as subsidies for energy are eliminated, market prices introduced, and other adjustments made, many facilities will become uncompetitive and shut down. The environment will benefit. Moreover, steps taken to make remaining operations more efficient by reducing resource consumption and making other process changes will yield additional environmental benefits.

Political forecasts about Eastern Europe have been notoriously wide of the mark in the last few years. But it is reasonable to believe that in heavily affected regions where forests are dying, babies are born prematurely, children are retarded, men and women are dying young, and the search for clean air takes people to distant areas, people will be willing to endure temporary unemployment and other economic dislocations for the promise of an economically and environmentally sustainable future.

ENVIRONMENTAL DEVASTATION IN THE SOVIET UNION

James Ridgeway

James Ridgeway is a columnist for the Village Voice

Moscow — A survey of the Soviet environmental situation reveals a country in desperate straits.

According to a mapping scheme developed by Soviet scientists, 16 percent of their country, containing a quarter of its population, is at environmental risk. Various data paint a somber picture: 40 percent of the Soviet people live in areas where air pollutants are three to four times the maximum allowable levels. Sanitation is primitive. Where it exists, for example in Moscow, it doesn't work properly. Half of all industrial waste water in the capital city goes untreated. In Leningrad, nearly half of the children have intestinal disorders caused by drinking contaminated water from what was once Europe's most pristine supply. An alternative source of clean water is now threatened by a joint Soviet-American tourist venture. Beaches along the Black, Azoz and Baltic seas are frequently closed because of pollution, and numerous rivers in the European part of the Soviet Union are off bounds because they are so filthy.

Since Stalin's time, the Soviet economy has been geared to establishing heavy industry, with little or no attention paid to environmental impacts. Most recently, the government has turned its attention to Siberia and the Soviet Far East. It has undertaken a colossal construction program, building hundreds of new cities, throwing oil and gas pipelines across the landscape (including the world's longest transcontinental gas pipeline, 2,700 miles from the far north toWwestern Europe) and constructing huge chemical complexes. While the Soviets have succeeded in building an industrial society, the country's economic malaise leaves it resembling an ungainly Third World nation, increasingly dependent on the sale of natural resources.

Most timbering of the forests is done with clear-cuts, with as much as 70 percent of the cut timber going to waste. According to one estimate, acid rain is killing more than 500,000 hectares of forests in northwestern Siberia.

An official Soviet report on the "State of the Environment" in 1988 pointed to metal poisoning in various cities: cadmium in Odessa, manganese in Dnepropetrovsk and Rostov. "The highest degree of atmospheric pollution in 1988 was observed in 68 cities," mainly in the Ukraine, in Kuzbass, the Caucasus, east Kazakhstan and near the Ural Mountains. "The most polluted reservoirs in the U.S.S.R. are along the rivers and lakes on Kola Peninsula and along the Bug, Dnyestr, Danube and Don rivers."

There is no such thing as an environmental impact statement in the Soviet Union, although there are informal efforts to establish something like one. And no one can be certain who is responsible for most of the polluting because so much of it is associated with secret military enterprises.

Soviet agriculture is especially vulnerable, with much of the topsoil eroding. Farmlands have been heavily doused with pesticides and fertilizers. In the western Ukraine, thousands of square miles of farm land remain contaminated from the Chernobyl nuclear accident.

The most celebrated case of pollution involves the inland Aral Sea . . . . Once the world's fourth largest inland body of saltwater, the Aral Sea has lost two-thirds of its original size since 1970 and is expected to vanish by the year 2000 because massive Stalinist irrigation projects have diverted the waters of rivers that once flowed into the Sea into enormous cotton spreads.

The Aral Sea fishing fleets are a thing of the past, with rotting hulls of the boats thrown up on a shrunken shore

From *Multinational Monitor,* September 1990, pp. 8-12. Reprinted with permission from *Multinational Monitor,* P.O. Box 19405, Washington, DC 20036. Individual subscription $25/yr.

of the sea which looks increasingly like a brackish mud puddle. The salt and silt from the dried up Sea blows across the cotton fields, requiring an immense effort to keep them in production.

The cotton plan was riddled with official corruption. State planners fiddled with the books to make it appear as if yields were increasing, thus providing the rationale for the government to keep pumping money into the project. Party officials, including Leonid Brezhnev's son-in-law, then stole the money.

Coal and natural gas: export substitution

While Europe has substantial coal deposits, it historically had no natural gas and had to import oil. After World War II, natural gas was discovered in the Netherlands, and then oil and gas were found in the North Sea. Eventually, with the introduction of tankers in which the gas was frozen and shipped from North Africa, more gas was brought to the continent. Natural gas made possible the revolution in the European economy, offering for the first time a clean, efficient fuel. And as the gas market grew, the Soviet Union came to play an integral role in the market. Probably more than any other single economic factor, natural gas tied the Soviet Union to Western Europe. And with mixed blessings.

As part of an immense effort to industrialize Siberia and the Far East, the Soviet Union has been developing large-scale deposits of hydrocarbons, most importantly natural gas. The overall plan has been to develop gas, and send it westward through new, often hastily and poorly constructed pipelines that provide gas to European Russia and its Warsaw Pact allies. This frees up other gas produced in the European part of the Soviet Union along with its imports from Iran for export to West Germany, Finland, Italy, Greece and elsewhere in Western Europe. As the Soviet economy worsened, this gas became ever more important as a source of hard currency. And as the need intensified for hard currency to buy imports to supplement declining agricultural production, gas that ought to have gone to Eastern Europe was instead sold to Western Europe, leaving Eastern Europe shackled to lignite coal as its only energy source. Burning coal resulted in more and more pollution, which, because of the prevailing winds, blew back and forth across Eastern Europe, into Scandinavia, and, most significantly, into the European parts of the Soviet Union — the Ukraine, the Baltic states and Byelorussia. If the export of energy helped pave the way for Western European prosperity, it also cost the Soviet Union heavily in environmental pollution and disease. In the Soviet Union itself, 25 percent of the atmospheric pollution comes from electric power plants. Now there is talk of increasing electricity sales to Europe, which, if created by coal, in all probability means aggravating the acid rain that already inundates part of the Soviet Union.

Taming the polluting state

Nikolai Vorontsov, chairman of *Goskompriroda*, the state committee for the protection of nature, described the uphill fight to get any sort of a grip on environmental pollution in a recent interview with the Institute for Soviet-American Relations. *Goskompriroda*, which was only started in 1988, employs just 450 people, compared with 600 in West Germany. The committee has only two computers, and a budget of 20 million rubles (less than $4 million) compared to the U.S. Environmental Protection Agency's budget of $5 billion. For the whole of the Soviet Union, which comprises one-sixth of the land mass of the planet, expenditures for environmental protection, besides those of *Goskompriroda*, amount to a mere 10 million rubles.

The amount of resources allocated to environmental causes may seem ludicrous, but money is not the primary issue. The big question in Moscow is whether the state, until now heedlessly pushing giant Stalinist development projects, can and will police itself. On the surface, the stated intent and the laws are not bad. The government has endorsed legislation to rehabilitate the Aral Sea and halted construction of nine half-built nuclear power plants. Beginning in 1991, polluting factories will supposedly be fined, with 85 percent of the money going to local governmental institutions for pollution control.

But the plan to fine polluting factories is a dubious endeavor. To begin with, the big polluters in heavy industry — chemicals and steel — are excluded. One power plant in Kazakhstan that installed special scrubbers had to cut its staff and lost 600 megawatts of generating capacity to pay for them. Other factories, which under the Soviet move towards a market economy are meant to be self-financing, will claim they cannot afford to provide similar controls. Still others will simply bribe inspectors. In poor parts of the nation, authorities prefer jobs to a clean environment. "The largest supply of artesian water in central Russia lies near the town of Pervomaisk," reports one area resident. "The town council is planning to construct a pesticide factory there, even though this will most probably endanger the water supply. Why? Because Pervomaisk is poor, and with the pesticide plant the town will get roads, schools and hospitals."

The government's management of the Chernobyl disaster does not bode well for its self-regulating capabilities. After Chernobyl, livestock thought to be contaminated from the accident were killed so their meat would not be eaten. But some of the meat was transported to distant parts of the country, mixed with other meat and made into sausage. This was possible because the Ministry of Public Health had secretly made temporary rules raising the limits for radiation concentrations in meat.

Taming foreign polluters

Foreign investment, centered around extractive industries and tourism projects, poses another dilemma. The most advanced projects are schemes involving multinational corporations, including prominent U.S. firms which have stood by the Soviets through thick and thin: Armand Hammer's Occidental Petroleum and Cyrus Eaton's Cleveland firm.

In western Siberia, Japanese, West German, U.S. and Italian firms are beginning work on five large petrochemical projects in the Tyumen area, which environmentalists insist is on its way to becoming one of the world's great waste dumps. The nearby Ob river has been reduced to an oily sluice. Pipeline accidents number 600 per year. Just about everyone in the Soviet Union hates the project, except, of course, the Stalin-era ministries that are pushing it ahead. In March 1988, the Soviet Union made a deal with Occidental, the Japanese firm Marubeni and two Italian outfits — Montedison and ENI — to build a gas and chemical plant near Guriev in the Caspian. The total cost is $6 billion and when it is finished the complex will be located near a gas-condensate plant at Astrakhan. The Astrakhan plant was equipped with a French-made

Trading in Timber

CONTAINING 40 PERCENT of the world's softwood supply, the forests of the Soviet Far East are the largest in the world. With U.S. companies poised to join Japanese, Cuban, North Korean and Bulgarian loggers, however, foreign interests may quickly decimate the Siberian forest.

One existing logging enterprise is the Tyndales Production Association in Tynda, Amur Oblast, which employs 7,000 North Koreans east of Lake Baikal. The joint venture does not involve currency exchange. In the Tyndales agreement, North Korea gets 39 percent of the timber cut and the U.S.S.R. gets 61 percent. The Soviet press has reported that the North Koreans have overcut tracts assigned to them, violated Soviet environmental laws and cut timber in unauthorized areas. Soviet authorities have imposed fines and sanctions, which the North Koreans have had to pay in timber because of the noncash nature of the agreement. Enforcement of environmental regulations has therefore led to cutting more timber, notes William Freeman of the U.S. Information Agency.

Environmental impacts have also been severe at Brusnichii, where North Korean logging operations have encroached on a legally protected nature preserve. The preserve director has led a public campaign to stop the logging, but government officials overseeing the timber industry have not responded to the appeals. The Soviet Union has given the agreement with North Korea precedence over local needs and state law, Freeman reports.

Japan is the largest non-Soviet bloc importer of Siberian timber, importing just under 15 billion board feet of timber, 19 percent of which comes from the Soviet Union. The United States and Malaysia each supply Japan with about one-third of its timber imports, but with international rainforest preservation efforts seeking to curtail further exploitation of Malaysian timber and to preserve the last remaining stands of old growth forest in the western United States, Japan may begin to rely more heavily on the U.S.S.R. to satisfy its timber appetite.

With the Soviet Union's increasing willingness to do business with multinational corporations, U.S. forest product companies have eagerly surveyed the U.S.S.R.'s massive forest. Companies such as forest products giant Louisiana-Pacific (L-P) recognize the potential to offset the dwindling supply of domestic U.S. timber with imports. L-P recently announced that it intends to import raw logs from Siberia to fill its Pacific Coast mills. L-P does not work directly with the Soviets, but through a Utah timber importer.

There are some obstacles to importing Soviet timber on a large scale, however. A shipment of Siberian pine destined for L-P, for example, was recently quarantined in San Francisco pending inspection by the U.S. Department of Agriculture. The USDA fumigated the timber to remove a pine nematode not found in the United States. The pine was then shipped to Eureka, Calif., for tests to determine if it would meet U.S. market standards and building codes for tensile and load bearing strength. A USDA spokesman said he did not believe the Siberian pine would satisfy the requirements of the U.S. market, although Bob Morris, an L-P spokesman, says he thinks it "holds promise." The viability of importing Siberian timber depends on the affordability of transport, developing plants to receive and treat imported Soviet timber and the inspection and fumigation process. The lumber, once milled, may be used for particle board or other non-structural uses and the sawdust will be burned in a power cogeneration facility at the mill site.

L-P intends to export only raw logs from Siberia, processing the timber outside the Soviet Union. One L-P company official told the Santa Rosa Press Democrat, "We're not interested in developing infrastructure for them."

Other U.S. paper corporations also hope to take advantage of the huge Soviet timber forests, but are positioning themselves at a slower rate than L-P. Georgia-Pacific Corp., the largest U.S. forest-products company, is a member of a trade consortium that includes RJR Nabisco, Ford Motor Co. and Johnson & Johnson. The consortium plans to establish joint ventures with the Soviet Union. Georgia-Pacific has denied plans to export unprocessed logs from the Soviet Union, and has so far limited its involvement to a ministry and consortium-sponsored Moscow trade show where it exhibited milling equipment. Other U.S. forest-products companies hoping to exploit the world's largest timber supply include Weyerhaeuser Corp. and James River Corp.

— Jay Lee

pollution control system, which did not work, necessitating the evacuation of nearby villages. Scientists predict the same thing will happen with the new projects, which have the potential to produce an accident comparable in scale to Chernobyl.

Occidental is involved in two other environmentally destructive projects. One is the Ventspils Petrochemical Complex in Latvia, which processes petroleum products for export. Enormous amounts of toxic and flammable chemicals from the complex swirl through the city and around the countryside. Occidental signed another deal to expand a chemical plant in the Ukraine that produces plastics. It already is a major polluter and each expansion is likely to make matters even worse. In both Latvia and Ukraine, independent or local reform groups have mounted serious, ongoing opposition to these foreign projects.

The Soviet populace is becoming wary of the likely environmental impacts of the multinational petrochemical corporations eager to invest in the U.S.S.R. "When perestroika began in the U.S.S.R., the country began to open doors wide for foreign companies," A.V. Yablokov, deputy chairman of the Supreme Soviet's Ecological Committee, told an international congress in Gothenburg last year. "Now the people in my country are beginning to understand that penetration of foreign firms and enterprises to the U.S.S.R. is connected with some dangers. Some people claim, and not without grounds, that many firms would like to develop their 'dirty' industries in our territory, such as ammoniac or pesticide productions."

"In a number of cases it occurred that Western chemical technology does not work in our land as it was pictured by advertisements, and chemical plants built according to Western projects proved to be sources of dangerous pollution of big regions, for instance near the northern coast of the Caspian Sea [the Astrakhan gas condensate complex]. A public campaign is developing now against construction of a tremendous oil-gas-chemical industrial complex in western Siberia. The construction had involved American and Japanese firms."

Expanding tourism can also be a problem. Leningrad is joining forces with the Cyrus Eaton Company to construct a hotel and entertainment complex northwest of the city. The site proposed for the complex sits atop an underground water reservoir which is a secondary source of drinking water for the dreadfully polluted Leningrad water supply. If Eaton builds the hotel, that water source will be permanently lost. In Leningrad, feeling against the project is so intense that 95 percent of the populace has come out in opposition.

Soviet environmentalism

In the Soviet Union, as in Eastern Europe, the environment has been a rallying cry for the political undergrounds that erupted in peaceful revolution in 1989.

The Soviet environmental movement originated in the early 1960s, when individual scientists questioned Nikita Krushchev's plans to sow the "virgin lands" of western Siberia with grain, pointing to the dangers of soil erosion. In 1966, conservatives and Russian nationalists appealed to the Party Congress to save Lake Baikal, the world's largest reservoir of fresh water whose existence was threatened by unchecked logging practices and two cellulose factories built to produce cord for bomber tires. A group of noted scientists formed a committee and Andrei Sakharov took up the cause directly with Krushchev. Then, a decade ago, a coalition of scientists protested plans to reverse the direction of rivers flowing north in Siberia and Central Asia so that they would instead spread their waters on the over-planted cotton plantations. Reestablished censorship in the late 1960s dampened the growth of the environmental movement, but, as Geoffrey Hosking points out in his account of environmentalism in *The Awakening of the Soviet Union*, what made its evolution especially difficult were the huge ministries charged with industrializing Siberia. *Gidroproekt*, in charge of hydroelectric projects, was part of Stalin's KGB secretive economic network, including the forced labor camps. Environmental concerns advanced as these different ministries came into competition with one another, for example, as the Ministry of Fisheries dis-

covered its catches declined on the lower reaches of streams dammed upstream.

In today's more open political climate, saving the environment is especially important to Russian nationalists as well as to independence groups across the country.

There are five large environmental groups in the Soviet Union, the largest of which is the Socio-Ecological Union, an umbrella group with 200 branches mostly in Russia itself (see "The State of Soviet Ecology," *Multinational Monitor*, March 1990). The Union believes change will come through political action. The other groups include the Ecological Society of the Soviet Union, which has ties with the far right *Pamyat*; the Ecological Union, which promotes solar and wind power and wants to use fines levied on polluters to clean up toxic wastes; and the All Union Movement of Greens, which is formally backed by the Communist Party's Youth organizations. People who participate in these groups often are also active in larger political organizations, such as the Popular Fronts, which were behind the rise of the reform politicians in Moscow, Leningrad and elsewhere. These reformers view environmental concerns as major issues.

A common past, a common future

Both the United States and the Soviet Union can trace the origins of environmental pollution to the industrial revolution and the ensuing reliance on hydrocarbons. Both countries experience horrendous pollution from coal-fired electric power plants and from projects aimed at industrializing their frontiers. These include, in the United States, diversion of the Colorado River, irrigation projects in California and the Southwest and the creation of the aluminum industry, and, in the Soviet Union, hydro-electric and water diversion schemes in Siberia and the Far East. The plans to reverse the flow of Soviet rivers is no more harebrained than Jim Wright's notorious scheme for damming up the western trench of the Rockies in Canada and piping the water through canals into the southwestern United States. In both countries, the modern environmental movement first took shape around the issue of water pollution during the mid-1960s—Lake Baikal in the Soviet Union, industrial pollution of the Raritan Bay, the Mississippi River and other bodies of water in the United States.

Today, the future of the environment can be tied to the often ignored race by both nations (perhaps more accurately, by corporations and ministries in both nations) to industrialize the last untrammeled part of the world in the far North. The Soviet Union exploits the Arctic in order to develop more gas for export to Europe. The United States is in search of more oil to maintain its wasteful energy system.

The Exxon Valdez spill is the most celebrated result of the race to tame the Arctic. But reports of Soviet exploitation are just as bad. For both countries, the most useful gift to future humanity would be a treaty preserving the Arctic along the lines set forth in Antarctica.

151

Resources: Land, Water, and Air

- Land (Articles 24–25)
- Water (Articles 26–27)
- Air (Articles 28–29)

The worldwide situations regarding scarce energy resources and environmental pollution have received the greatest amount of attention among members of the environmentalist community. But there are a number of other resource issues that demonstrate the interrelated nature of all human activities and the environments in which they occur; these issues may ultimately be of greater significance than whether consumers in modern nations can continue to operate energy-intensive lifestyles. One such issue is that of declining agricultural land. In the developing world, excessive rural populations have forced the overuse of lands and sparked such a shift into marginal areas that, today, the total availability of land is decreasing at an alarming rate of two percent per year. In the developed world, intensive mechanized agriculture has resulted in such a loss of topsoil (millions of tons per year in the United States alone) that some agricultural experts are predicting a decline in food production. Other natural resources, such as minerals and timber, are declining in quantity and quality as well; in some cases they are no longer usable at present levels of technology. The overuse of groundwater reserves has resulted in potential shortages beside which the energy crisis pales in significance. And the very productivity of the Earth's environmental systems—that is to say, their ability to support human and other life—is being threatened by processes that derive at least in part from energy overuse and inefficiency and from pollution. To make matters worse, there is a feeling that both the public and private sectors, including individuals, are continuing to act in a totally irresponsible manner with regard to the natural resources upon which we all depend. In "The 23rd Environmental Quality Index: The Year of the Deal," the editors of *National Wildlife* provide their annual report on the environmental crisis. An overview of the systems of land, water, and air, and the biosphere is presented in the context of what *National Wildlife* refers to as an administration in Washington "performing a precarious juggling act" by announcing conservation initiatives that seemed to mark an end to an era of environmental indifference and then cutting deals that indicated that little had changed after all. Among the many examples of environmental problems that remain unsolved, "The Year of the Deal" cites protection efforts for wetlands and endangered species, urban pollution and toxic waste disposal, continuing water pollution on a large scale, continuing timber harvesting in ancient forest stands, unchanging and unresponsive energy policies, and continuing topsoil erosion.

In the *LAND* section of this unit, two articles deal specifically with the issues of marginality and exploitation. In "A New Lay of the Land," by Worldwatch Institute researcher Holly B. Brough, takes a broad look at the issue of the uneven distribution of arable land that has pushed poor farmers into deserts and forests, speeding environmental breakdown. Many of the problems produced by this marginality might be reduced or eliminated with reallocation of land and shifts in land use. Reallocation or land reform strikes at the very heart of the problem of land marginality by creating small farms for many rather than large farms for a few, thereby stimulating rural economies and relieving poverty. Unfortunately, land reform has not been widely used in international development efforts, largely because of the entrenched power of the landed elite in many Third World countries. Brough suggests that if the environments of these regions are to be saved, land reform is absolutely necessary. Of all the marginal resources, the one that is most subject to overuse and abuse is soil, the resource upon which agricultural production depends the most. Problems of soil erosion and soil pollution are often the result of ignorance about the nature of the complex ecosystem within the soil itself. In "U.S. Farmers Cut Soil Erosion by One-Third" author Peter Weber claims that the massive soil erosion that occurred on U.S. farmlands in the 1970s has been cut drastically by the passage of the Food Security Bill. This bill removed the most erosive (most marginal) land from cultivation and required farmers to apply conservation techniques to other farmland.

The second subsection of the unit focuses on *WATER* and in this subsection, two articles deal with water quantity and water quality. Coastal zones are rarely defined as marginal locations and coastal resources are seldom marginal in nature. But increasing coastal populations have placed pressures on the use of coastal lands that have created zones of marginality in what were, only a short time ago, the world's most productive regions. "Holding Back the Sea" discusses the quantitative aspects of coastal water, especially in the context of the potential of rising sea levels as a consequence of global warming. Worldwatch Institute researcher Jodi L. Jacobson begins with the statement of a fact that is not as obvious as it probably should be: much of human society is defined by the world's coastal zones, and the boundary between land and water delineates much more than just the availability of beachfront property. Rather, it defines the availability of land for human settlement and agriculture, the economic and ecological productivity of tidal estuaries, the configuration of harbors and bays used in commercial shipping, and the availability of freshwater supplies in coastal communities. A rise in sea level, precipitated by global warming, would represent an environmental threat of unprecedented proportions, and Jacobson describes both the consequences of even minimal sea level increases and the strategies that might be adopted to deal with them. The second subsection article centers on global rather than regional water issues and approaches water issues from a quantitative rather than a qualitative sense. In "Water, Water, Everywhere, How Many Drops to Drink?" environmental writer William H. MacLeish notes that, while shortages of oil get the most publicity, it is shortages of water that represent the greatest real and potential shocks to human social and economic systems.

There are, of course, areas of the environment other than soil and water that suffer from the operation of the principle of resource marginality. When the shift is made to lower grade metallic ores, for example, greater applications of technology are necessary to extract these materials, leaving more residue to pollute the air and requiring the expenditures of greater and greater amounts of energy in the extraction process. In *AIR*, the final subsection in this unit, two articles deal with the most critical of the problems that characterize the global atmosphere: continuing accumulation of greenhouse gases and the concomitant potential for increasing atmospheric heat. In the first of these articles, environmental scientist S. Fred Singer of the University of Virginia points out that scientists are virtually unanimous in their belief that greenhouse gases are increasing but that the increase is contributing to global warming is still in question. Scientists also disagree on the magnitude and impact of the global warming trend. Singer adopts a more cautious perspective. He agrees that the available scientific evidence does point to a gradual increase in global temperatures. However, much of this evidence also suggests that relatively small increases in temperature might be more beneficial than harmful for such human activities as food production. In a counterpoint article, author Jon R. Luoma claims that even the small changes forecasted by Singer could have dramatic environmental impact. In "Gazing Into Our Greenhouse Future," Luoma states that much of the global warming theory is based on accepted scientific knowledge, although questions always arise when human activities are figured into scientists' equations. Basing his opinion on interviews with prominent global theorists from the National Center for Atmospheric Research, Luoma contends that significant steps to reduce emissions of greenhouse gases should be taken immediately in order to avoid further fueling of the greenhouse furnace.

There are two possible solutions to these and other problems posed by the use of increasingly marginal and scarce resources. One is to halt the basic cause of the problem—increasing population and consumption. The other is to provide incentives and techniques for the conservation and management of existing resources and for the discovery of alternative resources to eliminate the demand for more marginal resources.

Looking Ahead: Challenge Questions

How have U.S. federal laws and regulations designed to protect the environment been undercut by "deals" made between government and industry? Can public alarm over environmental concerns make a difference in convincing elected and appointed leaders to stiffen environmental laws?

How does the use of marginal farmland speed the deterioration of local and regional environments? In what ways can the process of land reform slow or reverse the trend toward use of more marginal lands?

How have U.S. farmers significantly reduced the amount of topsoil lost from their fields to erosion? Is there a relationship between the subsidies farmers receive from the federal government and their willingness to conserve their soil resources?

What environmentally protective choices are available to coastal communities that face the twin threat of increasing sea levels resulting from global warming and increasing coastal pollution from the concentration of human activities in the littoral zones of the world?

What are some of the most critical global problems concerning water quality and quantity, and how can new technologies and conservation measures be developed to reduce the dimensions of the global water crisis?

What are the major differences between public concern and scientific knowledge regarding the potential for global warming that results from increasing combustion of fossil fuels and other agencies that increase the levels of greenhouse gases in the atmosphere?

What are some of the uncertainties about the future impact of global temperature increase on social and economic systems? Are there reasons to enact strict legislation to control greenhouse gas emissions beyond the obvious concerns over global climate change?

23rd ENVIRONMENTAL QUALITY INDEX

ILLUSTRATIONS BY RICHARD WALDREP/EUCALYPTUS TREE STUDIO

THE YEAR OF THE DEAL

BEFORE the shadow of the Iraqi crisis fell across the future, 1990 had seemed on the surface like an unusually benign year for the environment. For one thing, it had been blessedly devoid of great environmental calamities. The evening news in this kinder, gentler year brought us no *Exxon Valdez* oil spill, no Chernobyl nuclear accident, no scenes of Midwest farm fields seared by apocalyptic drought. There were oil spills and serious contaminations and heat waves, to be sure, but none achieved the stature of a national disaster.

Instead, from Earth Day in the spring to election day in the fall, there was a great chorus of voices singing the praises of conservation and environmental responsibility. As the song lingered on, it was no longer played only by hard-core activists; corporate and political America had joined in. Being an environmentalist—or at least sounding like one—had become good business and good politics.

What happened to change the nation's environment in 1990 took place not in the forests, waterways and skies, but in board rooms and legislative chambers. Public relations people assured the country that what resulted from those conferences was good for citizens and the environment, but the end results of the Year of the Deal left many Americans wondering. The Bush Administration found itself performing a precarious juggling act. It announced conservation initiatives that seemed to mark an end to the indifference of the Reagan Era, but often went on to cut a deal that indicated little had changed after all.

One promise was protection for the country's wetlands, but the deal was filled with loopholes. Revision of the Clean Air Act started out with an impassioned promise to purify the air in America's cities, but ended up in some instances looking more like a fire sale of marked-down concessions and lowered standards. And the list kept growing. A vow to clean up the Great Lakes was accompanied by a sharp cut in the clean-up budget. The promise of a hard-hitting, comprehensive national energy policy by the end of the year evaporated into assurances that by December we would know the options.

The friendly environmental rhetoric of 1990 was as soothing as the band music on the deck of the slowly sinking *Titanic*. The dealmakers, meanwhile, were busy selling the lifeboats. On the following pages, *National Wildlife*'s twenty-third annual Environmental Quality Index recounts many of the gains and losses of the past year.

From *National Wildlife*, February/March 1991, pp. 33-40. Copyright © 1991 by the National Wildlife Federation.

WILDLIFE

Last June, in a long-awaited move, the U.S. Fish and Wildlife Service (FWS) officially listed the northern spotted owl as a threatened species. After a long controversy, the protective mantle of federal law fell across the remaining 2,000 pairs of owls and their ancient forest habitat in the Pacific Northwest. The listing of the owl was vindication for conservationists, who considered the bird's status a barometer of the health of the region's irreplaceable old-growth woodlands (see "Forests").

The victory lasted only four days. On June 26, the Bush Administration announced a new deal for the owl. The White House rejected plans developed by an interagency committee, headed by scientist Jack Ward Thomas, for protecting the owl's habitat and rewrote the plans purportedly to better balance economic and biological considerations—in other words, to permit more logging of old-growth forests. Then came a recommendation that the law be changed to make economic disruption grounds for witholding protection for a declining species. It was a serious blow to wildlife protection.

In response, National Wildlife Federation President Jay D. Hair said the proposal to weaken the law "should be recognized for what it is: a call to enshrine the almighty buck as the determining force over all life. The value of biodiversity, like that of cultural diversity, isn't readily

Although the Mt. Graham red squirrel has endangered species status, construction of an observatory could signal its demise.

reducible to just a cost-benefit analysis."

The previous month, Secretary of the Interior Manuel Lujan, Jr. had declared to a Colorado newspaper that the Endangered Species Act was "just too tough an act. We've got to change it." He was especially incensed by postponement of a $500-million reservoir project in southwestern Colorado. Construction was scheduled to begin in May, but was delayed indefinitely by a FWS ruling that the project threatened survival of the endangered Colorado squawfish. Lujan's objections to that ruling overlooked the fact that the project would amount to spending some $6,500 for every irrigated acre in a region where farmland typically sells for only a few hundred dollars an acre.

Lujan also objected to the obstacles faced by developers of a $200-million astrophysical observatory on Mount Graham in southeast Arizona. The site is a unique ecological environment for a number of plants and animals, including the last 100 or so surviving Mount Graham red squirrels. Although the FWS approved construction of the observatory in 1988, the General Accounting Office concluded in June that the approval wrongly ignored evidence that development on the site

---

Despite administration promises, wetlands and endangered species protection efforts come under repeated attack

---

"posed an unacceptable risk to the red squirrel's survival." A subsequent FWS call for more study was overruled, and with special dispensation from Congress to override the Endangered Species Act, construction of the observatory began.

Meanwhile, biologists were discussing a new report by the World Resources Institute predicting that if trends continue, one-fourth of all plant and animal species existing in the mid-1980s would be extinct in 25 years. This "biological impoverishment" of the Earth, said the report, is proceeding 100 times faster than it would without human influence, and is accelerating due to deforestation.

There was also distressing news regarding other North American wildlife in 1990. Off the coast of New England, commercial fisheries stocks continued to decline from overfishing at alarming rates. "The stocks are at their lowest levels since we've been keeping records," said U.S. Office of Fisheries Management deputy director David Crestin last spring.

Another beleaguered group, North America's ducks, remained nearly unchanged in numbers from 1989—still one of the smallest population counts on record. Estimated breeding population was 31.3 million, the third lowest recorded count and 22 percent below an average

With its habitat continuing to vanish, the nation's population of breeding ducks was among the lowest on record in 1990.

established between 1955 and 1989. "These numbers underscore the pressing need to save remaining wetlands," said FWS Director John Turner.

In 1989, Congress passed the North American Wetlands Conservation Act, which provides a stable source of funding to help save and restore wetlands. And last year President Bush embraced a pledge of no net loss of U.S. wetlands. This meant no wetlands could be drained, filled or converted by farmers or developers unless equivalent wetlands were created or restored. Implementation of that pledge was outlined in a memo between officials of the Environmental Protection Agency (EPA) and Army Corps of Engineers.

Environmentalists, full of praise for the policy, awaited formal approval. But the Bush Administration first delayed it, then softened it considerably. In the end, a no-net-loss policy did not apply where "minimal" disruption occurred, a reservation environmentalists claimed fatally weakened the policy, since most wetland losses are small increments. Expressing outrage, NWF's Jay Hair declared that "the administration's actions don't square with the President's promises."

AIR

THE news from New York City was ominous: After nearly two decades of struggling toward compliance with clean-air standards, new figures showed the city was losing ground on all fronts. For years, progress had been spotty. Along with 100 other U.S. cities, New York has never achieved federal standards for safe levels of ozone or carbon monoxide. Yet the lines on the charts had been pointed in the right direction. But figures released last year showed they were all reversing course.

New York's failure was no surprise to Connecticut. Officials there insisted that, because of pollution from New York City, they could not meet clean-air standards even if they stopped all manufacturing and traffic in their state.

The implacable enemy is the automobile. Although the nation's fleet is cleaner and more efficient than it was in 1970, it is so much larger that it is still degrading air quality. In 1970 there was one car for every 2.5 American citizens; by 1988 there was one for every 1.7 people. One result: According to figures released last year by the American Lung Association, air pollution from motor vehicles is responsible for at least $40 billion in annual health-care expenditures in this country.

Such glum news underscored more than ever the need to reduce air pollution. Yet nowhere was the administration's art of the deal more starkly revealed in 1990

More than half of all Americans live in areas where the air is not healthful to breathe. Auto pollution is a big factor.

than in its handling of the long-overdue overhaul of the Clean Air Act. Having garnered environmentalists' praise early on last year for announcing a redesign of the legislation, the President backed away.

First, he sent to Capitol Hill legislation that did not meet his own announced goals. Then he insisted on a crippling caveat: The law must not cost industry and consumers more than $21 billion a year.

To its credit, however, the Bush Administration succeeded where the Reagan Administration never ventured: It got a bill through the Senate in April and the House the following month. Both versions contained the nation's first regulations of emissions linked to acid rain, and the first tightening in 20 years of standards for tailpipe emissions. But as the legislation entered the final rounds of negotiation between the House and Senate, the dealmaking threatened to overshadow, and even negate, the progress. The President said he would veto the final version as too expensive if it contained a provision for unemployment and retraining benefits for those who lose jobs due to the bill's other provisions. However, such a veto was avoided after lawmakers agreed to dramatically reduce the scope of the provision

While urban pollution problems mount, U.S. lawmakers finally pass a new version of the Clean Air Act

in question. The bill was finally signed into law last fall.

Senate Majority Leader George Mitchell of Maine, long-time champion of clean-air legislation, insisted the bill that emerged from some behind-the-scenes sessions was far better than the existing law, and better than no legislation at all.

Meanwhile, the EPA was striking its own deal with nine major companies. The firms agreed to cut emissions of cancer-causing substances into the air at 40 chemical plants in the next two years. There was concern, however, over how such reductions would be tracked. In a study released last summer, the National Wildlife Federation noted that major industries are failing to reduce toxic air pollution at many facilities. Instead, they are reporting only "phantom reductions," using creative accounting techniques in their emission reports to the EPA. "What we found is that the public can be lulled into a false sense of security by declining toxic numbers that don't reflect the truth," said

Federation toxicologist Gerald Poje.

Similar dealmaking tainted the United States' international air-pollution partnerships, as well. When the leaders of the seven most industrialized nations convened their annual summit meeting in Houston last year, many of them expected

An NWF study found that major industries are failing to reduce toxic-air emissions, reporting only "phantom reductions."

to follow up on their firm environmental resolutions of 1989. Then, addressing the problems of the environment at their Paris summit, the seven heads of government resolved to take firm actions to reduce emissions implicated in predictions of global warming. President Bush declared the Paris accord to be a "watershed."

In Houston one year later, Bush, at the urging of his chief of staff, John Sununu, refused to be a part of any agreement to limit the so-called greenhouse gases. All the summit countries except the United States had announced plans to stabilize emissions by the turn of the century, and then-West German Chancellor Helmut Kohl proposed an agreement to reduce emissions by 25 percent in 15 years. His views were later supported by a new study by the International Panel on Climate Change, which found unanimity among experts that a warming is on the way and the consequences may be serious.

But the U.S. position, articulated by Sununu, was that this time there would be no deal. The proposal to limit greenhouse gases, he said, "does not understand not only our growth needs, but the growth needs of developing countries."

WATER

WITH the struggle to enact a new Clean Air Act seemingly almost over, EPA Administrator William Reilly last summer announced his agency's next top priority: cleaning up the Great Lakes. The EPA was criticized earlier in the year for failing to meet the goals of the 18-year-old Great Lakes water quality agreement between the United States and Canada. Despite significant improvements in the quality of municipal and industrial sewage, the lakes still resemble—in the words of Richard A. Liroff, director of the Conservation Foundation's Great Lakes Project—an "ecological slum."

According to a Conservation Fund study, the entire Great Lakes basin, home to 35 million Americans and Canadians, "is deteriorating from the impacts of toxic chemicals and other stresses." Among the stresses, according to a new Coast Guard study: an average of 425 oil spills and 75 chemical spills every month. Conservationists are particularly concerned about the long-term effects of such potent toxics as PCBs, DDT and dioxin.

Because of their persistence, these toxic chemicals accumulate in the food chain as they move upward from aquatic life to predators. Bald eagles that relocate to the vicinity of the lakes, according to a National Wildlife Federation study, begin to experience reproductive failure within as little as three years, and accumulate PCB

Bald eagles in the Great Lakes region are experiencing reproductive problems due to toxic contamination of the lakes.

levels six times higher than elsewhere. Similar buildups are blamed for reproductive failures in trout. Evidence, said Liroff, prompts concern that the contamination "may lead to developmental effects in chil-

dren," including premature births and impaired cognitive, motor and behavioral development.

During the 1980s, federal funds for Great Lakes cleanup declined sharply, a reflection of the Reagan Administration's position that the lakes are a regional resource to be maintained by the bordering states. In a major turnabout, however, Reilly called them "a national treasure" that is "absolutely crucial and unique in the world."

However, while Reilly was talking about a unified effort to deal with the situation, the Bush Administration was recommending cuts in budget appropriations for Great Lakes research and cleanup. According to the Northeast-Midwest Institute, a coalition of federal lawmakers, proposed cuts amounted to $131 million.

Meanwhile, the EPA was under fire for its lack of effort in another realm: drinking water. Charging that the agency is doing little to protect some 30 million Americans from exposure to potentially

As U.S. officials turn their attentions to Great Lakes pollution, drinking water threats surface elsewhere

contaminated drinking water, the National Wildlife Federation last summer sued the EPA for relaxing rules that require states to comply with the Safe Drinking Water Act. The NWF cited a Centers for Disease Control study, which found that between 1986-88, 24 states and Puerto Rico reported dozens of outbreaks of acute illness due to contaminated water. The lawsuit was supported by an August General Accounting Office report, which concluded that the nation's water safety system is plagued by unreliable statistics on contamination, serious underreporting of violations and falsification of data.

The news was better for freshwater lakes (other than the Great Lakes) in the eastern half of the nation. As the National Acid Precipitation Assessment Program (NAPAP) prepared its final report for Congress, after ten years of studies, its major finding was that acid rain damage is widespread but not as bad as expected. Of all of the U.S. lakes exposed to the effects of acid rain, NAPAP researchers

found just 4 percent to be acidified to the point that little can live in them. Another 5 percent were acidic enough to threaten certain species.

Throughout much of the western half of the country, authorities were more concerned in 1990 about the health—and

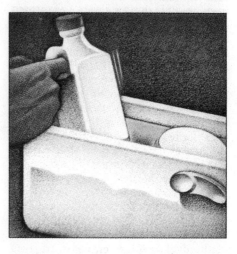

Many U.S. communities have implemented strong water-saving tactics, but in most of the country, conservation is lagging.

quantity—of their dimishing underground water supplies. Aquifers in many areas, including the vicinities of San Antonio and San Jose in California's Silicon Valley, are not only being polluted but also are being pumped faster than they can recharge naturally. Water experts say some areas of the high plains east of the Rocky Mountains will exhaust their groundwater supplies by the turn of the century.

In response, many communities in areas suffering from drought are belatedly implementing water conservation strategies. Last spring, for instance, Los Angeles initiated a program to cut water use in the city 10 percent below 1986 levels. Other cities are promoting use of more efficient plumbing standards and providing assistance in leak detection.

"In places where drought is a threat, the public perception of water conservation is at about the same level as where energy conservation was in this country in the mid-1970s," observed Edward Osann, director of NWF's Water Resources Division. "People are beginning to think about it systematically, but the nation as a whole has a long way to go in using water supplies more wisely."

FORESTS

THE heated controversy over the northern spotted owl and the Endangered Species Act tended to obscure what the fight was really over: the rapidly dwindling, irreplaceable ancient forest ecosystem of the Northwest.

In July, the head of the National Forest Service, F. Dale Robertson, confirmed what environmentalists had been saying and timber industry spokesmen denying for years—that only 10 percent of the ancient forests of the Northwest remain, and that half the survivors are scheduled for harvesting within 50 years. He added that the calculation did not mean the forests are being destroyed, as the centuries-old trees are being replaced with seedlings.

Many scientists believe that cutting down a biological treasure like the ancient forests of the Northwest before we understand its complexities is self-destructive. Researchers have found, for example, that these forests contain old-growth Pacific yew trees, the bark of which is the source of the drug taxol, effective in treating ovarian and other cancers. The yew requires 100 years of growth to reach a size useful for producing taxol.

Since there are no laws protecting endangered forests, conservationists have had to rely on those protecting the habitat of endangered species, such as the spotted owl. Because of its need for space and for old trees, the owl is a barometer of the health of the ecosystem.

Despite the focus on the spotted owl, the Northwest's ancient forest controversy is really over the loss of a vital ecosystem.

Meanwhile, for similar reasons, conservationists in the Southeast fought to protect a surrogate of their own. In all but 4 of the 26 national forests in the South, populations of red-cockaded woodpeckers

are declining or extinct. Like the owl, the woodpecker nests only in old, living trees. As a result, the Forest Service prohibited clear-cutting within three-quarters of a mile of any woodpecker's nest—a ruling that has put off-limits a million acres of federal forestland.

Loggers counterattacked last year with lawsuits and a campaign accusing environmentalists of trying, in the words of one representative, to "close national forests down to logging." But conservationists argue the loss of the woodpecker could seal the fate of 11 other species that in one way or another depend on the bird.

A National Research Council study released last summer took the Forest Service to task for concentrating its research on tree production when, in the words of study-committee chairman John Gordon of Yale University, "The number one job today is creating a livable environment." The report said U.S. forestry research is inadequate, underfunded and outdated.

Response from the Forest Service to the

Controversy centers on ancient forests, but federal timber harvest reductions signal overall policy shifts

unusually harsh criticism was immediate. Deputy chief for research Jerry A. Sesco announced that the agency would indeed change its ways. "We are going to do everything we can to heed the report's recommendations," he declared. "What we've got to do is embrace a much larger view of what a forest is," treating and studying it as a refuge of biological diversity and an indicator of the health of the planet. The response, marveled scientist Gordon, "is nothing short of a revolution."

Indeed, the revolution had already begun. At mid-year, the Forest Service had unveiled a new master plan calling for a substantial reduction of logging in national forests—from the current average annual cut of 12.2 billion board feet to 10.8 billion board feet by 1995. The plan would halve clear-cutting in national forests, and would put greater emphasis on wildlife protection and recreation.

The timber industry reacted to the pressures for reduced logging by predicting a 20 percent increase in the cost of lumber

and a consequent rise in housing prices beyond the reach of thousands of people who would otherwise buy new homes. "We are squeezing 65,000 families out of affordable housing," said one industry representative, "to provide bird houses for spotted owls." National Wildlife Federa-

Timber cutting will be reduced markedly in national forests in the years ahead, according to a new federal master plan.

tion forester Frances Hunt countered that the cost of saving "an incredibly unique and incredibly sensitive ecosystem" amounted to an increase in the average new-home payment of $13 a month.

Though the focus of much controversy about forestry issues, national forests are the source of only about 20 percent of the annual U.S. timber harvest; private woodlots provide about half the harvest and company-owned land the rest. In the private sector, timber companies face a squeeze far more formidable than that of wildlife habitat: land developers.

The situation is especially acute in New England, where, ironically, forests have been proliferating. In Vermont and New Hampshire, where there was once three times as much farmland as forest, the reverse is now true. But the most willing buyers in recent years have been developers, who subdivide the land. The trend poses problems for land-use planners and conservationists, and prompted studies last year by federal and state agencies. Said one Vermont official: "We're going to have to find new ways of conservation that include both public and private land and don't necessarily involve buying it up."

ENERGY

ERHAPS most surprising about the sudden destabilization of the Middle East and of the world's oil economies was that anyone should have been surprised. The effects of a move such as Iraq's lightning August invasion of Kuwait had been predicted for some time. To cite but one example, Louisiana Senator L. Bennett Johnston, chairman of the Senate Energy and Natural Resources Committee, warned in April that rising imports, declining U.S. production and the emerging Persian-Gulf oil monopoly posed a serious threat to this country's economy and security.

When the crisis came, the immediate effects were no less surprising: Within days, the price of crude oil doubled and the price of gasoline went up by 20 percent. The Bush Administration had immediate diplomatic and military responses to the invasion, but had little to say about its energy policy implications. Indeed, the most significant proposal in energy policy to deal with the changed world was made by Interior Secretary Lujan. "There is no reason," he insisted after the invasion, "why we should not explore in the Arctic National Wildife Refuge."

The long-standing argument about whether to probe the fragile Alaskan refuge for what might be the country's largest untapped oilfield had been tipped firmly in nature's direction by outrage over the *Exxon Valdez* oil spill. The trend

Opening the arctic wildlife refuge to oil development, say conservationists, won't solve the nation's long-term energy needs.

was encouraged by new findings announced in February, confirming that the refuge's Coastal Plain is a favorite denning site for pregnant polar bears, and suggestions that interference could cause a pop-

ulation decline in the internationally protected species. But the administration hoped that fears about oil shortages would override worries about wildlife.

The impact of the latest oil shock was eased, at least at the outset, by the presence of a cushion created after the bitter experiences of 1973 and 1979: the 590-million-barrel strategic oil reserve, bought by the U.S. government and stored in salt caverns in Louisiana. Capable of replacing oil imports for six months, the reserve was a factor in the early days of the crisis even though it went untapped; its presence prevented panic buying.

The loss of oil from Iraq and Kuwait made worse a situation that was already becoming gloomy. During a year that saw demand for petroleum rising implacably— by 800,000 barrels per day among Pacific Rim countries alone—production in this country and the Soviet Union was declining just as irreversibly.

With U.S. domestic production plunging, options for satisfying America's pe-

A new Middle East crisis creates a predictable response, but not changes in U.S. energy policy

troleum addiction appeared increasingly limited. But environmentalists pointed out that the solution to the oil dilemma is not to increase drilling but rather to use supplies more efficiently. Better energy conservation, they said, is the country's greatest untapped reserve. Indeed, studies confirm that using the same quantity of energy, other industrialized nations such as Japan and Germany produce almost twice as many goods as the United States.

One area where vast amounts of oil could be saved is in car fuel efficiency. But last year, the administration actively opposed a bill, sponsored by Nevada Senator Richard Bryan, that would require automakers to increase gas mileage of all cars sold in this country by 40 percent by the year 2001. Administration opposition helped kill the measure, which would have saved 2.8 million barrels of oil a day.

Meanwhile, two months before the Iraqi surprise, President Bush declared a moratorium until the turn of the century on offshore oil and gas drilling near the West

Coast, Florida and in the Georges Bank area of the North Atlantic due to disturbing questions about the environmental impacts of such drilling. Crafting a deal that the administration hoped would work for everybody had taken 16 months, a task force and considerable staff time. Oilmen

A presidential moratorium on offshore oil and gas drilling in most U.S. coastal waters was hailed by environmentalists.

were presented with a delay, not a ban, and left with areas near Santa Barbara and North Carolina's Outer Banks where drilling could possibly proceed.

Embracing the decision, environmentalists also called for a strong U.S. energy policy based on conservation measures and increased development of solar power and other renewable sources to wean the country off its oil dependency. Currently, renewables supply only about 8 to 10 percent of U.S. energy.

However, a comprehensive national energy policy was not to be seen in 1990. Energy Department Secretary James Watkins continued his work on crafting such a policy, begun in 1989. But as the year progressed and the deadline for submission of the final report neared, reports surfaced that the administration had cut another deal: The final report would contain no policy recommendations for the President, only a series of options. To observers both from industry and environmental groups, the deal meant another delay in making difficult choices about such issues as the government's role in encouraging conservation, the future of nuclear power and the need for energy taxation.

WORSE   BETTER

SOIL

AFTER years of reports about consistent high losses to erosion, there was good news in 1990 regarding the nation's topsoil. According to the Worldwatch Institute in Washington, D.C., America's farmers have reduced soil erosion by more than one-third since 1985, largely because of new programs initiated by the federal Farm Bill passed that year.

During the previous decade, farmers had lost to wind and water erosion 5 pounds of topsoil for every pound of grain they produced—an estimated total of 1.6 billion tons per year, according to Worldwatch. The 1985 Farm Bill offered farmers $49 per acre per year to plant grass or trees on land most likely to erode. It also required farmers with erodible land to develop soil-conservation plans or lose federal subsidies. By last year, there were 33 million acres in that Conservation Reserve Program, and the rate of soil loss had been reduced to about one billion tons annually.

That remains a staggering loss. And there are worries about the future of the program. The Agriculture Department has eased many of its standards and failed to enforce others in response to farmers' pleas of economic hardship; there is no sign yet of a willingness to go beyond the first, easy steps mandated by the 1985 law toward controlling erosion. Still, those steps have been taken.

American farmers are not only trying

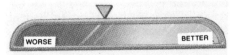

According to the Worldwatch Institute, the nation's farmers have cut soil erosion by more than a third since 1985.

harder to keep their soil, but they are increasingly reluctant to douse it with chemicals. One half of the farmers in a recent poll expressed concern about the dangers of using too many agricultural chemicals

on their land. There was ample cause for concern; for one thing, the EPA has found pesticide traces in the groundwater supplies of 26 states.

Another factor affecting farming practices is the increasing resistance of many insect pests to chemical pesticides. Even though some 430 million pounds of pesticides are applied yearly to U.S. crops— a tenfold increase since World War II— crop damage appears to have increased. One example is corn, which, according to Cornell University entomologist David Pimental, "is the single largest user of insecticides in the nation." Forty-five years ago, corn grew in rotation with other crops and without insecticides. Today, despite massive use of insecticides on corn, losses from insects have increased from 3.5 percent to 12 percent.

One result of such changing attitudes can be found in the sales charts of the nation's chemical companies; for them, a 40-year boom is over. Fertilizer prices at mid-year 1990 were off by more than 10

Topsoil erosion is down thanks to recent programs, and many farmers are rethinking use of chemicals

percent. What's more, the number of pesticide manufacturers and dealers was in steady decline, and per-acre use of chemicals was down (although total sales of such chemicals were still growing because of increased acreage).

The chemical companies have been predicting that reduced chemical use would mean increased prices for grain and produce. "People in Hollywood and Georgetown can afford a $3 tomato," said Gary Myers, president of the Fertilizer Institute, "but what about the rest of the consumers?" However, some farmers trying techniques of sustainable agriculture and integrated pest management were finding that the results can be profitable. "Farmers who are wanting to convert," insisted Texas agriculture commissioner Jim Hightower, "are able to make as much money as high-chemical, conventional farmers."

Meanwhile, the successes of the 1985 Farm Bill made its revision during 1990 a matter of intense interest to more than the farmers whose livelihoods it was to regu-

late. The dealmaking, summarized at one stage in a 2,363-page congressional document weighing 6 pounds, was watched closely by environmentalists, who were concerned about the fate of some proposed conservation and pollution-control measures in the new law.

The increasing resistance of insects to chemical pesticides has led many farmers to seek safer methods for fighting pests.

At one point, a Texas Republican and a New York Democrat introduced a popular amendment that would have denied federal farm subsidies to rich farmers. The proposal was gaining a lot of support. However, opponents of the measure pointed out that the exclusion of rich farmers from benefits would drive them out of the farm program and remove their large holdings from the program's environmental controls and incentives. The amendment was overwhelmingly defeated.

The bill that eventually won approval by both House and Senate offered no major changes in agricultural support programs. But it did break new ground in environmental protection, particularly in regard to pesticide control and forest and wetland protection programs.

Though budget cuts threatened passage of some measures of the bill, environmentalists remained hopeful. "Congress must understand that the public's desire for ensuring environmental protection should be an integral part of the nation's farm policies," observed NWF legislative representative Steven Moyer. The new bill also included a ban on export of farm chemicals that are illegal in the United States.

WORSE ▼ BETTER

QUALITY OF LIFE

THE quality of life that Americans experienced in 1990—Year One of what many people call the Decade of the Environment—depended to a great extent on where they lived and what they knew.

Residents of Chattanooga, Tennessee, for example, had reason to feel pretty good. A citizens' movement there that spans two decades has helped turn one of the nation's worst environments into one of the best. While toxic air emissions are still a problem, other air pollutants that once were so bad they ate holes in nylon stockings on women's legs are now well within federal standards for safe breathing. And a riverfront that was degraded by abandoned industrial sprawl is now graced by a multi-million-dollar park.

In northeast Detroit, on the other hand, the picture was not so rosy. Residents there suffered last year under a constant drizzle of oily soot, coming from a steel plant operating in what Wayne County officials described as "gross violation" of pollution standards.

Living in New England, according to one study of relative environmental health, was in general more wholesome than living in the South. The Institute for Southern Studies in North Carolina applied 35 indicators, taking into account everything from pollution to politics, and concluded last spring that Vermont enjoyed the most healthful environment, Alabama the least.

Though recycling programs are growing, Americans still recycle or compost only about 13 percent of their solid waste.

While six New England states ranked among the top ten, a spokesman for the study described the South as "the nation's biggest waste dump."

All across the country, however, the liv-ing was uneasy in 1990 as more information about environmental dangers made its way into the consciousness of more and more Americans. They found themselves confronted not only by seemingly abstract worries such as global warming and atmospheric ozone depletion, but also by threats to many of their fundamental needs: Throughout last year, for instance, warnings proliferated about the contamination of meat, fish, eggs, milk, fruit and even bottled water from toxic chemicals and other substances.

Amid the worries, people were looking for ways to help turn things around, to make a contribution to the health of their planet. A book on the subject was a runaway best seller. Increasingly, citizens were recycling and composting their garbage instead of shipping it off to local landfills. But while recycling efforts have doubled in the United States since 1960, they still account for only 13 percent of the nation's flow of solid waste. And while many people were intially excited about new local

---

As more information about environmental dangers surfaces, Americans are showing increasing concern

---

recycling programs, their enthusiasm waned as they learned that the materials they painstakingly separate often wind up in a landfill anyway because there is not yet a market for them. "People want to do the right thing," said one Midwest recycling consultant, "but they don't know what the right thing is."

Two states took the initiative last year to help solve that dilemma. Wisconsin and Massachusetts became the first states to pass laws banning disposal of a broad range of recyclable materials, from newspapers to plastic containers. The laws are intended to promote investment in recycling plants by ensuring a steady supply of raw materials. Landfill and incinerator operators are subject to stiff fines (as much as $25,000 a day in Massachusetts) if they improperly dispose of any materials covered by the bans.

Throughout last year, there were other positive signs of environmental concern in the political arena, from California's Big Green initiative—the toughest set of en-vironmental regulations proposed anywhere—to Sarasota, Florida's referendum for a total moratorium on new construction. But such measures seemed to have little positive impact on the lives of those Americans who suffer most from environmental contamination.

One recent study found that half of all U.S. blacks and Hispanics live near a closed or abandoned hazardous dump site.

"If we seek to know where pollution's threat to survival is now greatest," noted NWF President Jay D. Hair, "we must acknowledge the disproportionate burden borne by this nation's poor."

A comprehensive, nationwide study conducted by the United Church of Christ Commission for Racial Justice found that more than half of all U.S. blacks and Hispanics live in communities that have at least one closed or abandoned hazardous waste dump site. What's more, researchers found that communities or companies looking for sites for health-threatening chemical or waste sites most often look in low-income minority neighborhoods where residents generally are less-informed and less organized than elsewhere.

"Black Americans are afflicted by twice the rate of cancer and are killed by asthma at three times the rate of whites," observed Hair. "Yet the person who needs a job most to hold his or her family together is the least likely to complain about chemical contamination of a workplace. This is one of the conservation community's greatest challenges in the years ahead: the necessity to forge an alliance between environmental and social justice."

5. RESOURCES

National Wildlife's annual Environmental Quality Index is a subjective analysis of the state of the nation's natural resources. The information included in each section is based on personal interviews, news reports and the most current scientific studies. The judgments on resource trends represent the collective thinking of the editors and the National Wildlife Federation staff, based on consultation with government experts, private specialists and academic researchers.

A NEW LAY OF THE LAND

Uneven distribution of arable land has pushed poor peasants into deserts and forests, speeding environmental breakdown. Finding a way to parcel good land might spell a break for threatened ecosystems.

HOLLY B. BROUGH

Holly Brough researches poverty and land-use issues at the Worldwatch Institute.

Fifty years after the tribes of Israel entered the Promised Land, their God declared a Jubilee celebration, when all land fortunes accumulated during those years were redivided among the original owners. Every half-century thereafter, the land was equitably distributed among the people. Today in most developing countries, the Promised Land looks more like a wasteland and a Jubilee is long overdue.

Land ownership in much of the developing world is highly skewed. While a few wealthy plantation owners and cattle ranchers monopolize fruitful land, thousands more peasants are pressed to the diminishing margins of soil and society. In Guatemala, for instance, a mere 2 percent of landowners claim 63 percent of agricultural land.

Worse, the world's growing contingent of landless peasants must rake fragile hillsides, torch rain forests, and plow near-deserts for their next meal, driving the environmental threats of global deforestation and desertification. Norman Myers, a British forestry analyst, estimates that landless peasants clear three-fifths of the 42 million acres of rain forest lost annually. Small farmers and their

livestock also pose a major danger to the third of the earth's land surface that is threatened with desertification.

Land reform strikes at the heart of these problems by creating small farms for many, rather than large farms for a few. More equitably distributing agricultural land can stimulate rural economies and relieve poverty. Bolstered by efforts to curb population growth and promote soil conservation, land reform could help stem environmental destruction in many countries.

Yet land reform has been relegated to the darkest recesses of the international development effort. Promising a dramatic transformation of the rural power structure, as it does, reform is an ambitious undertaking more easily ignored than initiated.

Indeed, the odds against land reform are hard to overstate. The power of landed elites and tenacious practical obstacles make redistribution an unpopular policy among government leaders—despite its appeal on election platforms. Resistance to reform is fierce, and often violent. Still, if the earth is to survive and its people prosper, land reform is indispensable.

The Lay of the Land

Across centuries and cultures, from the Incan empire to the feudal estates of pre-revolutionary France, land has been inequitably divided between haves and have-nots. As feudalism crumbled in Europe, however, its pattern re-emerged overseas. Many of the rural environmental problems facing developing countries today trace their roots to colonialism.

Colonialism spread European immigrants sparsely over fertile land in Africa, South America, and Asia, and pushed out the native majority. In Algeria, 20,000 Europeans seized 6 million acres of alluvial, water-blessed land, leaving 630,000 Algerians with 12 million acres of parched, thin-soiled earth. Spanish and Portuguese colonizers of Latin America recreated the sprawling estates of their mother countries, relegating native Indians to surrounding uplands.

Export crops have taken over where colonialism left off. Large farms devoted to bananas, cotton, sugar, and beef cattle have virtually monopolized the rich soils of Central America, according to Jeffrey Leonard, vice president for environmental quality at the World Wildlife Fund — Conservation Foundation in Washington, D.C. In the African Sahel, peanut and cotton plantations line the irrigated valleys of the southern Senegal and Niger rivers, pressing small farmers north into areas of sparse rainfall and poor soils.

Population growth further cuts into peasant holdings. Generations of dividing land among heirs has left mere patches of soil to family members. In Tamil Nadu, India, marginal holdings (less than 2.5 acres) shot up from 73 percent of farms in 1972, to 82 percent 10 years later. The remaining 18 percent control 76 percent of the land, but much of this is in plots not much larger than five acres. Complicating matters, properties are fragmented over time. Tilling scattered bits of land drains a farmer's time and investment, often ruling out conservation.

Land concentration figures highlight the legacy of colonialism and the impact of export-based agriculture. In El Salvador, where 2 percent of landowners hold 60 percent of the land, almost two-thirds of rural families have little or none. White farmers, who comprise less than 1 percent of Zimbabwe's people, command 39 percent of all land. Although statistics in Asia are less dramatic, 3 percent of Filipino landowners control a quarter of the country; 60 percent of rural families either can't survive

on the little land they own or own no land at all.

Forced off good land by plantations and wealthier farms, many peasants return to work as wage-laborers. During critical planting and harvest times they must neglect their own scraps of earth, again precluding proper soil husbandry.

Those left behind try to wrench the greatest yield from already deficient soil. As one labor leader from the National Association of Farm Workers in El Salvador laments: "What does a poor *campesino* [peasant farmer] do? He starts working what little land he does have year after year without leaving it fallow to recuperate. Soon the land doesn't produce and becomes barren. The soil is gone." Within the memories of *campesinos* in the Honduran highlands, fallow periods have plummeted from an average of 15 to 20 years, to 0 to 7 years.

Without rest, soil loses the shallow seam of organic matter that contains crop-feeding nutrients. Soil structure gradually collapses, reducing the earth's ability to hold rainwater and resist erosion. The process only accelerates on hillsides.

Ambiguous rights to land place the environment in double jeopardy. For tenants unsure whether they will till the same earth next season, good soil husbandry makes little economic sense. Moreover, since some laws give tenants ownership of land they improve, landlords may protect their holdings by evicting tenants ambitious enough to try conservation.

Squatters likewise have no incentive to preserve the soil. According to a World Bank study of four provinces in Thailand, untitled farmers build "bunds"—raised earth walls that catch soil and water—at half the rate of farmers with clear land titles. Getting a bank loan to finance soil conservation is nearly impossible for those without a land title to put up as collateral.

As a last resort, cornered by poverty and tiny plots of infertile land, peasants flee to the frontier—forest or desert—often with their government's blessings, and try to scrape a living from smoldering tree ashes or shifting sands.

Life on the Edge
"We ate the forest like a fire. There was no forest left. We pulled the stumps and cut the limbs. We thought that after a year of this, it would get better, but it sent us downwards."

These words, spoken by an Ethiopian farmer uprooted from his ancestral home, epitomize the plight of the world's swelling ranks of landless peasants. Today in Honduras, 200,000 families, or about 35 percent of the rural population, subsist on forest lands; in Thailand, a million squatter families have cleared nearly one-fifth of forest reserves for farming.

Both migrant peasants and the indigenous tribes they displace suffer from the move to the forests. First, the indigenous practice of shifting slash-and-burn agriculture—a sustainable cycle of clearing, cropping, fallowing, then reclearing forest land—is interrupted as more landless flood the forest. Encroachment on tribal lands has forced the Palaw'ans of the Philippines to reduce fallow time from eight to two years.

Migrants adopt a far more destructive form of shifting agriculture. They farm cleared land until crop yields deteriorate with the soil, then move into the woods. Seventy percent of Peruvian farmers who migrate to the Amazon do not settle even for a few years, but continuously open up new land.

During the mid-1980s, shifting agriculture in Indonesia cleared away about a million acres of forests each year, believes Soedjarwo, Indonesia's former minister of forests. Logging felled less than one-fifth as much area during the same period. But forestry practices encourage peasant migration by building roads that the land-hungry follow.

In Latin America, cattle ranches trail in the wake of peasants, subsuming the abandoned land. Ranchers in Central America give settlers short-term leases on uncleared frontier, then move in their cattle once the leases are up. Expanding ranches constantly push the peasant frontline further into the forest.

Government laws also lock squatters into destructive ways. Ecuador's Institute for Agrarian Reform and Colonization, which has encouraged settlement of remote areas, grants title to forest plots only if at least half the land has been cleared. Similar laws exist in countries from Malaysia to Nicaragua. To secure this promised land, migrants clear as much as they can. A study of settlers in Peru found that squatters cut forests at twice the rate of legal tenants—2.7 acres a year versus 1.3.

States have colonized fragile ecosystems in lieu of real land reforms to relieve pressure on overcrowded land and pacify angry peasants.

Brazil's Amazon colonization scheme is perhaps the most notorious, but it is not alone. Bolivia, Ecuador, and Peru have similar programs, while Ethiopia has been relocating farmers from the war-torn north for years. An Indonesian program has resettled 722,000 Javanese peasant families on forest-blanketed outer islands since 1970, while at least 574,000 families have migrated on their own.

The environment has paid in full for these programs. From 1964 to 1985, the Amazon lost 7.4 million acres to Ecuador's colonization efforts, and Indonesian settlers leveled 500,000 to 700,000 acres of forests a year during the 1980s alone.

Burning forests causes soil to irretrievably lose its fertile organic matter, already at low levels in rain forests, where 90 percent is stored in vegetation. Repeated burning eradicates sensitive native flora, allowing weeds to invade. And, with the absorptive topsoil gone, water runs off, taking more earth with it. The western highlands of Guatemala annually lose as much as 16 tons of soil per acre. In India, tree and plant destruction allow 7 billion tons of soil to wash into the sea each year, including 6 million tons of plant nutrients.

Downstream, flooding increases and water supplies are depleted. The flood-prone area of Honduras has doubled in the last 10 years from 50 to 100 million acres, damaging crops, roads, bridges, and dams. Silt fills reservoirs, like the one supplying Honduras's capital of Tegucigalpa, while eroded hillsides, now impervious to water, cannot replenish groundwater supplies.

In the fragile arid and semi-arid regions of Africa, many of the same factors are at work. The Jebel al-Awilya Dam in Sudan's Western Nile province irrigates tenant farms of 5 to 18 acres along the river's basins. Farmers excluded from the project have been pushed to sandy uplands and a belt consisting of dunes and clay soil, where the combination of population and poor soils has forced them to forsake fallow periods and cultivate more land simply to feed themselves. According to Sudanese geographer Anwar Abdu, formerly stable dunes in this area have begun to spread.

The problem is aggravated by drought. Threatened by dry spells, peasants pull as much marginal land into production as possible in the hope that at least some will yield crops. In western Sudan, the cultivation of millet (a staple crop for the poor) has ex-

A Common Tragedy

For all the ecological ravages attributable to inequitable land distribution, the collapse of common property ownership can be equally destructive. From African pastoralists controlling access to watering holes, to Indian villagers monitoring tree cutting in local forests, traditional forms of common resource management have minimized human impacts on fragile resources and ecosystems.

Forests, grasslands, fisheries, and water all constitute common properties, subject to use but not ownership. But contrary to the familiar "tragedy of the commons" scenario, where each user exploits a resource for private profit, common property systems are not "free-for-alls." Recognized authorities control access and preserve the resource.

Among the Berber herders of Morocco, land susceptible to degradation was carefully monitored by each sub-tribal group's *amghar n'tougga* (chief of grass). Vested with authority to decide when and where to move livestock, the chief of grass would close grazing areas seasonally to allow the trampled range to regenerate.

Forest resources are protected similarly. In the Indian hill village of Silpar in Uttar Pradesh, each family is informally responsible for reporting illegal tree cutting to the village council. The violator must pay a fine, half of which goes to the family that turned in the transgressor.

As vital as common properties are for the earth, they are equally important to the poor. Village studies in the dryland regions of India by N.S. Jodha, formerly of the International Crop Research Institute for the Semi-Arid Tropics in Hyderabad, India, reveal that wood, water, and fodder from common properties supply peasants with half their income during drought years.

Sadly, common properties are being besieged from all sides—by population growth, cultural changes, and state policies. First and foremost, however, the resources themselves are shrinking from privatization and nationalization.

Privatization usurps grazing areas, especially those reserved for dry seasons. The Il Chamus herders of Kenya have seen one of their dry-season grazing areas reduced by 26,000 acres since the 1940s, as small farmers encroached on their range. The hungry peasants, in turn, had been edged off land further south by the expansion of commercial farms. The Il Chamus have been forced to graze their herds in swamps, trampling the surrounding soil and vegetation.

Elsewhere in Africa where farms have seized dry-season grazing land, herders are pushed onto poorer soil when it is most susceptible to disruption. Multiplied across the globe, these local cases of livestock overcrowding have led to intense land degradation. According to the United Nations, overgrazing has degraded 7.7 billion acres—about 80 percent of the world's dry lands.

State-sponsored privatization also undermines common property regimes. The indigenous Indian community of Pasu Urcu in eastern Ecuador abandoned its fallow system in the 1970s after the land reform agency informed them that uncultivated land could be claimed by nearby peasants. Their sustainable cycle of shifting cultivation was severed.

In India, state land reforms have distributed traditional common properties, much of it to well-off farmers who had illegally encroached on the areas. Jodha's studies show that common property land has been nearly halved in size since 1950. More and more people foraging smaller areas for wood and fodder spell ruin for these ecologically marginal areas.

While privatization directly attacks common resources, nationalization sabotages the systems that manage them, obscuring lines of authority over the resource. In the Sudan, the Unregistered Land Act of 1970 gave the state title to rural land, undermining the power of village sheikhs to control tree cutting. When people aren't answerable to a single authority, a common property can swiftly be opened to all—and the environment loses out.

Preserving common property systems can begin by protecting the resource. Since April 1988, the Colombian government has ceded nearly 30 million acres of the Amazon to indigenous tribes for collective and inalienable ownership. Jodha calls for a complete moratorium on conversion of common property, which would give users legal leverage to protect it. Careful research at the local level can determine how nationalization or privatization plans would hurt common property.

Ultimately, however, common property is more than a resource. Careful attention must be paid to the communities that manage such properties, and the forces undermining these communities.

panded northwards 120 miles beyond the accepted growing region into delicate Saharan soils. Since these soils are most fragile during drought, erosion problems are compounded.

Herculean Challenge

The benefits of effective land reform are indisputable. Farmers who own their land are better stewards of it. Small family farms absorb more labor than mechanized plantations—employing family members who might otherwise head for forests or job-scarce cities. While sheer numbers preclude settling all peasants who need land, the new income that beneficiaries earn can increase demand for goods and services that others might supply.

Contrary to what might be expected, reforms also increase agricultural output. Except for slips directly after land transfers, supported small farmers have proven themselves more productive than large farms. Small owner-operated farms in Taiwan produce 50 percent more than sugar plantations in the Philippines.

In the dozens of countries where land reforms have been undertaken, however, reality has seldom met the promise (see Table 1). In most cases, a fraction of landless families have benefited. Measured against the enormous obstacles facing land reforms—resistant landowners, uncommitted leaders, unorganized peasants, and sketchy land records—these results are not surprising. Land reform is a risky proposition, and there is much room for error.

Imposing limits on farm size is a prime method for freeing private land for redistribution. Large landowners, however, can dodge such ceilings. In Mexico, estate owners held on to their property during land reforms in the 1930s by registering it under *prestanombres* (name lenders), who were either children or sympathetic neighbors. Incomplete rural land records are easily altered to this purpose.

Idle or "unproductive" estate lands are theoretically ripe for the taking. Almost invariably, farms use land in inverse proportion to their size: the larger they are, the more land is left alone. Although some land is fallow, more often estate owners let land lie because their incomes do not depend on agriculture. Their economic lifelines run to city businesses, while farms are mere investments against inflation. Brazil, with 9 million land-

Table 1.

History of Land Reforms, Selected Countries

Taiwan	Prior to reform, United States and Japan conducted detailed land ownership survey. Reform (1949-1953) slashed rent, and sold public lands and farmland to tenants. Local Farm Tenancy Committees comprised of tenants, officials, and landlords settled disputes and enforced the law.
China	Post-revolution reform (1950) created cooperatives but allowed peasants small private plots. Since then, a legal rollercoaster has formed communes and eliminated private plots (1958), reinstated private plots (1961), given peasants "use certificates" to land with no rights to sell or lease (1980), and in 1990 reclaimed property given to peasants.
Kenya	1954 Swynnerton Plan consolidated land fragments and gave title to farmers. By 1978, the reform had affected nearly half of the land.
Zimbabwe	From 1980 to 1989, the state settled 52,000 households onto farms abandoned at independence and public land. State legally owns settlement land (mostly of poor quality) and users cannot sell or inherit it.
Nicaragua	1979 reform seized land held by former ruler Anastasio Somoza and his cronies for state farms and established cooperatives from underutilized large farms. State gave credit, education, and health care to peasants. President Violeta Chamorra is trying to restore cooperative and state farmland to former owners.
Mexico	Continual reform since 1917 has created group farms (*ejidos*), which cannot be sold or rented to non-heirs. After 1940s, state invested heavily in large commercial farms immune to reform, forcing the struggling *ejido* farmers to seek low-paying agricultural work elsewhere.
Chile	A fairly successful reform was reversed after military coup in 1973. More than a quarter of reform land was restored to landowners, and reform cooperatives were divided into private holdings.

Sources: William Theisenhusen, ed., Searching for Agrarian Reform in Latin America (Winchester, MA: Unwin Hyman Press, 1989); Roy L. Prosterman, et al., eds., Agrarian Reform and Grassroots Development: Ten Case Studies (Boulder, CO: Lynne Reinner Publishers, 1990); John P. Powelson and Richard Stock, eds., The Peasant Betrayed (Washington, D.C.: CATO Institute, 1990).

less households, hoards an estimated 82 million idle acres of farmland.

When it comes to seizing this land, however, political opponents to reform have defined "unproductive" ambiguously in the law. Without a clear definition, land liable for expropriation cannot be taken. Landowners in Ecuador and Honduras have slipped through this loophole.

Then there are money questions. Crushing foreign debts and fear that farm productivity will plummet compel countries to exempt export cropland from reforms. Politicians opposed to the 1969 reform in Peru arranged generous compensations for former landowners, partially tying the hands of the government because of limited funds.

Even where landowners surrender land, they give up only their poorest acres. Of the 62 million acres of land parceled out under Mexico's reform in the 1960s, only 10 percent was arable.

"Land to the Tiller" programs, designed to give tenants title to their rented land, often complement legislation that limits farm size, but they present their own hurdles. The programs instead can compel owners to evict tenants preemptively. For victims of such practices, justice is often elusive. Bureaucratic red tape can also bog down land titling, as in the case of El Salvador.

Since it is obviously easier to transfer land to those already living on it, the truly landless also lose out from "Land to the Tiller" programs. Peru's land reform of 1969 did little for seasonal workers and *minifundistas* (marginal farmers) of the highlands.

Land titling—giving formal deeds either to reform beneficiaries or illegal squatters—is another dimension of land reforms. Though often well-supported (the World Bank underwrites projects in northeast Brazil and Thailand), titling frequently excludes women, who shoulder 70 percent of the farming burden in Africa and 50 percent in Latin America. Women need title to obtain loans for seed and fertilizer. Titling also impinges on common properties.

Where all other obstacles are surmounted, supporting small-scale farmers with credit and advice is the final challenge. After doling out vast amounts of land in the 1930s, Mexico's government abandoned the struggling new landowners in the late 1940s, underwriting northern plantations producing for export instead. Peasants unable to survive on their land were forced into low-paying labor on these very plantations.

The fundamental issue that underlies all the barriers to land reform is summed up by Demetrios Christodoulou, formerly a researcher with the U.N. Food and Agriculture Organization: "Agrarian reform is about power." In rural societies, owning land means wielding power; surrendering it means sharing that power. Few landlords surrender willingly.

Yet despite the odds, some countries have succeeded at land reform. Japan, South Korea, Taiwan, and China all pushed through comprehensive reforms in the wake of wars or in the midst of political and social upheaval. In Korea, the landed aristocracy had been weakened during World War II and tenants were already refusing to pay rent before the reform. In Kerala, India, an organized, literate peasantry and a political party responsive to their demands pushed through a strong land reform in the 1970s against stiff landlord opposition.

Peasants today are pushing against similarly entrenched power bases opposed to land reform. It promises to be a long struggle.

Up from the Roots

Grass-roots efforts may be the best hope for putting land reform back on national agendas, and such movements appear to be increasing. Since 1989, landless peasants in Paraguay have undertaken about 90 land invasions—occupying and demanding rights to private and public land. Although the peasants are constantly threatened with military repression, invasions continue. In Brazil, which has a long, brutal history of peasant land struggles, demonstrations and land occupations persevere, spearheaded by the National Landless Workers Movement.

Although everywhere confronted with political intransigence and landlord violence, the poor are not fighting alone. In Tamil Nadu, a group called LAFTI (Land for the Tillers Freedom) is bringing land to the poor through bank loans, legal aid, and pressure on landlords. Land dividends are small, but as one woman said, "The one acre of land is precious for my family."

To Brazilian peasants, who have gained land through occupations and legal battles, the National Development Bank in Rio Grande do Sul is extending credit without demanding land as collateral. Thus, during the first uncertain years of farming, peasants are freed from worrying about foreclosure.

Since 1981, World Neighbors, a development group based in Oklahoma City, Oklahoma, has been teaching soil and water conservation methods to Filipino farmers working tiny plots of steep, rugged land on Cebu Island. These methods have eliminated the need to cut forests for fresh land. Also, in Guerrero state, Mexico, the peasant-run Coalition of Collective Ejidos (cooperative farms) provides a range of services from credit to women's workshops. More than 1,000 peasants have raised their incomes through this program.

By strengthening, educating, and empowering the poor, these efforts are critical to reform. But without support at the national and international level, they cannot succeed. Foreign donors can further set the stage for reform by buttressing the work of grass-roots groups and related outside organizations. With financial and technical support, governments or private groups also can gather data vital to implementing reforms: the amount of farmland available, beneficiary groups, acres-per-family ratios, costs, etc.

More directly, donors can earmark aid funds for land redistribution. Ensuring that targeted money is actually used to purchase land rather than just title beneficiaries of

colonization, however, is essential if such efforts are to work. Typically, the World Bank and the U.S. Agency for International Development have supported the latter programs.

For the environment's sake, legal codes and their interpretation need changing. As Alvaro Umano, Costa Rica's former minister of natural resources, contends, "We need to address the problem of the law that says to get possession of land you have to show improvements, and to show improvements you have to remove the forest cover." In short, colonization schemes that level forests and destroy common property are not reform.

Finally, the end of the Cold War bodes well for land reform. Donors and developing states now may be less likely to perceive radical land reform as a Communist conspiracy and more as a policy capable of addressing the needs of both the environment and the people.

Still, the difficulties of reform cannot be glossed over. There are no quick fixes. Building the conditions and coalitions for reform carefully, and putting the issue back on the global agenda whenever possible, are the first steps. With these prerequisites in place, today's landless—and the world—may look forward to a jubilee year, and a healthier promised land.

U.S. FARMERS CUT SOIL EROSION BY ONE-THIRD

PETER WEBER

"If Moses had foreseen what was to become of the Promised Land after 3,000 years and what was to become of hundreds of millions of acres of once good lands … in China, Korea, North Africa, the Near East, and in our own fair land of America … [he might] have been inspired to deliver another Commandment to establish man's relationship to the earth."

So concluded W.C. Lowdermilk, one of the founders of the Soil Conservation Service (SCS), after his globe-spanning fact-finding mission in 1938-39. From older civilizations he was looking for insights into how to repair damage in the Dust Bowl and the gullied South. Except for pockets of stewardship, he found 7,000 years of destruction.

Fifty years after Lowdermilk's seminal studies, environmentalists and soil conservationists seem to have hit on the kind of formula that eluded him. Now embodied in the 1985 Farm Bill (a.k.a. the Food Security Act), their ideas have cut excessive soil erosion by one-third in the last five years, and would exceed that achievement in another five years if it were fully implemented.

Even with the SCS still on the job in the mid-1970s, soil erosion was again climbing to record levels, this time because farmers motivated by high grain prices were plowing up highly erodible grassland and continuously monocropping without any fallow period. For every pound of grain they harvested, farmers squandered five pounds of topsoil.

Those startling erosion rates were the impetus for change. The legislation that eventually emerged in the 1985 Farm Bill established a two-stage strategy that first removes the most erosive land from production and then requires farmers to apply soil conserving practices to erosive farmland still in cultivation. The philosophy behind the legislation has it that farmers receiving federal payments and other subsidies should in return conserve the nation's soil resources.

Farmers have responded to the farm bill by reducing excessive soil loss from 1.6 billion tons per year to 1 billion tons. (Excessive soil loss is defined as the amount of soil that erodes above new soil formation, which averages four tons to the acre each year.)

The crown jewel of the bill's two-stage strategy is the Conservation Reserve Program (CRP), which now encompasses 34 million acres. It was created to protect the most-threatened cropland and help farmers meet the program's soil conservation requirements. Under the CRP, farmers receive around $49 per acre in annual rental payments for planting their most erosive land—land that should probably never have been cropped in the first place—in grass or trees for 10 years.

In four years, the CRP has slashed excess annual erosion by nearly 600 million tons, bringing the annual per-acre erosion rate on this land down from roughly 21 to 2 tons.

The second stage of the program required farmers with highly erodible land to develop an SCS-approved erosion control plan by the end of last year. By 1995, these farmers have to follow through on their plans, or they lose their eligibility for subsidies. So far, an estimated 1.5 million farmers have worked with SCS to develop plans to stabilize the soils on 134 million acres. When all of the conservation plans are in place, they have the potential to cut excess soil erosion by another 700 million tons per year—this on some of the most productive farmland under plow.

More than likely, though, this part of the program will not reach its full potential. There are a number of factors working against it, the first of which is the U.S. Department of Agriculture's (USDA) decision to lower the program's soil loss goals to permit erosion that's in excess of natural soil formation. Under pressure from farmers, the agency justified less stringent

standards by citing the economic hardships farmers now face.

Second, farmers have avoided doing the more time-consuming and expensive, but very effective, anti-erosion practices of contour farming and terrace building in favor of much less expensive but marginally effective practices, such as leaving stubble on the field after harvest. Soil conservationists doubt farmers can meet the goals set in their conservation plans if they don't go to extra lengths.

Third, and perhaps of most concern in the long run, a rising demand for grain could recreate the record erosion levels seen in the 1970s. World grain scarcity, caused by continued global population pressures, could push grain prices high enough to encourage farmers to return to mono-cropping and other intensive practices, which in turn would cause increased erosion. Rising grain prices at a certain point could

entice farmers to forgo their federal price supports, crop insurance and other subsidies, meaning they could ignore conservation if they chose.

Once that decision is made, farmers could also bring CRP land back into production without a conservation plan, either breaking or waiting until the end of their 10-year contracts.

A repeat of the erosion-crazy 1970s seems unlikely in light of farmers' dependence on federal programs, but, nonetheless, the USDA reports that around one-third of the enrolled farmers don't plan on implementing their conservation plans until the 1995 deadline nears, in hopes that higher prices will allow them to exit federal programs. The agency estimates that conservation plans will cut back excessive erosion by only 300 to 500 million tons, well below their potential.

Despite its weaknesses, the 1985 Farm Bill is a piece of landmark legis-

lation, distinguishing the United States as the first major food-producing nation to make soil conservation a national priority. Other countries with more serious erosion problems than the United States, of which there are many, would do well to adopt elements of the bill.

Yet, even in the United States, excessive soil erosion will continue. Aside from minor adjustments, the farm bill now before Congress contains no provisions to strengthen or expand on the erosion control provisions in its predecessor. For now, USDA is resting on its laurels and waiting to see how the conservation plans unfold.

Given present attitudes and past practices, one wonders whether a future W.C. Lowdermilk will make "our own fair land of America" just another stop on the tour of eroded and degraded Promised Lands.

HOLDING BACK THE SEA

Rising sea levels in response to global warming represent
an environmental threat of unprecedented proportion.
Coastal communities face two fundamental choices:
retreat from the shore or fend off the sea.

Jodi L. Jacobson

Jodi L. Jacobson is a senior researcher at
the Worldwatch Institute, 1776 Massachusetts Avenue, N.W., Washington, D.C.
20036. This article is excerpted from *State of
the World 1990*, which is available from the
Futurist Bookstore.

Quick study of a world map illustrates an obvious but rarely considered fact: Much of human society is
defined by the planet's oceans.
And the boundary between land
and water determines a great deal
that is often taken for granted, including the amount of land available for human settlement and
agriculture, the economic and
ecological productivity of deltas
and estuaries, the shape of bays
and harbors used for commerce,
and the abundance or scarcity of
fresh water in coastal communities.

For most of recorded history, sea
level has changed slowly enough
to allow the development of a social
order based on its relative con-
stancy. Global warming will radically alter this. Increasing concentrations of greenhouse gases in the
atmosphere are expected to raise
the earth's average temperature between 2.5° C and 5.5° C over the
next 100 years. In response, the
rate of change in sea level is likely

to accelerate due to thermal expansion of the earth's surface waters
and a rapid melting of alpine and
polar glaciers and of ice caps. Although the issue of how quickly
oceans will rise is still a matter of
debate, the economic and environmental losses of coastal nations
under most scenarios are enormous. One thing is clear: No coastal
nation, whether rich or poor, will
be totally immune.

Accelerated sea-level rise, like
global warming, represents an environmental threat of unprecedented proportion. Yet most discussions of the impending increase
in global rates obscure a critical
issue: In some regions of the world,
the local sea level is already rising
quickly. Egypt, Thailand, and the
United States are just a few of the
countries where extensive coastal
land degradation, combined with
even the recent small incremental
changes in global sea level, is contributing to large-scale land loss.
These trends will be exacerbated in
a greenhouse world.

Low- to middle-range estimates
by the U.S. Environmental Protection Agency (EPA) indicate a
warming-induced rise by 2100 of

anywhere from a half-meter to just
over two meters. A one-meter rise
by 2075, well within the projections, could result in widespread
economic, environmental, and social disruption. G.P. Hekstra of
the Ministry of Housing, Physical
Planning, and Environment in the
Netherlands asserts that such a rise
could affect all land up to five
meters in elevation. Although only
a small percentage of the world
total — about 3% — this area encompasses one-third of global cropland and is home to a billion people.

As sea level rises, coastal communities face two fundamental
choices: retreat from the shore or
fend off the sea. Decisions about
which strategy to adopt must be
made relatively soon because of the
long lead time involved in building
dikes and other structures and because of the continuing development of coasts. Yet allocating
scarce resources on the basis of unknown future conditions — how
fast the sea will rise and by what
date — entails a fair amount of risk.

Protecting beaches, homes, and
resorts can cost a country with a
long coastline billions of dollars,
money that is only well spent if cur-

From *The Futurist*, September/October 1990, pp. 20-27. This article is excerpted from *State of the World, 1990*, published by
W. W. Norton & Company. Copyright © 1990 by Worldwatch Institute. Reprinted by permission.

The coastline of an island in the Ganges Delta is badly damaged and eroded by a deadly typhoon and tidal wave. Scientists predict more-powerful hurricanes and storm surges with a warmer atmosphere.

rent assumptions about future sea level are borne out. Assessing the real environmental costs is difficult because traditional economic models do not reflect the fact that structural barriers built to hold back the sea often hasten the decline of ecosystems important to fish and birds. Moreover, protecting private property on one part of the coast often contributes to higher rates of erosion elsewhere, making one person's seawall another's woe.

Global Changes, Local Outcomes

A higher global average temperature can alter sea level in several ways: The density can decrease through the warming and subsequent expansion of seawater, which increases volume. The volume also can be raised by the melting of alpine glaciers, by a net increase in water as the fringes of polar glaciers melt, or by more ice being discharged from ice caps into the oceans.

The slight variations in global climate of the last 5,000 years are responsible for correspondingly small fluctuations in sea level, on the order of 1–10 centimeters every century. Over the past 100 years, however, global sea level rose 10–15 centimeters (4–6 inches), a somewhat faster pace. Scientists continue to debate the cause of this rise, many arguing that there is no evidence that it is due to human-induced warming, while others are not so sure.

Uncertainties abound on the pace of all the possible changes expected from global warming. The most immediate effect will probably be an increase in ocean volume through thermal expansion. The rate of thermal expansion depends on how quickly ocean volume responds to rising atmospheric temperatures, how fast surface layers warm, and how rapidly the warming reaches deeper water masses. The pace of glacial melt and the exact responses of large masses such as the Antarctic shelf are equally unclear. Over the long term, however, glaciers and ice caps will make the largest contribution to increased volume if a full-scale global warming occurs.

Over the past five years, a number of scientists have estimated the possible range of greenhouse-induced sea-level rise by 2100. Gordon de Q. Robin projects an increase of anywhere from 20 to 165 centimeters. Computations by other scientists yield projections as high as two to four meters over the next

5. RESOURCES: Water

110 years. Widely cited EPA estimates of global mean sea-level rise by 2100 range from 50 to 200 centimeters depending on various assumptions about the rate of climate change. Most models do agree that initial rates of increase will be small relative to the much more rapid acceleration expected from 2050 on. In any case, even the low range of estimates portends a marked increase over the current global pace.

What is important about the sea-level rise expected from global warming is the pace of change. The rate expected at the global level in the foreseeable future — one meter by 2075 is certainly plausible — is unprecedented on a human time scale. Unfortunately, with today's level of population and investment in coastal areas, the world has much more to lose from sea-level rise than ever before.

Lands and Peoples at Risk

Intense population pressures and economic demands are already taking their toll on deltas, shores, and barrier islands. Rapid rates of subsidence and coastal erosion ensure that many areas of the world will experience a one-meter increase in sea level well before a global change of the same magnitude.

The ebb and flow of higher tides will cause dramatic declines in a wide variety of coastal ecosystems. Wetlands and coastal forests, which account for most of the world's land area less than a meter above mean tide, are universally at risk.

According to EPA estimates, erosion, inundation, and saltwater intrusion could reduce the area of present-day U.S. coastal wetlands up to 80% if current projections of future global sea level are realized. The Mississippi Delta, the Chesapeake Bay, and other vital estuaries and wetland regions would be irreparably damaged.

No one has yet calculated the immense economic and ecological costs of such a loss for the United States, much less extrapolated them to the global level. Yet as global sea level rises, these problems will surely become more severe and widespread in ecosystems around the world.

Flood victims in Bangladesh are rescued in boats and rafts. With sea-level rise, floods such as this could become commonplace in Bangladesh.

U.S. AGENCY FOR INTERNATIONAL DEVELOPMENT

Experts Downgrade Sea-Level Forecast

Some scientists have changed their predictions about sea-level rise, based on updated information. "We have revised rather drastically our best estimates of how much global sea level will rise due to greenhouse warming," says Mark F. Meier of the University of Colorado at Boulder, who in 1985 chaired a National Research Council committee investigating changes in sea level.

In 1985, Meier's committee predicted a sea-level rise of about one meter by the year 2100, based on the information available then. Meier now says that a rise of only one-third meter can be predicted before 2100, perhaps by midcentury.

These predictions are wide ranging in time frame and in effect, due mainly to uncertainty over how Antarctica will be affected in a warmer world, according to Charles R. Bentley of the University of Wisconsin–Madison, who, along with Meier, spoke at a recent meeting of the American Geophysical Union. Climate models indicate that greater snowfall in a warmer Antarctica would actually build up the ice sheets there and help lower global sea levels, but effects in other regions should counterbalance this and result in a moderate increase in global sea levels, says Bentley. Meier and his colleagues say that, at the extreme, ocean levels could climb by 0.7 meter, or they could actually fall by 0.1 meter.

The scientists agree that much remains unknown about both global warming and its effect on ocean levels. "Our understanding of the system is not very good at the moment," Meier concluded.

Source: "Predictions Drop for Future Sea-Level Rise" by Richard Monastersky. *Science News*, December 16, 1989. Science Service, Inc., 1719 N Street, N.W., Washington, D.C. 20036.

Land subsidence is a key issue in the case of river deltas, such as the Nile and Bengal, where human activities are interfering with the normal geophysical processes that could balance out the effects of rising water levels. These low-lying regions, important from both ecological and social standpoints, will be among the first lost to inundation under global warming.

Under natural conditions, deltas are in dynamic equilibrium, forming and breaking down in a continuous pattern of accretion and subsidence. Subsidence in deltas occurs naturally on a local and regional scale through the compaction of recently deposited river-borne sediments. As long as enough sediment reaches a delta to offset subsidence, the area either grows or maintains its size. If sediments are stopped along the way, continuing compaction and erosion cause loss of land relative to the sea, even if the absolute level of the sea remains unchanged.

Large-scale human interference in natural processes has had dramatic effects on both relative rates of sea-level rise and on coastal ecosystems in several major deltas. Channeling, diverting, or damming of rivers can greatly reduce the amount of sediment that reaches a delta, as has happened in the Indus, the Mississippi, and many other major river systems, resulting in heavier shoreline erosion and an increase in local relative water levels.

Worldwide, erosion of coastlines, beaches, and barrier islands has accelerated over the past 10 years as a result of rising sea level. A survey by a commission of the International Geophysical Union demonstrated that erosion had become prevalent on the world's sandy coastlines, at least 70% of which have retreated during the past few decades.

Increased erosion would decrease natural storm barriers. Coastal floods associated with storm surges surpass even earthquakes in loss of life and property damage worldwide. Apart from greater erosion of the barrier islands that safeguard mainland coasts, higher seas will increase flooding and storm damage in coastal

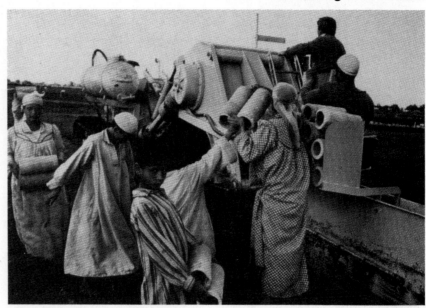

WORLD BANK
Workers lay tiles for drainage of irrigated land in the Nile Delta in Egypt. Dams used in irrigation have greatly decreased the amount of sediment reaching the delta, causing land subsidence and encroaching salt water.

areas because raised water levels would provide storm surges with a higher base to build upon. And the higher seas would decrease natural and artificial drainage.

A one-meter sea-level rise could turn a moderate storm into a catastrophic one. A storm of a severity that now occurs about every 15 years, for example, could flood areas that are today only affected by truly massive storms once a century.

In Bangladesh, storm surges now reach as far as 160 kilometers upriver. In 1970, this century's worst storm surge tore through the countryside, initially taking some 300,000 lives, drowning millions of livestock, and destroying most of Bangladesh's fishing fleet. The toll climbed higher in its aftermath. As the region's population mounts, so does the potential for another disaster.

Studies indicate a dramatic increase in the area vulnerable to flooding in the United States as well. A one-meter rise would boost the portion of Charleston, South Carolina, now lying within the 10-year floodplain from 20% to 45%. A 1.5-meter rise would bring that figure to more than 60%, the current area of the 100-year floodplain. As a result, "once-a-century" floods would then occur on the order of every 10 years. Sea-level rise will also permanently affect freshwater supplies. Miami is a case in point. The city's first settle-

ments were built on what little high ground could be found, but today most of greater Miami lies at or just above sea level on swampland reclaimed from the Everglades. Water for its 3 million residents is drawn from the Biscayne aquifer, which flows right below the city streets. That the city exists and prospers is due to what engineers call a "hydrologic masterwork" of natural and artificial systems that hold back swamp and sea.

Against a one-meter rise in ocean levels, Miami's only defense would be a costly system of seawalls and dikes. But that might not be enough to spare it from insidious assault. Fresh water floats atop salt water, so as sea levels rise the water table would be pushed nearly a meter closer to the surface. The elaborate pumping and drainage system that currently maintains the integrity of the highly porous aquifer could be overwhelmed. The higher water table would cause roads to buckle, bridge abutments to sink, and land to revert back to swamp. Miami's experience would not be unique. Large cities around the world — Bangkok, New Orleans, Taipei, and Venice, to name a few — face similar prospects.

WORLD BANK

Workers finishing construction on the Tarbela Dam on the Indus River in Pakistan. While the dam helps control flooding downstream, it also decreases the natural flow of sediment, resulting in greater coastal erosion in the Indus Delta.

"Almost without exception, the prognosis for these vulnerable, low-lying countries in a greenhouse world is grim."

Most Vulnerable, Least Responsible

The social and environmental costs of sea-level rise will be highest in countries where deltas are extensive, densely populated, and extremely food-productive. In these countries, most of which are in the Third World, heavy reliance on groundwater and the completed or proposed damming and diversion of large rivers — for increased hydropower and agricultural use, for flood control, and for transportation — have already begun to compound problems with sea-level rise. Almost without exception, the prognosis for these vulnerable, low-lying countries in a greenhouse world is grim.

The stakes are particularly high throughout Asia, where damming and diversion of such systems as the Indus, Ganges-Brahmaputra, and Yellow rivers have greatly decreased the amount of sediment getting to deltas. As elsewhere, the deltas reliant on these sediments support sizable human and wildlife populations while creating protective barriers between inland areas and the sea. Large cities, including Bangkok, Calcutta, Dhaka, Hanoi, Karachi, and Shanghai, have grown up on the low-lying river banks.

These heavily populated areas are almost certain to be flooded as sea-level rise accelerates.

The U.N. Environment Programme's 1989 global survey represents the first attempt to analyze systematically the regions most vulnerable to sea-level rise. Ten countries — Bangladesh, Egypt, The Gambia, Indonesia, the Maldives, Mozambique, Pakistan, Senegal, Surinam, and Thailand — were identified as "most vulnerable." These 10 share many characteristics, including the fact that they are, by and large, poor and populous.

UNEP estimates based on current population size and density show that 15% of Bangladesh's land area, inhabited by 15 million people, is threatened by total inundation from a primary rise of up to 1.5 meters. Secondary increases of up to three meters would wipe out over 28,500 square kilometers (20% of the total land area), displacing an additional 8 million people.

Pressures to develop agriculture have quickened the pace of damming and channeling on the three giant rivers — the Brahmaputra, the Ganges, and the Meghna — that feed the delta. As a result, subsidence is increasing. This situation is being made worse by the increasing withdrawal of groundwater.

In the Nile Delta of Egypt, extending from just east of the port city of Alexandria to west of Port

Said at the northern entrance of the Suez Canal, local sea-level rise already far exceeds the global average due to high rates of subsidence.

At least 40% of Indonesia's land surface is classified as vulnerable to sea-level rise. In terms of both size and diversity, the country is home to one of the world's richest and most extensive series of wetlands. Here, too, population pressures are already threatening these fragile ecosystems.

A one- to two-meter rise in sea level could be disastrous for the Chinese economy as well. The Yangtze Delta is one of China's most heavily farmed areas. Damming and subsidence have contributed to a continuing loss of this valuable land on the order of nearly 70 square kilometers per year since 1947. A sea-level rise of even one meter could sweep away large areas of the delta, causing a devastating loss in agricultural productivity for China.

Paying by the Meter

Assuming a long-run increase in rates of global sea-level rise, societies will have to choose some adaptive strategies. Broadly speaking, they face two choices: fight or flight.

Along with the intensified settlement of coastal areas worldwide over the past century has come a belief that human ingenuity could tame any natural force. As a result, people have been inclined to build closer and closer to the ocean, investing billions of dollars in homes and seaside resorts and responding to danger by confrontation.

Nowhere in the world is the battle against the sea more actively engaged than in the Netherlands. Hundreds of kilometers of carefully maintained dikes and natural dunes keep the part of the country that is now well below sea level —

more than half the total — from being flooded.

The Dutch continue to spend heavily to keep their extensive system of dikes and pumps in shape and are now protected against storm surges up to those with a probability of occurring once in 10,000 years. But the prospect of accelerated sea-level rise implies that maintaining this level of safety may require additional investments of up to $10 billion by 2040.

Large though these expenditures are, they are trivial compared with what the United States would have to spend to protect its more than 30,000 kilometers (19,000 miles) of coastline. Preliminary estimates by EPA of the total bill for holding the sea back from U.S. shores — including costs to build bulkheads and levees, raise barrier islands, and pump sand — range from $32 billion to $309 billion for a one-half-meter to two-meter rise in sea levels.

Many countries have made vast investments reclaiming land from the sea; witness the efforts in Singapore, Hong Kong, and Tokyo. Political pressures to maintain these lands through dikes, dams, and the like will be high, but political support for subsidizing coastal areas may be undercut by competing fiscal demands over the long run. With increasing competition for scarce tax dollars, property owners in the year 2050 may find the general public reluctant to foot the bill for seawalls.

Moreover, what may seem like permanent protection often turns out to be only a temporary measure. While concrete structures may divert the ocean's energies from one beach, they usually displace it onto another. And by changing the dynamics of coastal currents and sediment flow, these hard structures interrupt the natural processes that allow wetlands and beaches to reestablish upland, causing them to deteriorate and in many cases disappear.

Beach nourishment is a relatively benign defensive strategy that can work in some cases. And comparing the costs and benefits illustrates that it is not usually as prohibitively expensive as other approaches. Sand or beach nourishment, for

Land where the Indus River has receded following floods in its delta. Floods damage low-lying delta regions, sometimes making the land unusable for agriculture.

WORLD BANK

example, can cost $620,000 per kilometer, but these costs are often justified by economic and recreational use of the areas. A recent study of Ocean City, New Jersey, found it would cost about 25¢ per visitor to rebuild beaches to cope with a 30-centimeter rise, less than 1% of the average cost of a trip to the beach. And the fact remains that most beach replenishment is temporary at best, indicating the need for continuous investment.

The legal definitions of private property and of who is responsible for compensation in the event of natural disasters are already coming into question. As sea-level rise accelerates, pushing up the costs of adaptation, these issues will likely become part of an increasingly acrimonious debate over property rights and individual interests versus those of society at large.

Site-specific studies of several towns in the United States suggest that incorporating projections of sea-level rise into land-use planning can save money in the long run. Projections of costs in Charleston, South Carolina, show that a

WORLD BANK

Construction of a new drainage ditch in Jakarta, Indonesia, to relieve the threat of flooding. Global warming and the resultant sea-level rise could seriously increase the threat of flooding in many low-lying coastal cities.

> "As millions of people displaced by rising seas move inland, competition with those already living there for scarce food, water, and land may spur regional clashes."

strategy that fails to anticipate and plan for the greenhouse world can be expensive. Depending on the zoning and development policies followed, including the amount of land lost and the costs of protective structures built, the costs of a one-meter sea-level rise may exceed $1.9 billion by 2075 — an amount equal to 26% of total current economic activity in this area. If land-use policies and building codes are modified to anticipate rising sea levels, this figure could be reduced by more than 60%. Similar studies of Galveston, Texas, show that economic impacts could be lowered from $965 million to $550 million through advanced planning.

Whatever the strategy, industrial countries are in a far better financial position to react than are developing nations. Debates over land loss may be a moot point in poorer countries like Bangladesh, where evacuation and abandonment of coastal land may be the only option when submergence and erosion take their toll and when soil and water salinity increase. As millions of people displaced by rising seas move inland, competition with

those already living there for scarce food, water, and land may spur regional clashes. Ongoing land tenure and equity disputes within countries will worsen.

Planning Ahead

An active public debate on coastal-development policies is needed, extending from the obvious issues of the here and now — beach erosion, river damming and diversion, subsidence, wetland loss — to the uncertainties of how changes in sea

level in a greenhouse world will make matters far worse. Raising public awareness on the forthcoming changes, developing assessments that account for all future and present costs, and devising sustainable strategies based on those costs are all essential.

Taking action now to safeguard coastal areas will have immediate benefits while preventing losses from soaring higher in the event of an accelerated sea-level rise. Limiting coastal development is a first step, although strategies to accomplish this will differ in every country. Governments may begin by ensuring that private-property owners bear more of the costs of settling in coastal areas.

A new concept of property rights will have to be developed. Unbridled development of rivers and settlement of vulnerable coasts and low-lying deltas mean that more and more people and property will be exposed to land loss and potential disasters arising from storm surges and the like. Governments

U.S. AGENCY FOR INTERNATIONAL DEVELOPMENT

Bengalese carry loads of earth in an effort to reclaim land and provide irrigation. Developing countries in low-lying areas are caught between two desperate needs: increasing the amount of arable land for food and keeping back rising ocean levels.

that plan over the long term to limit development of endangered coasts and deltas can save not only money, but resources as well. Wherever wetlands and beaches are not bordered by permanent structures, they will be able to migrate and reestablish further upland, allowing society to reap the intangible ecological benefits of biodiversity.

Of course, protection strategies will inevitably be carried out where the value of capital investments outweigh other considerations. But again the key is to plan ahead. As the Dutch discovered, more money can be saved over the long term if dikes and drainage systems are planned for before rather than after sea levels have risen considerably.

A cap on problematic dam-building and river-diversion projects in large deltas would lessen the ongoing destruction of wetland areas and prevent further reductions in sedimentation, thereby minimizing subsidence as well.

Additional money is needed to do more research on sea level globally and regionally. Funds are needed to support studies of beach and wetland dynamics, to take more-frequent and widespread measurements of global and regional sea levels, and to design cost-effective, environmentally benign methods of coping with coastal inundation.

The majority of developing nations most vulnerable to sea-level rise can do little about global warming independently. But they have a clear stake in reducing pressures on coastal areas by taking immediate actions. Among the most important of these is slowing population growth and, where necessary, changing inequitable patterns of land tenure in interior regions that promote coastal settlement of endangered areas. Furthermore, the governments of Bangladesh, China, Egypt, India, and Indonesia, to name just a few, are currently promoting river development projects that will harm delta ecosystems in the short term and hasten the date they are lost permanently to rising seas.

The issue of how to share the costs of adaptation equitably may well be among the hardest to resolve. Industrial countries are responsible for by far the largest share of the greenhouse gases emitted into the atmosphere. And no matter what strategies poorer nations adopt to deal with sea-level rise, they will need financial assistance to carry them out. The way industrial countries come to terms with their own liability in the face of accelerated sea-level rise will play a significant role in the evolution of international cooperation.

WATER, WATER, EVERYWHERE, HOW MANY DROPS TO DRINK?

William H. MacLeish

William H. MacLeish's current environmental book, "The Gulf Stream: Encounters with the Blue God," was preceded by "Oil and Water" and will be followed by "The Day Before America" (Houghton Mifflin). Mr. MacLeish was editor of Oceanus, a publication of Woods Hole Oceanographic Institute, from 1972 to 1982. His magazine writings include "The Silver Whistler," a memoir about his father, Archibald MacLeish, and "Ways to Beat the Noxious '90s."

CROCKER POND IS A KETTLE HOLE —an oval depression left by the melting of buried glacial ice—in upper Cape Cod. I had been in it as many times as I could during a sweltering weekend in mid-July. Between immersions, I joined a couple of hundred people being poached in the summer heat wave and listened to discussions of a subject I thought I knew well—the meaning of water. Hour by hour, I came to understand the limitations of my hydrologic literacy.

Now, packed to leave, I lie on Crocker's north bank and let new learning mix with old. The grass under me is hydrated carbon; the blades defy gravity because of the water in them. I am 70% water, and my blood, if iron were replaced by magnesium, would be twin to the sea.

I, the grass, all earthly flora and fauna, are—to steal a phrase from the eminent Russian scientist Vladimir Vernadsky, animated water. My kind collapses quickly without it, unlike fungi and other microbes and certain seeds that can survive a drought of decades, even centuries.

In the Middle East, for example, oil may appear to be the geopolitical resource of 1991, but water soon will be. Iraq's water supplies are threatened by Turkey's water diversion upstream on the Euphrates and Tigris rivers, despite plans for "peace pipelines" to ease shortages. Egypt is almost entirely dependent on the the world's longest river, the Nile, yet it is the ninth country along its banks in line for its water. With growing populations and varieties of needs, it seems likely that water battles, if not wars, may increase in future.

Yet we pay water little mind, especially in the moistness of a New England summer.

Humans have been withdrawing water from rivers, lakes, and wells, altering its properties and returning it since the first tribes camped by the first fords. Such human impact was small until the arrival of agriculture and settlement some 10,000 years ago. Even then, most anthropogenic effects (those caused by human activities) had to do with local pollutions from body wastes and attendant pathogens. The gradual intensification of irrigation in the more arid parts of the world resulted in the deposit of salts and other harmful compounds in soils and subsoils, and these soluble deposits often found their way to flowing water.

They were later joined by farm fertilizers— first natural manures and, in this century, millions of tons of synthetic plant foods high in nutrients like nitrogen. Nitrogen compounds can cause populations of algae and other plant life to explode, overshooting the carrying capacity of host waters. Then decaying vegetable matter uses up dissolved oxygen, often killing resident fish. This process of overfertilization leading to oxygen deprivation is known as eutrophication.

THE WORST INSULT

In this century our most egregious insult to water has been our own population explosion—from about 1.6 billion in 1900 to a projected 6 billion by 2000. In the rush of numbers, two basic constants remain: Experts suggest that the average human drinks half a gallon (64 oz.) of water a day more or less, depending on climatic and other conditions,

From *World Monitor*, December 1990, pp. 54-58. Copyright © 1990 by William H. MacLeish. Reprinted by permission.

such as physical exertion. (Soldiers in the Saudi desert require six gallons, as we have lately become aware.) And the average human produces about six cups of excreta per day, most of which mixes with water at some point. Guaranteeing the purity of what goes in and restoring purity to what comes out is a source of growing concern in cities from San Diego to Singapore and in hundreds of thousands of villages on every continent.

Adding to what might indelicately be called this ordure of magnitude is the waste that goes with the commercial development we take for granted as an integral part of modern life. Elevated living standards float on oceans of industrial residues, as thousands of chemical and mechanical processes go on stream each year, many in the name of convenience. Manufacturers of paper products, chemicals, petrochemicals and refined petroleum products, textiles, and foods are among the most important generators of impurities and sometimes poisons in the richer, industrialized regions of the world—and, increasingly, the lower income regions. Brazil, for example, is the fifth largest consumer of pesticides.

Dioxins, PCBs, lead, cadmium, and other toxic agents are showing up in rivers and lakes around the world. The United States, despite problems with the residues of its industrialization, has instituted one of the world's most thoroughgoing campaigns against pollution. Many of the so-called "point" sources of poisons entering the US water supply—factory waste pipes, for instance—have been identified and ordered into compliance.

But, as industrial effluents improve in purity, municipal authorities are discovering many of the same poisons in individual home septic tanks. House and home may sparkle from shower grout to shining sink, but they are not the temple to purity portrayed in TV commercials.

DIMENSIONS OF THE PROBLEM

Global monitoring of water problems is still far from complete. But a sampling of worldwide press reports and surveys by the United Nations Environment Program suggests the dimensions:

• Some rivers in Latin America carry a fecal coliform load that is more than a thousand times beyond that set for safe drinking water.

• Up to 90% of Europe's rivers show a buildup of nitrogen compounds, mostly from fertilizers.

• Acid rain from burning high sulfur coal and oil without corrective systems degrades water quality in North America, Europe, and China.

• New York City, which for decades has boasted of one of the cleanest water supplies in the world, is now faced with installing expensive systems to filter out pollutants from housing developments encroaching on upstate rivers and reservoirs.

• Heavy metals have accumulated in industry-crowded rivers like the Rhine. Levels have declined in the last two decades as a result of pollution controls. But enough remain to leave Coleridge's Rhine query of 1828 still pertinent:

The river Rhine, it is well known,
Doth wash your city of Cologne;
But tell me, nymphs! what power divine
Shall henceforth wash the river Rhine?

(It was exactly 30 years earlier, incidentally, that Coleridge wrote what might qualify as the perfect slogan for water quality campaigners: "Water, water, everywhere,/ Nor any drop to drink.")

THE EAST'S ECOLOGICAL RUIN

And what will rinse away the fouled waters of Eastern Europe and Eurasia? Communist regimes denied environmental problems for so long that the ecological ruin there has come as a shock to the rest of the world. Drinking water in parts of Poland and Czechoslovakia is hazardous to health, and some rivers are too filthy for even industrial use. Baikal, the deepest freshwater lake in the world (and largest lake in volume, discounting the saltwater Caspian Sea), has been degraded by effluents from Soviet industry on the shore.

Some of the most congested landscapes in the world are found along stretches of coastline: the lowlands of Europe, northeastern, southeastern, and southwestern United States, central Japan, eastern China, Bangladesh, and southeastern India, for example.

Early this year, the Group of Experts on the Scientific Aspects of Marine Pollution (GESAMP) released its latest report. The open sea, the scientists said, is still relatively clean, though "man's fingerprint is found everywhere in the oceans." Where our weight bears heaviest is along the continental margins of the world, the coastal zones that yield most of the world's harvest of marine fish. The damage is concentrated near large cities and industrial centers or down current from the mouths of severely polluted rivers.

Industries along many of these industrialized coasts, like those along inland rivers, are increasingly being required to treat their wastes before discharging them to the sea. The horror stories of mercury poisoning in Japan and other instances of seaborn chemicals devastating coastal dwellers are rare today, but they could easily return as industries locate new plants in areas of lax regulation.

The GESAMP scientists conclude that "at the end of the 1980s the major causes of immediate concern in the marine environment on a global basis are coastal development and the attendant destruction of habitats, eutrophication, microbial contamination of seafood and beaches, fouling of the seas by plastic litter, progressive buildup of chlorinated hydrocarbons [such as slow-degrading pesticides], especially in the tropics and subtropics, and accumulation of tar [from oil seepage and spills] on beaches."

n Arcata, California, wastewater receives its final processing in two attractive man-made marshes, which also serve as habitats for ducks and other wildlife.

Sandra Postel, who reports on freshwater problems and policies for the Worldwatch Institute in Washington, is worried by the number of "exhausted rivers, falling water tables and shrinking lakes" she sees worldwide. One has only to look at the mean trickle of the Colorado River as it enters the Gulf of California to take her point. Before the dams came, and the irrigation projects and the thirsty cities far beyond its watershed, the river rushed the Gulf at full volume.

MINING WATER

To support its commitment to farming what were once considered marginal agricultural areas, the United States is in essence mining water. So are nations such as Jordan and Israel, which tap subterranean aquifers with deep wells. The great Ogallala Aquifer under America's Great Plains is being pumped, primarily in the Texas Panhandle, where it has been depleted by 25% of its pre-agricultural age size. The Soviets have so abused the Aral Sea, by diverting river water which feeds it, that its surface area has been reduced by 40% in the last 30 years, and hopes for its recovery are slight. In China and India, demands for irrigation are confronting demands for drinking water in heavily urbanized areas, and crops are likely to be the losers.

New approaches are needed in the dry American West, and they are forthcoming.

One of the most promising involves what America uses most often, the market. Trading in water rights was popular in the western US before federal water programs were voted. Those programs, with their expensive subsidies, may not survive in present form during a period of budget deficits.

TRADING WATER

Today water trading often involves urban districts looking for new supplies doing deals with farmers or land companies with excess supplies to sell. Prices can get exorbitant; enough water to meet the needs of a typical home in the Denver area for two years can cost upward of $6,000. Rural areas can end up ecologically degraded by exporting too much of their stream flows. And the bartering does nothing to correct the underlying problems of regional water shortages.

Fortunately, some of these drawbacks are beginning to be addressed:

• The Metropolitan Water District of Southern California, facing severe urban drought, has agreed to finance rural water conservation projects in the area—fixing leaks in canals, for example. The water supplies it receives in exchange are relatively cheap, and the irrigation of cropland through which the canals flow remains unaffected.

• The Rocky Mountain Institute (RMI), a leader in energy and water efficiency research, reports that Santa Monica, California, has installed low-flow toilets in one-quarter of the city's homes, saving almost a million gallons a day. The $2.8 million program, which offered free installation or rebates to cover installation costs of the low-flow toilets and shower heads, was financed by a conservation incentive fee (participants were exempt from this surcharge on their water bill), general water revenues, and money from the water wholesaler. That eases demand not only on supply (reducing per capita consumption for indoor use from 80 gallons to 50 gallons per day) but also on another, often forgotten component of municipal water systems— treatment of wastewater, which in this article is defined as any "used" water pumped or diverted for industrial, agricultural, or household purposes. The less water consumed in home and factory, the less sent down the line to the sewage plant.

SAVING WATER

• RMI says a single hotel in Michigan saved three-quarters of a million dollars over eight years simply by using state-of-the-art water-saving showerheads, faucets, and toilets. These thrifty devices are increasingly popular in Scandinavia, Germany, and other parts of Europe where resource efficiency is highly regarded. The US lags by comparison. It's too early to forecast whether Saddam Hussein's power grab will jolt Americans into re-embracing such old Yankee homilies as "Waste not, want not" and "You'll never miss the water 'til the well runs dry."

Americans' attitudes toward oil and water are strangely parallel. If there isn't an oil shock or persistent drought in sight to reinforce thrifty slogans, much of the population reverts to the understandable largess of the Opening of the West, when people were few and resources plentiful. Americans tend to regard both oil and water as renewable, though oil is not, and clean water may become decreasingly so if consumption and pollution proceed unabated.

The amount of industrial poisons in wastewater can be greatly reduced by switching to more benign substances in manufacturing and by recycling. Metal plating firms in New England have found that it makes economic sense to recover and reuse some of the dangerous pollutants they once dumped down the drain.

Throughout the world, people need to understand the water limits. Farmers in India can be taught more efficient methods of irrigation. House-holders in Los Angeles can learn not to throw cleaning fluids and cosmetics down the drain. If their water bills ascend to cost-accounting reality, consumers will go shopping for toilets that use less than two gallons per flush rather than seven.

Human wastes and the diseases they can carry still remain one of the world's great waterborne scourges. Countries that can afford to do so have built treatment systems that reduce the hazards. But most sewage plants are so costly that governments have a hard time maintaining them, let alone upgrading and expanding them. The plants often contribute some pollution on their own—chemicals like aluminum salts and chlorine, used in treatment—and they produce large amounts of sludge, often of sufficient toxicity to make disposal a difficult problem.

There are alternatives, and one of the most

promising combines innovative engineering with natural systems. In Arcata, California, communal wastewater receives its final processing in two attractive man-made marshes, which also serve as habitats for ducks and other wildlife. Denham Springs, Louisiana, runs its processed sewage through a series of shallow pits that have been lined, carpeted with stones, filled with water, and planted with lilies and other plants.

SOLAR AQUATICS

A related experiment in ecological engineering is a technique called solar aquatics. It uses less land than the synthetic marshes, works better in cold climates, and is more adaptable to urban application. Right now, a pilot program is ending its first year of operations in a specially designed greenhouse set up within sight and smell of the conventional sewage treatment plant that serves Providence, Rhode Island.

There a large greenhouse equipped with sophisticated venting and heating devices (the latter for winter use) houses 48 1,300-gallon translucent plastic tanks arranged in four trains. The tanks support different mixes—engineered ecosystems, really—of snails, crayfish, algae, bacteria, and plants. These combinations take the dank effluents of the city and turn them into clear water.

Solar aquatics, or rather the man who developed it, was what brought me to Crocker Pond. I first met John Todd and his wife, Nancy, when they were building the New Alchemy Institute a few miles from Crocker Pond. Now, here they were, with a new organization to run—the Center for the Protection and Restoration of Waters—and with a water-treatment technique of global promise. The walls of old farm buildings at the center were covered with Todd's visions of urban landscapes built around water gardens that delight the eye while detoxifying waste.

I left the kettle hole reluctantly, the cool of pond water still on my skin. Then I reached the highway, and the sweat started. With it came dribbles of memorized data: 5 trillion gallons of waste from industries in the US dumped into coastal waters each year; leaking landfills, poisonous septage, all in this one country; the global threat of climate change and with it shifts in regional precipitation patterns. Too much, I thought. Too much to handle.

That, of course, is the habitual response to the habitual presentation of environmental problems, one in which the data of disaster are the only frame of reference. There are always other dimensions, other perspectives. One came to me on the road. A speaker at the conference had quoted from Thoreau. The thought could just as easily come from John Todd or the hundreds of others around the world committed to the restoration of rivers, aquifers, and seas.

"He who hears the rippling of water in these degenerate days will not utterly despair."

Precisely.

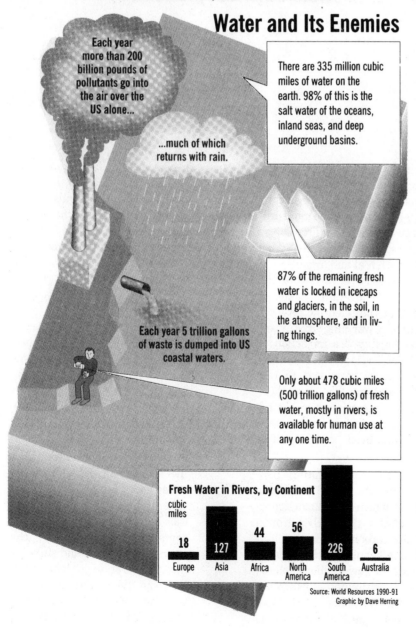

Water and Its Enemies

Each year more than 200 billion pounds of pollutants go into the air over the US alone...

...much of which returns with rain.

There are 335 million cubic miles of water on the earth. 98% of this is the salt water of the oceans, inland seas, and deep underground basins.

87% of the remaining fresh water is locked in icecaps and glaciers, in the soil, in the atmosphere, and in living things.

Each year 5 trillion gallons of waste is dumped into US coastal waters.

Only about 478 cubic miles (500 trillion gallons) of fresh water, mostly in rivers, is available for human use at any one time.

Fresh Water in Rivers, by Continent

cubic miles

Europe	Asia	Africa	North America	South America	Australia
18	127	44	56	226	6

Source: World Resources 1990-91
Graphic by Dave Herring

Global Climate Change: Fact and Fiction

Although greenhouse gases have been increasing in the atmosphere, the climate record does not show the temperature increase predicted by climate models.

S. Fred Singer

S. Fred Singer is professor of environmental sciences at the University of Virginia and served as the first director of the U.S. weather satellite program. His most recent book is Global Climate Change, *published by Paragon House.*

Greenhouse warming has emerged as a major issue of the 1990s. The easing of international tension with the Soviet Union could make greenhouse warming a leading foreign policy issue, along with other global environmental concerns.

Wide acceptance of the Montreal Protocol, which limits and rolls back the manufacture of chlorofluorocarbons (CFCs), has encouraged environmental activists—at conferences in Toronto (1988), The Hague (1989), and Geneva (1990)—to call for similar controls on carbon dioxide from fossil-fuel burning. They have expressed disappointment with the White House for not supporting immediate action.

But should the United States assume leadership in a campaign that could cripple its economy, or would it be more prudent to assure first, through scientific research, that the problem is both real and urgent?

An objective review of the best available data makes it clear that the scientific base for greenhouse warming is too uncertain to justify drastic action at this time. There is little risk in delaying policy responses to this century-old problem, since there is every expectation that scientific understanding will be substantially improved within a few years.

Instead of initiating panicky, premature, and ineffective actions, which would slow down but not stop the further growth of global carbon dioxide, the United States should use the same resources—a few trillion dollars, by some estimates—to increase economic resilience so that we can apply specific remedies as necessary. This is not to say that prudent steps cannot be taken now. Indeed, many kinds of energy conservation and efficiency increases make economic sense even *without* the threat of greenhouse warming.

The scientific base: The IPCC report

The scientific base for greenhouse warming includes some facts, lots of uncertainty, and just plain ignorance—clearly, more observations, better theories, and more extensive calculations are required.

There is a consensus about the increase in so-called greenhouse gases in the earth's atmosphere as a result of human activities in recent decades. There is some uncertainty about the strength of sources and sinks for these gases, or, to put it differently, about their rates of generation and removal. There is major uncertainty and disagreement about whether this increase has caused a change in the climate during the last 100 years.

There is also disagreement in the scientific community about changes predicted as a result of further increases in greenhouse gases. The models used to calculate future climate are not yet good enough. As a consequence, we cannot be sure whether the next century will bring a warming that is negligible or a warming that is significant.

Finally, even if there is global warming and associated climate changes, it is debatable whether the consequences will be good or bad. Likely, the effects would be mixed.

The policy issue now is whether the nearly 30 percent increase in carbon dioxide, mainly since World War II, calls for immediate and drastic action. Taking account of increases in the other trace gases that produce greenhouse effects, we have already gone halfway to an effective doubling of atmospheric carbon dioxide above the pre-industrial level. Regardless of what anyone does, this cannot be reversed in our lifetime, and according to the prevailing theory we are faced with an inevitable temperature increase of about 1.5°C. The climate record, however, does not support the theory.

Advocates of carbon dioxide control base their case on the *Policymakers Summary* and a more complete scientific report on greenhouse warming released in June 1990 by a working group of the Intergovernmental Panel on Climate Change (IPCC). Formed as an ad hoc group under UN auspices, the IPCC purports to present a "scientific consensus" about the imminent danger of major global warming stemming from the use of fossil fuels.

No such consensus exists—not when dozens of scientists disagree, individually and collectively. The IPCC summary—not accurately reflecting the report itself—ignores valid scientific objections to computer models that predict a significant global warming. It plays fast and loose with the historical climate data (which clearly disprove the current greenhouse theory), and it is silent about other human activities thought to promote climate cooling.

The IPCC summary puts a spin on the report's major conclusions, which can only serve to mislead nonscientist decision makers earnestly seeking answers to global problems—problems that may not even be real.

For example, its claim of "certainty" that "there is a natural greenhouse effect which keeps the earth warmer than it would otherwise be" sounds ominous, but it is on par with revealing, in hushed tones, that the earth is round.

After all, the greenhouse effect has been known for over a century and studied intensively for decades. Without it there would be no life on earth. And to claim with certainty that human activities—like the emission of carbon dioxide and other gases—"will enhance the greenhouse effect" is not particularly startling either, unless one can show that the additional warming is significant.

Yet these two IPCC conclusions, issued without any further qualifications, suggest to the unwary a future in which the American heartland is turned into a desert by near-continuous droughts, lashed by frequent hurricanes, or—worse still—flooded by raging oceans as sea level rises to unprecedented heights.

For the past few years, such images have been assiduously promoted by environmental groups and many politicians, and reported uncritically by news media anxious to hype disaster. These fantasies are not based on scientific fact and are not even supported by the IPCC report.

Lack of scientific consensus

One should be suspicious about any claim of scientific consensus, particularly in a field as complicated as climate change. In fact, there is constructive tension between those who rely primarily on theory (global climate modelers) and a growing number of climate experts who put weight on empirical evidence. The latter object to the IPCC's treatment of climate data, which—contrary to what is claimed—do not back up but rather contradict the IPCC's computer-model predictions.

Global climate models try to simulate what is happening in the real atmosphere-ocean system. However, they are severely limited by the capacity and speed of the computers.

Even more restrictive is our limited knowledge of the relevant atmospheric and ocean processes: heat transfer in the atmosphere, details of cloud formation, and vagaries of ocean current, to name but a few crucial uncertainties.

Even within these severe limitations, the half-dozen or so global models now available give rather divergent results, predicting an average global warming, sometime in the next century, ranging from 1.5° up to 4.5° C.

None of the computer models include probable offsetting cooling effects from human activities. For example, Virginia climatologist Patrick Michaels has developed strong evidence that sulfur emission from fossil-fuel burning increases the number of cloud condensation nuclei and clouds' reflecting power; and Joyce Penner of the Livermore National Laboratory offers evidence that smoke from increased biomass burning will similarly increase cloud cover, reducing solar warming.

The models diverge even further when trying to predict regional changes rather than global averages. For example, while all models predict more global evaporation (and more rainfall) as a result of warming, some predict a drier, others a wetter United States. One model result yields a Sahara as wet as Scotland.

The only way to validate the models is by "hindcasting"—applying the models to the temperature trends of the past 100 or so years and comparing their predictions with the available record. Here the models fail miserably.

For example, one would expect that half of the global warming of 1.5° to 4.5°C should have occurred by 1990, in view of the measured increase in atmospheric greenhouse gases. Indeed, by allowing maximum error one can discern a temperature rise of 0.5°C since 1880, barely approaching the lower limit of model calculations.

The IPCC report makes much of this "agreement" but fails to mention the fact that there has been hardly any net temperature increase over the last fifty years, yet most of the greenhouse gas increase in the atmosphere has occurred since then.

"Global" temperature-trend curves—and there are several—are stitched together from records of land stations and ships. The best quality records, from the more numerous stations in the Northern Hemisphere, clearly show a long-term cooling between 1938 and 1975, which greenhouse advocates either ignore or ascribe to natural climate fluctuations. This begs the question as to why one should not similarly explain the sharp increase observed between 1975 and 1980 as a natural fluctuation.

Many climatologists would ascribe the pre-1938 warming in the record to a recovery from the Little Ice Age, the climate cooling between about 1500 and 1850 that had an adverse economic impact on European food production. As Arizona State University climatologist Robert Balling points out, this interpretation wipes out any warming attributable to greenhouse effects.

The most reliable and complete climate record for a large land area is for the United States. The data show a general cooling since 1940 and certainly no appreciable recent warming. The hottest years on record were in the 1930s, not in the 1980s.

An absence of warming since 1978 is also shown by newly published satellite observations. They provide the only truly *global* air temperature record, undisturbed by land effects such as urban heat islands. Contrast this negative satellite result with the model-projected increase of 0.2° to 0.3°C per decade.

The IPCC summary finesses

A differing view of global warming

Greenhouse warming entered the public consciousness during the particularly hot 1988 summer as a result of testimony presented before a U.S. Senate committee by James Hansen, director of NASA's Goddard Institute for Space Studies. Hansen stated that after carefully studying the global temperature record data he could say with a "high degree of confidence" that there was "a cause and effect" relationship between the greenhouse effect and the "observed warming."

While Hansen and some other scientists have continued to see evidence of greenhouse warming in the climate record, computer-based global climate models have become the main source of claimed scientific support for the hypothesis Although the global climate models have never been validated, global warming has become a "hot" international issue. A media release by the United Nations Environmental Programme (UNEP) on the occasion of World Environment Day, June 5, 1991, warned that "the world is committed to unprecedented warming" and that "inaction . . . could quite literally cost us the Earth." It refers to the Intergovernmental Panel on Climate Change (IPCC) report as "scientific consensus" and calls for "immediate and drastic reductions, of the order of 60-80 percent in carbon dioxide emissions."

The UNEP release lists a number of catastrophes resulting from "vast changes in climate": savage and more frequent storms, severe flooding "as seas move inland," drops in crop yields, disruption of marine ecosystems, disease and famine, and tens of millions of environmental refugees.

—*S.F.S.*

Another Ice Age coming?

S. Fred Singer, Roger Revelle, and Chauncey Starr

Global temperatures have generally been declining since the dinosaurs roamed the earth some 70 million years ago. About 2 million years ago, a new "ice age" began—most probably as a result of the drift of the continents and the build-up of mountains. Since that time, the earth has seen 17 or more cycles of glaciation, interrupted by short (10,000 to 12,000 years) interglacial or warm periods. We are now in such an interglacial interval, the Holocene, that started 10,800 years ago. The onset of the next glacial cycle cannot be very far away.

It is believed that the length of a glaciation cycle, about 100,000 to 120,000 years, is controlled by small changes in the seasonal and latitudinal distribution of solar energy received as a result of changes in the earth's orbit and spin axis. While the theory can explain the timing, the detailed mechanism is not well understood—especially the sudden transition from full glacial to interglacial warming. Very likely an ocean-atmosphere interaction is triggered and becomes the direct cause of the transition in climate.

The climate record also reveals evidence for major climatic changes on time scales shorter than those for astronomical cycles. During the past millennium, the earth experienced a "climate optimum" around 1100 A.D., when Vikings found Greenland to be green and Vinland (Labrador?) able to support grape growing. The "Little Ice Age" found European glaciers advancing well before 1600 and suddenly retreating starting in 1860. The warming reported in the global temperature record since 1880 may thus simply be the escape from this Little Ice Age rather than our entrance into the human greenhouse.

Excerpted from "What to Do about Global Warming: Look Before You Leap," *Cosmos*, vol. 1, no. 1, (1991): 28-31.

Roger Revelle is professor of science and public policy at the University of California at San Diego. He is a former director of the Scripps Institution of Oceanography.
Chauncey Starr is president emeritus of the Electric Power Research Institute and former dean of engineering at the University of California at Los Angeles.

this obvious disagreement by blaming it, again, on natural variability of climate. It cautions that "the unequivocal detection of the enhanced greenhouse effect from observations is not likely for a decade or more."

In an interview published in *World Climate Change Report* (May 1990), IPCC chairman and University of Stockholm meteorologist Bert Bolin said, "It is impossible to distinguish the [greenhouse] change from natural [climate] variability."

Yet environmental-political activists insist on immediate controls on greenhouse gas emissions.

Why no greenhouse warming?

The disagreement between *detailed* model calculations and the climate record is even worse than just described. Livermore Laboratory meteorologist Hugh Ellsaesser and others have examined the climate record for model-predicted "fingerprints" that can distinguish greenhouse warming from natural climate warming.

Among the findings:
● Models predict greatest greenhouse warming at the highest latitudes, but the north polar zone has actually cooled.
● Models predict more rapid warming of continents than oceans—and therefore of the Northern Hemisphere than the Southern—but the data show the opposite.
● Models predict that greenhouse warming will increase with altitude in the tropics, and decrease with altitude in the polar regions, but again there is disagreement with observations.

In view of this lack of validation, Ellsaesser concludes that present computer models cannot be relied on to predict future climate change. That does not mean that models will not improve, but this improvement cannot happen overnight. Accurate prediction will require faster computers, and, above all, a better understanding of physical processes in the atmosphere and ocean.

Why *hasn't* there been evidence for greenhouse warming in response to the increase in greenhouse gases? In addition to the mitigating effects of cloud cover increase due to the burning of fossil fuels and biomass, the most important greenhouse gas is not carbon dioxide but water

5. RESOURCES: Air

vapor—which is not directly controlled by human activities.

None of the existing computer models incorporate the role of water vapor properly. As Ellsaesser and Massachusetts Institute of Technology professor Richard Lindzen have pointed out repeatedly, water vapor could act in a way that overcomes the warming effects of the greenhouse gas increase. The IPCC report, apparently starting with the conclusion that the climate record supports computer predictions, chooses to ignore this crucial issue.

Ellsaesser and Lindzen are among several prominent scientists interviewed for the British TV production "The Greenhouse Conspiracy." The show makes the scientific case against the IPCC report and against current global warming hysteria generally. Unfortunately it has not yet been shown on U.S. television even though many programs holding an alarmist view have been aired.

Impacts of climate change

But assume the most likely outcome—a modest general warming of less than 1°C in the next century, mostly at high latitudes and in the winter. Is this necessarily bad?

One should perhaps recall that only a decade ago when climate cooling was a looming issue, government-funded economic studies arrived at a huge national cost associated with such cooling. More to the point perhaps, actual climate cooling, experienced during the Little Ice Age or in New England's famous "Year Without a Summer" in 1816, caused large agricultural losses and even famines.

If cooling is bad, then a modest warming should be good, it would seem—provided the warming is slow enough so that adjustment is relatively cost free. Crop varieties are available that can

benefit from higher temperatures with either more or less moisture.

U.S. Department of Agriculture scientist Sherwood Idso and other agricultural experts agree that as atmospheric carbon dioxide increases, plants will grow faster and need less water. The

warmer night temperatures, found in the analysis of U.S. climate data, translate into longer growing seasons and fewer frosts. And increased global precipitation should also be beneficial to plant growth.

Keep in mind, finally, that year-to-year changes at any loca-

THE CLIMATE RECORD

Average Annual Temperatures

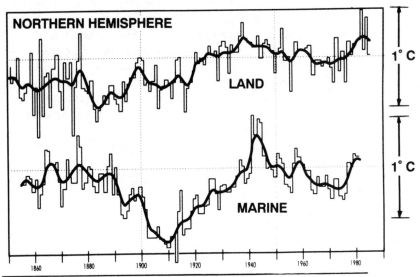

SOURCE (LAND): JONES, T.D., ET AL. "NORTHERN HEMISPHERE SURFACE AIR TEMPERATURE VARIATONS 1851-1984." J. OF CLIMATE AND APPLIED METEOROLOGY, VOL. 25, PP. 161-79, 1986.
SOURCE (MARINE): FOLLAND, C.D., ET AL. "MARINE TEMPERATURE FLUCTUATIONS, 1856-1981." NATURE, VOL. 310, PP. 670-3, 1984

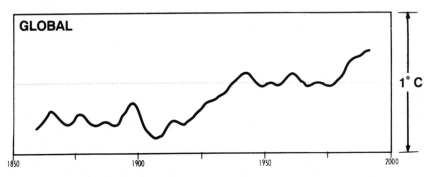

IPCC REPORT. CAMBRIDGE UNIVERSITY PRESS, 1990.

Reading the climate record requires decisions about what data to use for compilations—at least until uniform and accurate long-term weather satellite data become available.
■ *Top:* **Separating the data from land and from sea produces two significantly different curves for the Northern Hemisphere. Note the major increases before 1938, the sustained cooling to 1975, and the sharp rise to 1980.**

■ *Above:* **The global average temperature curve, as used in the IPCC Policymakers Summary (1990), shows temperature variations quite different from the consistent temperature rise predicted due to the increase of humanly produced greenhouse gases. By contrast, the U.S. temperature record, which is corrected for urban heat island effects, shows highest annual averages in the 1930s rather than in the 1980s.**

tion are far greater and more rapid than what might be expected from greenhouse warming. Nature, crops, and people already are adapted to such changes.

It is the *extreme* climate events that cause the great ecological and economic problems: crippling winters, persistent droughts, killer hurricanes, and the like. But there is no indication from modeling or from actual experience that such extreme events would become more frequent if greenhouse warming ever becomes appreciable, despite the media claims.

On the contrary, climate models predict that global precipitation should increase by 3-11 percent and polar temperatures should warm the most, thus reducing the driving force for severe winter weather.

And what of predictions of rising sea level and catastrophic flooding as glaciers melt? Such fears are often discussed in the tabloids. Mountain glaciers and the polar ice caps certainly contain enough ice to raise sea level by 100 meters. Conversely, during recent ice ages enough ice accumulated to drop sea level 100 meters below the present value.

But these are extreme possibilities. Tidal gauge records of the past century suggest that sea level has risen modestly, about 0.3 meter (1 foot). But the gauges measure only *relative* sea level, and many of their locations have dropped because of land subsidence.

Besides, the locations are too highly concentrated geographically, mostly on the U.S. East Coast, to permit global conclusions. The situation will improve greatly, in the next few years as precise absolute global data become available from a variety of satellite systems.

In the meantime, satellite-radar altimeters have already given a surprising result. As reported recently by NASA scientist Jay Zwally, Greenland ice sheets are gaining thickness—a net increase in the ice stored in the cryosphere and an inferred *drop* in sea level—leading to predictions with a reduced rate of sea level rise.

The available evidence, then indicates that even if significant warming were to occur in the next century, the net impact might well be beneficial.

GAZING INTO OUR GREENHOUSE FUTURE

*With supercomputers, scientists attempt to untangle the knot of knots:
the workings of Earth's atmosphere.*

Jon R. Luoma

America's Temple to Science sits, spectacular and imperious, all alone above Table Mesa, up a serpentine road from the university town of Boulder, Colorado. The Temple, more properly known as the headquarters of the National Center for Atmospheric Research, constructed in apparent self-tribute by the National Science Foundation, is at once futuristic (Woody Allen shot much of the 22nd Century spoof Sleeper here) and—all quarried stone and soaring towers—oddly cathedralish. Craning one's neck up at the NCAR headquarters from its parking lot below, one is struck nearly reverent. Introibo ad altare dei.

One bright autumn afternoon, in the glistening lobby of the Temple to Science, I strode beside Steve Schneider, on the way from his cluttered office to his Volvo outside. No, that's not right. I didn't quite stride. Schneider, a sort-of human vortex who makes tag-along visitors feel like they've been thrust into a renegade revolving door, was striding. I, notebook in hand, was asking questions and intermittently trotting. I was nearly keeping up. When Hollywood comes back to the Temple to produce its epic film about global warming, it might as well go ahead and put Schneider himself in the role of the brave super-scientist. If anything, he seems too Hollywood to be true: energetic, charismatic, as polished as an anchorman, with lean good looks right out of central casting: James Woods without the angst.

One of NCAR's—and Schneider's —key missions is to untangle the knot of knots: the inhalations and exhalations of Gaia, the workings of the planet's atmosphere. Schneider and his colleagues here at the Temple are among the world's leaders in trying to untangle the knot. And oh, it is tangled. Is the Earth really going to heat up? How much? How fast? To what effect?

NCAR is one of a handful of top purchasers of high speed supercomputers in the world. (In roughly the same league as the cryptic spooks in the National Security Agency.) Supercomputer power is measured in "gigaflops," or billions of calculations per second, meaning that within about fifteen minutes a supercomputer can do as much arithmetic as a crack mathematician with a calculator, working full time, can do in—oh—roughly one ice age. And it is in part through such muscular electronic number-nuking technology that NCAR hopes to untangle the knot. Schneider, when he's not on the international symposium tour showing slides to scientific colleagues and unwashed media types, spends his time at the NCAR Temple thinking about global atmospheric chemistry and all manner of arcane gobbledygook having to do with these supercomputers, in particular something called Global Circulation Models, or GCMs.

This article will get around to telling you more about GCMs than you probably want to know. To Schneider and men and women of his ilk, they matter a great deal when it comes to gazing into what may well be our greenhouse future. But to me, the way Schneider treats guests in his car has more to do with the heart of global warming. As he drove, just a tad too fast, down the mesa's serpentine road toward a picnic in Boulder he had arranged for a squadron of visiting scientists, Schneider noticed that I wasn't wearing my seatbelt.

"I'd like you to fasten it," he said, with a sideways glance, and then a grin. "I'm a risk aversive person. Especially when the ratio between cost and risk is so great."

IN THE HEARTS AND MINDS of the American news media, and hence the American public, global warming became real in the hot, droughty summer of 1988. An illustration of a sweating family in a bell jar appeared on the cover of a national newsmagazine, and newspapers everywhere carried features on the topic.

But then, beginning late in 1989, global warming appeared suddenly to be a more dubious proposition. A major article in the business magazine *Forbes* attempted to debunk what it called "The Greenhouse Panic." Newspapers ran features suggesting that scientists had begun to reverse their course, or that global warming wasn't going to be all that bad, or that it might be good, or that even the experts were just plain confused.

Is it possible that global warming isn't a real threat after all—merely hazy speculation? Apparently the Bush Administration thinks so. At a major world climate conference held in the Netherlands late in 1989, President George Bush's science adviser, Allan Bromley, abjectly refused to sign a resolution that would have called for nations at least to stabilize emissions of "greenhouse gases" such as carbon dioxide (CO_2), methane (CH_4), and chlorofluorocarbons ($CFCs$). And again in 1990, at the Second World Climate Conference in Geneva, the United States balked, refusing to commit to specific cutbacks in emissions, even though a call for cutbacks had the support of most other industrialized nations and seven hundred scientists at the conference had declared that the

As the World Warms

GRAPHICS BY GARY COX FOR AUDUBON

Much of the global warming theory is based on accepted scientific knowledge. As solar radiation, or heat from the sun, approaches the Earth, it is either (A) absorbed by the planet, (B) absorbed by the atmosphere, or (C) reflected back into space by the atmosphere. Some of the heat that penetrates the atmosphere is radiated back toward space through (D) evaporation or in the form of (E) infrared radiation from the Earth. In the phenomenon known as the green-house effect, (F) carbon dioxide and other gases, along with (G) water vapor and clouds, help to maintain the global average temperature by (H) trapping some of this heat in the atmosphere.

Questions arise when human activities figure into the equation. In recent years the amount of carbon dioxide in the atmosphere has increased from emissions of CO_2 from factories, planes, and vehicle exhaust, and through burning of fossil fuels to heat homes and run vehicles. Deforestation also contributes to the buildup by releasing CO_2 held in trees. Some twenty other greenhouse gases, including chlorofluorocarbons, methane, nitrous oxide, and ozone, are also accumulating. Many scientists believe global warming is occurring because this buildup of gases causes (H) more heat to be radiated back to Earth, raising the average temperature.

Storms and a rise in sea level cause flooding.

Changes in temperature and decreases in rainfall wreak havoc on agricultural areas.

Unless regional rainfall increases with temperatures, deciduous, mountainous and rainforest areas are at risk.

Boreal forests are stressed as a result of rapid temperature rise.

191

time to act is now. The United States' stated reasons for foot-dragging come straight from the Ronald Reagan, Anne Burford, James Watt environmental bible: The issue needs more research.

But does it? The answer is yes, the issue really does need far more research. There is a conundrum, however: While most scientists readily concede that there is no absolute proof that the world will become drastically hotter, many experts, including Schneider, point to two decades of evidence that, absolute proof or not, the risk appears to be extraordinarily high. If we wait to deal with the problem while scientists look for a conclusive answer, the hothouse century may already be upon us.

My own look into the global warming story suggests that we are careening into the future not with a sober climatologist in a Volvo, but at extraordinary speed with a carbon dioxide-, methane-, and CFC-intoxicated teenager in a Yugo with loose steering. Probably on the wrong side of the expressway. And the driver appears to be nodding off. Maybe, just maybe, we'll survive this drunken joyride. But maybe we'd at least better think about fastening our seatbelts.

SCHNEIDER IS HAPPY TO disabuse anyone who will listen of a misconception or two. "The greenhouse effect," he says, "is not a 'controversial' theory. It is probably the best established principle in atmospheric science, and it has been well established for over one hundred years."

The greenhouse effect, indeed, is responsible for warmth and life on Earth. It works like this: High in the atmosphere molecules of certain trace gases, principally carbon dioxide, allow heat radiation from the sun to filter through to the Earth, while also trapping near the Earth the infrared heat energy that is attempting to escape into space. Just like the panes of glass in a greenhouse, these gases keep the lower atmosphere of the Earth far warmer than it would otherwise be. Without them, in fact, temperatures on Earth would roughly approximate the refrigeration of Mars.

No one, not even White House Chief of Staff John Sununu, can dispute that. And no one can credibly dispute that levels of greenhouse gases

Negative Feedbacks

Positive Feedbacks

in the atmosphere are steadily rising. The most compelling evidence of that comes from the CO_2 monitoring gadgets atop Mauna Loa, on the big island of Hawaii, far from any industrial or major urban pollution source. Continuous monitoring on the mountaintop since 1958 has shown a steady increase in atmospheric CO_2, rising from 315 parts per million to about 350 ppm today, an 11 percent increase in only three decades.

Further, no one can dispute that adding more of these greenhouse gases to the atmosphere should cause the Earth to warm up significantly, all other things being equal. But there's the rub. All things are not equal.

It is possible, just maybe, that the Earth has enough built-in self-regulating resistance to atmospheric alteration that the planet will not be devas-

*Some "happy Pollyannas" are relying on **negative feedbacks** to provide resistance to the buildup of carbon dioxide, and so stave off global warming. They believe that (A) exploding algae populations in a warming ocean will absorb more carbon dioxide, (B) green plants will respond by stepping up photosynthesis and thus absorbing more CO_2, and (C) a warmer Earth will produce more cloud cover, reflecting more heat back into space.*

*Equally likely is the possibility of **positive feedbacks**, which complicates the scenario. A warmer ocean could (A) add more water vapor to the atmosphere, while a warming of the permafrost could (B) release more methane, as well as (C) decrease the amount of heat reflected back into space. A warmer climate might induce us to use more (D) air conditioning, further increasing CO_2 emissions.*

tated. That resistance comes from phenomena called "negative feedbacks." And the notion that negative feedbacks will greatly moderate the effects of greenhouse gases provides the entire basis for argument by those who deny that global warming is an actual threat.

Just what are these negative feedbacks? For one, trees and other green plants may respond to an abundance of carbon dioxide in the atmosphere by photosynthesizing at a higher rate, and in the process use up much of the excess carbon dioxide. Similarly, exploding populations of algae phytoplankton in the warming oceans may take in much more carbon dioxide; or a warming may produce more bright, white clouds, which will reflect more solar heat back into space.

The hitch is that there are also "positive feedbacks." For instance, enormous amounts of methane are believed to be trapped in the Arctic permafrost. Warming the permafrost may release some methane, which will cause more warming and release more methane, and so on. Shrinking ice sheets may mean that the Earth reflects less heat back into space. And a warmer climate may induce more air conditioning, causing more fossil fuel burning, causing more CO_2 emissions, causing more heat...

In any case a few scientists, pro-business policy-makers, chamber of commerce types, and other happy Pollyannas are relying almost entirely on negative feedbacks to balance, like magic, not only the straightforward warming itself but all the positive feedbacks as well. Although they can't come close to proving that they're right, maybe we'll get lucky.

OH, FOR THE GOOD old days when environmental poisoning was binary, on and off, black and white. Noxious fumes either made one sick or they didn't. DDT either caused eagles' eggshells to thin, or accumulated in mothers' milk, or it didn't. And there were straightforward ways to measure it. Today's scientific prognosticators, however, who are trying to determine just how serious a problem global warming might be, face surprising limitations.

Researchers attempting to untangle the climate knot, to sort out the positive and negative feedbacks, can't do much by way of standard lab work. They are, in fact, left with only two

categories of investigatory tools. One is history. The other is math and the Global Circulation Model.

History has been of some help. Researchers studying air bubbles trapped in glaciers have been able to verify that CO_2 levels have risen about 25 percent as the world industrialized over the past century. As part of a remarkable project at the Soviet Vostok ice station in Antarctica, Soviet scientists have, over a period of years, drilled about a mile into a glacier to remove an ice core. The core is an accumulation of snow and ice layers that extend back in time, like the rings of an enormously ancient tree, for 160,000 years—to a time before the last glaciation began. Recorded in the trapped air bubbles is the atmosphere's chemical history. The study has confirmed that never in that enormous stretch of time have atmospheric CO_2 levels been as high as they are today.

But are those gas levels having a warming effect? Worldwide, the answer appears to be yes, sort of. On a global average, according to calculations by laboratories in both Great Britain and the United States, temperatures have indeed risen over the last century, although only by about one degree Fahrenheit.

Even evidence of that modest increase, however, has been challenged. There's no assurance, after all, that thermometers were properly calibrated in decades past. And in some cases, cities have grown up around village thermometers, with masses of heat-retaining concrete threatening to skew temperature readings. Or thermometers have been moved from an urban center to an airport.

Moreover, temperatures always have and always will fluctuate. Whole clusters of years can be cooler or warmer than normal, greenhouse effect or no greenhouse effect. The challenge for statisticians is to try to discern a strong warming trend amidst all the normal fluctuation—to be like a sonar operator on a submarine, trying to sort out true signals from the background noise. It's not an easy distinction to make. After reaching a peak in the 1940s, the entire globe began a minor cooling trend that lasted for more than twenty years. Then, between the mid-1970s and the mid-1980s, although there was an unprecedented warm-up worldwide, temperatures in Europe actually cooled by about one-half de-

gree. And in 1989, Thomas Karl of the National Oceanic and Atmospheric Administration (NOAA) reported that no significant historical warming trend had occurred at all in the United States in the past century.

But it appears that Karl's findings may be noise amidst a stronger signal —the United States represents only about 1.5 percent of the globe's surface. Other recent studies that have attempted to compensate for the growth of "urban heat islands" continue to point to a temperature increase as carbon dioxide levels have increased.

The increase of a single degree noted by the British and American laboratories, to be sure, is not much. But most scientists looking for a correlation between rising greenhouse gas emissions and rising temperatures find reassurance, if you can call it that, in the apparent trend.

Mathematical modeling has its own host of problems. Essentially, a computer inhales all of the known data about climate, interactions between chemicals in the atmosphere, physical factors such as the location, size, and mass of land versus oceans, and thousands of other factors—the configuration of icefields, the exchange of heat between oceans and air, the metabolism of plankton in the sea and all the plants in a forest, projected human population growth, and industrialization. Such a Global Circulation Model then proceeds, through billions of calculations, to develop an atmosphere in a bottle—or, in this case, an atmosphere on a computer printout.

If the GCM works, a researcher should be able to alter one factor, say the power of radiation from the sun, and see precisely how it would affect temperature, rainfall, and winds around the planet in four dimensions, including both space—latitude, longitude, and altitude—and time. To gaze into the greenhouse future, modelers would simply tweak their GCM atmosphere by adding hypothetical amounts of CO_2 or other gases. (This, deliciously, is referred to as "perturbing the model.") If the GCM is a valid one, it should show in detail what happens to the world's climate as the concentration of greenhouse gases increases.

It's a great idea. The problem is that it really doesn't work very well, just yet. In fact, according to Schneider's recent book, *Global Warming*, it was only in the past decade that modelers delighted themselves by proving, with their zillion-dollar computers and sophisticated models, that winter is colder than summer! (Actually not a completely frivolous notion. If the models can be proven to work for such gross and easily observed phenomena, it greatly raises confidence in their ability to peer into the future.)

When it comes to focusing on finer details, however, the problems faced by the computer modelers are monumental. Models "are marvels of mathematics and computer science," argues Reid Bryson of the University of Wisconsin, "but rather crude imitators of reality." Bryson, a sharp critic of those who suggest signs of global warming are evident from current computer models, says, "The models are still very primitive. What it really gets down to is that this whole hubbub about global warming is theoretical. The theory may be correct. But where's the empirical evidence that it's correct and complete?" The models, according to Bryson, are still too unsophisticated to present an adequately accurate picture of the future.

Modeling and predicting a single physical and chemical reaction is no big problem—not much more complicated than calculating the trajectory of a billiard ball rebounding from a bank on a pool table. But modeling the planet's entire atmosphere is more like attempting to calculate the trajectories of a few billion racquetballs in a planet-sized room, where millions of balls are colliding, rebounding, glancing off each other and all six surfaces, all at different speeds, trajectories, and spins. It's worse than that. Some of the speeds and spins are unknown. Some of the balls are oddly balanced. Worse, some are invisible.

Although they are complex and techno-wondrous, present models can still provide only a coarse description of future events. None are able to resolve their mathematical focus tightly enough, for example, to take fully into account the apparent negative feedback of the net cooling effects of clouds. (This is complicated by the fact that scientists still aren't yet dead certain whether clouds actually cause cooling or warming.) Conversely, none have fully taken into account the positive feedback hypothesis forcefully advocated by scientist George Woodwell of the Woods Hole Research Center.

According to Woodwell's analysis, increased soil temperatures will greatly increase the activity of decomposer bacteria in soils, which would release vast amounts of CO_2, a process that former U.S. Environmental Protection Agency scientist Daniel Lashof has called a "sleeping giant" in the greenhouse debate.

To make matters worse, there is some dispute about whether even a supercomputer vastly faster than anything now available could adequately model the planetary climate, even if every bit of important data could be located and entered. Pointing to "chaos theory," some theorists have speculated that the globe's complex climate might have several equilibrium states, randomly variable even with the same inputs, as if a cookbook recipe randomly produced a birthday cake one time and liver and onions the next.

SO IF WE DON'T know what's going on, why should we worry, and why should we act?

Steve Schneider is passionate on the topic. "Despite the uncertainties, there is very little disagreement that large changes in the global climate are possible and even probable," he argues. "And even an increase of one or two degrees is not trivial."

Indeed, even the current lower limit for how much the planet will warm in the next three to four decades—three to four degrees Fahrenheit—suggests great changes: surges in sea level from thermal expansion of the oceans, with dire effects on wetlands and low-lying coastal regions, and vast alterations in global rainfall and drought patterns, with devastating impacts on agriculture and endangered species.

Evidence from both the historical and modeling perspectives suggests that adding greenhouse gases to the atmosphere will force the climate to heat up. None of the most sophisticated GCMs indicates that the temperature will remain the same if we continue to perturb the air around us. None of the historical data suggests that the globe is cooling or maintaining status quo.

The extent to which the world may warm up has always been presented by modelers as a range, from a worst case to a better case. The worst case was that the Earth would warm by as much as nine degrees by the middle of the next century. By 1990 many scientists ap-

peared to be moving toward a consensus that the extent of the warming may well be closer to the better case—a total warming of "only" three or four degrees by about 2050. And even those who had suggested that negative feedbacks might squeak us through unscathed now agree that some warming appears inevitable. Even Massachusetts Institute of Technology Professor Richard Lindzen, who has long been one of the most vocal skeptics, conceded at a January 1990 symposium that some warming now appears inevitable.

"How much uncertainty is enough to prevent policy action based upon a climate model?" Schneider once asked his assembled colleagues. "The dilemma rests...in our need to gaze into a very dirty crystal ball. But the tough judgment to be made is precisely how long to clean the glass before acting on what we think we see inside."

Is there still a possibility that the scientists are missing something, that there's yet another negative feedback hiding in nature that will neatly balance the perturbation? That would be nice.

More than twenty years ago, ecologist Barry Commoner suggested that perturbing natural ecosystems was analogous to opening up the back of a fine watch and jamming a screwdriver into the intricate works. Doing so leads to only a few possible outcomes. There is always the chance that the screwdriver will slip through and do no harm. There is even the chance that the screwdriver will knock some mechanism that was misaligned back into place. Common sense, however, suggests that perturbing the inner workings of the watch will virtually always do harm.

ANTI-REGULATION advocates are already howling about the potential economic horrors from a large-scale effort to control global warming. "Under a global warming scenario, EPA would become the most powerful government agency on Earth, involved in massive levels of economic, social, scientific, and political spending and interference," screeched writer Warren Brookes in *Forbes* magazine in 1989. Taking action to prevent global warming, he gloomily suggested, "could well spell the end of the American dream for us and the world." (The nuclear power industry, to the contrary, appears to believe that global warming may be the best thing since Enrico Fermi, since concerns about burning fossil fuels could lead the world to reconsider the atom.)

The Bush Administration has even raised the specter of global injustice, suggesting, in a finessed bit of Double-think, that America's justification for not agreeing to greenhouse gas limits was concern about the effects on energy production, and hence economic development, in the Third World. This, even in light of the fact that the Bangladeshes and Indonesias of the world are threatened with horrendous flooding, drought, and other costs from a warming created, most of all, by energy piggery in the industrialized nations, most notably this one.

Meanwhile, evidence abounds that, at the very least, some significant steps could be taken to slow global warming while simultaneously providing actual economic and political benefits to the United States. Schneider, in fact, has in recent years been promoting a "tie-in strategy," suggesting that we at least take actions which would "provide widely agreed societal benefits even if the predicated [global warming] did not materialize."

Consider, for example, that our major economic competitors, Germany and Japan, now use about half the energy per unit of Gross National Product that we in the United States do, with correspondingly lower greenhouse gas emissions. Consider that we in the United States alone, on our little 1.5 percent of the planet's surface, with roughly 5 percent of the planet's population, produce some 26 percent of its carbon dioxide.

Consider that between 1973 and the early 1980s, facing the economic penalties of an energy crisis driven by the stranglehold on oil by an international oil cartel, Americans quietly slashed energy consumption by 25 percent, to the extent that we're now saving some 13 million barrels of oil every day—and the carbon dioxide associated with burning that oil. (If we'd kept on saving energy at that rate, we wouldn't have needed any oil from the Persian Gulf after 1985, according to Amory

---

1990: WARMEST YEAR ON RECORD

Last January scientists in the United States and Great Britain dropped a grenade into the global warming debate when they simultaneously released data showing that 1990 had been the warmest year ever recorded.

A report from NASA's Goddard Institute for Space Studies on data collected from about 2,000 sites showed a worldwide average temperature of 59.8 degrees Fahrenheit, the highest in the 140 years such records have been collected. A study by the British Meteorological Office exactly corroborated the NASA study, with both showing an average warming of .09 degrees for the year. The previous record was set only in 1988.

Any given year has the potential to be the hottest, but scientists point to a more disquieting fact: The dubious 1990 record comes after a string of extraordinarily warm years. Globally, since 1980, the world has suffered six of the seven warmest years on record.

Skeptics in the scientific community continue to point out that temperatures always have and always will fluctuate, and that the new data is more likely statistical "noise," rather than a real trend.

James Hansen of the Goddard Institute disagrees. "In our opinion, the case for a cause and effect relationship between global warming and the greenhouse effect is become harder to deny," he said in his report on the data.

For Rafe Pomerance, a global climate specialist with the World Resources Institute, the new evidence suggests that it is time for politicians to "reverse the burden of proof." He points out that, if indeed global warming is in progress, absolute proof will be exceedingly slow in coming. "But this is more evidence that it is prudent to assume we're going to have global warming, and that we should act accordingly." —J.R.L.

---

and Hunter Lovins of the Rocky Mountain Institute.) Most of these savings came without either vast pain or vast effort but merely enlightened response

to economic pressures by business, institutions, and industry.

Consider also that a host of energy experts now firmly believe that further cuts in fossil fuel emissions through the use of existing energy-efficient lighting, heating, and motor technologies could greatly reduce energy consumption, and hence greenhouse gas emissions. Merely providing the incentives to switch America's businesses and homes to existing energy-efficient light bulbs could cut electrical consumption in the United States by as much as 12.5 percent, and save literally billions of dollars in the process.

Providing incentives for such alternative energy sources as solar photovoltaics could, like efficiency improvements, carry the positive multiple-whammy of abating global warming while simultaneously providing a host of other benefits: reducing our dependence on politically and economically unstable foreign fuels; abating acid rain and other forms of air pollution; avoiding oil drilling and disastrous spills in sensitive areas; and helping to wean our dependency on the whims and armies of oil-controlling megalomaniacs.

Likewise, slowing global deforestation and accelerating reforestation would increase the capacity of forests to consume carbon dioxide, while helping to ease an extinction crisis facing many of the planet's animal and plant species, particularly in the tropics. At the same time, slowing deforestation would vastly reduce CO_2 additions to the atmosphere from burning trees and slash.

None of which would solve global warming. But it would put on the brakes. "Slowing down the system would be a major victory," Schneider insists. "If we could 'only' cut emissions in half, we'd buy a great deal of time to study global warming and come up with better solutions. And if all our hypotheses are wrong—if it turns out to be an infrared herring—we will still have improved problems like acid rain, urban air pollution, and overconsumption of fossil fuels."

In other words, the least we could do is fasten our seatbelts.

Biosphere: Endangered Species

- Plants (Articles 31–33)
- Animals (Articles 34–35)

One of the greatest tragedies of the conservation movement is that it began after it was already too late to save many species of plants and animals from extinction. In fact, even after concern for the biosphere developed among resource managers, their effectiveness in halting the decline of herds and flocks, packs and schools, or groves and grasslands has been limited by the ruthlessness and efficiency of the competition. Plants and animals compete directly with human beings for living space and for other resources such as sunlight, air, water, and soil. As the historical record of this competition in North America and other areas attests, since the seventeenth century settlement has been responsible—either directly or indirectly—for the demise of many plant and wildlife species. In addition to habitat destruction, other factors that have contributed to the decline of animal, bird, and plant populations include the pollution of waterways, unnecessary overkills, the inability of indigenous animals to compete with domestic stock, and competition between wild plants and their crop cousins. It should be noted that extinction is a natural process—part of the evolutionary cycle—and is not always created by human activity. But human actions have the capacity to accelerate a natural process that might otherwise take millennia.

In the closing years of the twentieth century, it has become ever more apparent that the problems will continue for the biosphere, and that the short-term benefits of human activities will continue to clash with the long-term values of environmental quality. Nowhere is this more clear than in viewing the implications of human activity for modifications in global biological systems, with attendant consequences for the interlocking web of all life on Earth.

In the lead article of the unit, consultant Otto T. Solbrig of the UN's Man and the Biosphere Program describes his biodiversity from a functional point-of-view. The tremendous variety in nature has always astonished and delighted humans, Solbrig notes, but the current rate of loss of species diversity seems sure to reduce this immense richness and seriously affect human welfare in the process. For all the importance of biodiversity to the health of the global system, relatively few scientists are studying biodiversity as a phenomenon. At a time when the conceptual and practical importance of biodiversity is becoming clear, there is a shortage of trained personnel and a lack of necessary funds with which to pursue research. Increased public awareness of humanity's role in biodiversity depletion is necessary if research efforts to learn more about biodiversity and how to maintain it are to go forward.

The most important component of any biosphere is the primary production component of living vegetation, and in the three articles in the *PLANT* subsection the issue of human impact on these most important biological systems is central. In the first article, authors Nigel J. H. Smith, J. T. Williams, and Donald L. Plucknett discuss the fate of the tropical forests. "Conserving the Tropical Cornucopia," an article based on a new book by Smith, Williams, and Plucknett, approaches tropical forest clearance from the standpoint of a serious problem that has been largely ignored: the shrinking population of plants in the wild, which includes the near relatives of many species that yield food, beverages, oils, latex, spices, timber, and fuel wood. More than just contributing to global climate change, regional soil erosion, or the loss of plants of medicinal value, tropical forest clearance is destroying enormous reserves of highly valuable "crops" of both subsistence and commercial value.

Deterioration of the world's forest reserves is far from just a problem of the tropical regions, however, and in the final two articles of the subsection, midlatitude forest zones are the subject of scrutiny. In "Timber's Last Stand" John C. Ryan of the Worldwatch Institute describes the need for a "New Forestry" that could help save a destructive industry from itself. Current commercial forestry practices rely on clear-cutting—the wholesale removal of all trees from a selected area. Ryan argues that this management strategy is shortsighted in both an environmental and an economic sense. A "new forestry" based on selective cutting and maintenance of forest diversity is necessary, according to Ryan, if the forest industry is to be continued, and if jobs of those involved in the timber industry are to be saved. The current confusion over the management of the world's mid-latitude forests, typified by the debate between clear-cutting and selective cutting proponents, exemplifies the crucial role that human management—or the lack of it—can play in the world's vegetative systems. The long-term picture for the stability of vegetative regions, either in the tropical or mid-latitude world, is certainly not a very optimistic one, given the direction of recent events. A similar point of view is expressed by Jeffrey L. Chapman, a research assistant to the Native Forest Council, in "Forests Under Siege." Chapman notes the inconsistency of an American public clamoring for preservation of tropical forestlands while allowing the rapid clearance of the temperate zone's last rain forests in the federally owned national forests of the Pacific regions of the United States. The environmental impact of lumbering in the old-growth forests of Alaska, Washington, and Oregon is no less serious than that of deforestation in the tropics. Unlike forest clearance in the tropics, much of which is for the purpose of clearing land for agriculture to grow food to supply impoverished and undernourished peasant farmers, the forest clearance in the United States is based on the timber's market value on the international market. These ancient forests, claims Chapman, provide us with clear water, clean air, wildlife habitat, and a tranquil sanctuary. The country should not let the search for profit lead it into the vegetative equivalent of the last buffalo hunt.

The prospect for the future of wildlife is not any better than that of the world's remaining native forestlands. Land developers destroy animal habitats as cities encroach upon the countryside. Living space for all wild species is destroyed as river valleys are transformed into reservoirs for the generation of hydropower and as forests are removed for construction materials and for paper. Toxic wastes from urban areas work their way into the food chain, annually killing thousands of animals in the United States alone. Rural lands are sprayed with herbicides and pesticides that also kill birdlife and small mammals.

In the first article in the *ANIMALS* subsection, environmental consultant Dick Pitman focuses on the competition between farmers and wildlife in Africa, home of much of the world's remaining great game herds. Much of the unique animal heritage of Africa has been and is in jeopardy because of the unchecked population growth on that continent and the rapid expansion of cultivated land that accompanies it. In Zimbabwe, president Robert Mugabe, proposed to make wildlife an agricultural option to complement crop production and cattle raising. Wildlife management would then be rationalized to bring economic benefits to the rural communities that engage in it, and game meat would be processed to supplement the beef supply in the local market. Such an agricultural option might well work in much of impoverished rural Africa, and the successful integration of wildlife management and agricultural policy could bring benefits to both humans and the animals who occupy the continent.

In "How the West Was Eaten," biologist and former park ranger George Wuerthner attributes the reasons for the growing list of endangered species in the American West to the competition between wildlife and the livestock industry. Rather than being a benign use of land, livestock raising has the potential to be enormously destructive in biological terms. Domesticated stock compete for the same vegetative resources required by herbivorous wildlife and, since their numbers are controlled by ranchers rather than environmental forces, domesticated stock are often overstocked and deplete the quality of the vegetative resources through overgrazing. Just as wildlife who compete directly with cattle and sheep are driven off the land by their human-aided competitors, so are predatory animals impacted by livestock ranchers who see dollar losses in every coyote, wolf, bear, mountain lion, or golden eagle. The irony of the "eating of the West" is that most of it takes place on federal, that is to say public, lands—belonging to all the people of the United States but managed as if it were the private fiefdom of the ranching industry.

The question of plant and animal extinction is much more than one of losing plants and animals that may make our lives richer in less-than-tangible ways by their very presence. Rather, the question is a broader one, relating to our own survival as a species. The stability of an ecological system depends upon its diversity. As diversity declines, so does stability. Simplified ecosystems are much more fragile than complex ones. We do not fully understand the role played by plants and animals in all the world's natural systems, but we do know that it must be an important one. What damages will be wrought to natural systems if the important natural component of native species is reduced or removed? We cannot halt the process of extinction but we can moderate it. We must moderate it if we hope to maintain our precious biosphere resource. The preservation of plant and animal species may even be necessary to our own continued existence, even if that preservation clashes with our own economic and social objectives.

Looking Ahead: Challenge Questions

What is meant by the term biodiversity, and how does it relate to the welfare of humans? How is biodiversity being affected by human activities?

What are some of the viable economic uses of tropical forest vegetation? How can tropical forests be utilized for economic purposes valuable enough to allow them to be saved rather than cut for agricultural land?

What is the "new forestry," and how does it relate to the maintenance of existing forest reserves? Are there connections between the concept of the "new forestry" and that of the sustainable ecosystem?

Why are native or old-growth forests in the western United States being cleared when the U.S. market does not require the timber?

How can the conception of wildlife as a crop or harvestable resource be applied to agricultural policy and wildlife management in Africa to help preserve that continent's threatened species?

What is the nature of competition between domesticated livestock and wildlife on public lands in the American West? Are there ways other than direct competition in which wildlife suffers from the presence of a livestock industry?

The Origin and Function of Biodiversity

Otto T. Solbrig

OTTO T. SOLBRIG is the Bussey Professor of Biology at Harvard University and a consultant on biodiversity with the Man and the Biosphere Program of the United Nations Educational, Scientific and Cultural Organization.

There is an enormous variety of plants and animals on Earth, and the high degree of species richness in tropical forests and coral reefs is a marvel. But even in environments that are relatively species poor, such as the ocean bottom, hundreds of kinds of microorganisms, plants, and animals flourish. This tremendous prodigality of nature has astonished and delighted humans and is ultimately the source of their sustenance. Yet, in using plants and animals for food and clothing, for building houses and for medicine, humans are endangering this immense richness of species. The current rate of loss of species diversity and the reduction in the genetic variety of crops and wild species could seriously affect human welfare.

People alter the biodiversity of the Earth in both direct and indirect ways. The use of renewable resources, for instance, often directly decreases species diversity. This is especially true in extractive industries such as forestry or fisheries, which tend to overexploit useful species and destroy unwanted ones. Agriculture and animal husbandry also destroy or modify the native biota. In addition, people indirectly change biodiversity by burning fossil fuels and biomass for energy, by altering hydrological patterns, by intentionally or accidentally introducing exotic species that reduce interregional biodiversity; and by destroying hedges, forest fringes, and fallow lands that provide habitat. A new and powerful human threat to species diversity is the release of toxic chemicals, such as lead, mercury, fluorocarbons, and chlorinated pesticides, into the atmosphere, soil, rivers, lakes, and oceans.

The task of describing, cataloguing, and explaining the diversity of living organisms belongs to biologists. But because the number of organisms on the planet is estimated to be 3 to 8 million or possibly higher,[1] and fewer than 1 million of these have been described and catalogued to date, this job is far from complete. The task is so formidable, in fact, that it may never be completed. Nevertheless, it is important to seek patterns in the distribution and abundance of species and to test hypotheses about the function of species diversity in ecosystems.

Today, scientists have only a very rudimentary knowledge of biodiversity. In many temperate and arctic areas, where there are comparatively few species, acceptable catalogues of the vascular plants and vertebrate animals do exist. There are also reasonable estimates of invertebrate animals and nonvascular plants, including fungi. Less known in these areas are soil organisms, bacteria, and viruses. On the other hand, in tropical regions, where there is a high degree of species diversity, the story is very different. There are no detailed lists of the plants and animals of many tropical countries. At best, there are very rough estimates of the sorts of animals, plants, and fungi that live in tropical regions. Most tropical insects, soil organisms, bacteria, and fungi have yet to be collected and described.

Marine species, both temperate and tropical, are probably the least known. The oceans have the highest diversity of animal and plant phyla, yet until recently, the deep sea was believed to be devoid of life. Today, scientists know the ocean floor has an abundant biota with more than 800 known species in more than 100 families and a dozen phyla. Ocean-bottom hydrothermal

CENTER FOR MARINE CONSERVATION—MICHAEL WEBER

From *Environment*, Vol. 33, No. 5, June 1991, pp. 16-20, 34-38. Reprinted with permission of the Helen Dwight Reid Educational Foundation. Published by Heldref Publications, 1319 Eighteenth St., N.W., Washington, DC 20036-1802. Copyright © 1991.

vents, such as the sulfide chimneys called "black smokers," have been found to contain at least 16 families of invertebrates that were unknown just five years ago.[2] Recently, an entirely new set of unicellular organisms called picoplankton, with cell diameters between 1 and 4 microns, was discovered. The productivity of marine systems may have been underestimated by 50 percent because scientists were ignorant of the role played by picoplankton and had no appropriate methods of measuring them.

Another serious problem is that scientists do not understand exactly how the diversity of genes, genotypes, species, and communities influences ecosystem function. Over the past 100 years, geneticists, taxonomists, evolutionists, and ecologists have accumulated much knowledge about diversity. The information gathered attests to the importance of diversity for the proper functioning of many organisms and ecosystems. However, a comprehensive, rigorous, and general theory of biodiversity is lacking. Because the threat to biological diversity is now great, scientists must learn how living systems are influenced by changes in diversity. Given the rapid pace of landscape transformation worldwide, there is some urgency in obtaining this information. Knowledge of biodiversity is also very important for evaluating the impact of global climatic change.

Fossil records show that drastic environmental change has been a major cause of species extinction.[3] Species losses also derive from mutual interaction, such as competition and predation. Also, because the environment is in a constant state of transformation, some species are always being lost. Some changes in the physical environment are cyclical and repeated, while others are less predictable. (However, even cyclical changes are subject to chance.) In any case, it seems reasonable that genetic diversity provides organisms and ecosystems with the capacity to recuperate after change has occurred. The scientific evidence for this hypothesis is not conclusive, however.[4]

Equilibrium in Nature

An ideal state in which every element

Natural selection probably allowed longer-necked individuals of this African ruminant to survive and reproduce when environmental changes killed others.

is in equilibrium, or "the balance of nature," cannot exist. The weather changes constantly; the diversity of plants and animals fluctuates; mountains erode; and lakes get silted in. Yet, this idea of a balance, of an equilibrium in nature, has persisted. Conservation managers, for example, endeavor to reduce disorder and create undisturbed environments. Those scientists who support the notion of a balance of nature maintain that ecosystems, although not in balance now, are moving constantly toward equilibrium. These scientists argue that ecosystems are prevented from attaining balance because of exterior forces, which are called disturbances. Storms, floods, pests, outbreaks, fires, and human-induced changes are all examples of disturbances that are presumed to keep ecosystems from reaching equilibrium.

The concept of the balance of nature gained credence in the early 18th century when Isaac Newton introduced very successfully the notion that nature could be explained in terms of a few simple laws. According to the Newtonian view, the world is formed by elements that are simple and that respond to regular and deterministic dynamics.[5] Today, physicists believe that nature is complex, not simple, and that the Newtonian view is insufficient as a general explanation of how the universe func-

tions. The universe, physicists are finding, is not in balance. Disturbances and irregularities of all sorts are no longer seen as aberrations but as integral parts of nature. Physicists are also discovering that at the very origin of the cosmos, at the so-called big bang, a singular, irreversible, and complex universe was produced. Similarly, ecologists have been observing and documenting that most, if not all, ecosystems are not in balance.[6]

But why should scientists be concerned with whether an ecosystem is in equilibrium? The answer to this question is crucial for understanding biodiversity. Systems not in equilibrium behave very differently than do systems in equilibrium. Their behavior can even appear strange and mysterious. For example, when the source of a disturbance is removed from a system that is near equilibrium, that system is expected to return to its previous state. When, in an undisturbed forest, a gap is created by a landslide or a falling tree, new seedlings grow in the gap and restore the equilibrium. After a few years, it is almost impossible to tell that there once was a gap. But when the source of change is removed from systems that are not in equilibrium, they do not return to equilibrium. Instead, they adopt a new state. For example, when land used for agriculture in the Amazon for-

6. BIOSPHERE: Endangered Species

est is abandoned, it grows into a grassland or a savanna and does not return to the original forest. Natural disturbances are a necessary part of the forest ecosystem's function. For example, without gaps creating disturbances, most forests cannot renew themselves. And fires, storms, and hurricanes are and have been a part of life since the beginning of the planet; without them, ecosystems could not function properly.

Knowing about the features of nonequilibrium systems is important for proper ecosystem management. Because a nonequilibrium ecosystem does not necessarily revert to its previous state when the convulsion is removed, it is often impossible for humans to return ecosystems to their original condition. Therefore, one set of management guidelines will not suffice for all situations. The previous history of changes and the present disturbance regime will determine the consequences of any management.

Origin of Diversity

There are two conditions that cause population diversity. First, new genotypes are constantly cropping up in a population through mutation, recombination, and related genetic phenomena and through immigration of individuals, their gametes, or their propagules. Second, diversity in the population is eliminated by natural selection and lost through emigration of individuals. Every genetic variation, from gene mutations to entire species, will disappear eventually. This loss can be a very fast process, or the variants can survive for a long time. Species that have survived for extended periods include horseshoe crabs, which have been around for 200 million years, and cockroaches, which originated even earlier, in the Carboniferous period. The speed at which new variations originate in relation to the rate at which they are eliminated determines the actual diversity of the system.

In the process of gene mutation, all heritable diversity ultimately arises at the molecular level. Gene mutations are chemical changes that take place in the composition of the DNA (deoxyribonucleic acid) molecule, the chemical substance responsible for heredity.

DNA is remarkable in that it shows complex chemical behavior not expected from systems in equilibrium. The most remarkable characteristic of DNA is its ability to regenerate itself, its "autocatalytic" behavior.

The DNA molecule consists of four kinds of repeating units called nucleotides, which can be arranged in any sequential order. DNA has the curious ability to maintain its physicochemical integrity regardless of the order of the four nucleotides. That is, any DNA molecule behaves like any other of the same length, regardless of the nucleotide content. This characteristic of DNA makes life feasible because if only one arrangement were possible, or if the chemical stability of one arrangement were significantly different from that of others, then all DNA molecules would be alike.

The order of the nucleotides uniquely determines the characteristics of the chemical products made by DNA. But the order of the nucleotides in a DNA molecule can change, and these changes are called mutations. The regular appearance of mutations gives life its diversity.[7] Mutated molecules of DNA reproduce their changes and make modified enzymes, which, in turn, make altered cells. Mutated cells result in modified organisms.

Thus, new mutations permit the evolution of new characteristics by changing the structure and/or function of enzymes and other proteins. The characteristics of the enzymes and proteins are determined by those of the DNA molecule, but not the reverse—the "information" flows only in one direction. This flow contrasts with that of an ordinary chemical reaction, in which

The continuing destruction of the dense forests between Arizona and Argentina could threaten the survival of the ocelot and, thus, the equilibrium of the entire subtropical ecosystem.

The Serengeti plain in Kenya and Tanzania, home to a wide variety of vertebrates, is threatened by rapidly encroaching agricultural activity and growing human populations.

there are forward and backward reactions.

The appearance of mutations is a random process. There is no way to anticipate or predict what mutation is going to occur, only that some will. It is natural selection, however, that determines whether new mutations get established or are eliminated. The processes of mutation and selection can be looked at in two ways. Mutation and selection might be random, independent processes that are not influenced by the characteristics of the system, in which case no active process of diversity maintenance exists. On the other hand, a certain degree of variety may be required for living systems to function properly. If so, there may be system feedbacks that modify the rates of mutation and selection. Below a certain threshold, a system may collapse because of insufficient variety. According to this last view, diversity is actively maintained in a system.

Speciation

Mutation and natural selection generate genetic and morphological variation within a lineage.[8] Yet what is most distinctive in a community is the coexistence of different species, each acting independently and not sharing its DNA

or mutations. How, then, does the great variety of species originate?

Speciation is the process that separates genetic variations into distinct units, or species. In speciation, the original population of organisms with similar genes, called a gene pool, is divided into two or more gene pools. Each of these new gene pools then acquires a unique set of characteristics (cellular, tissue, organ, and organismic) through mutation and selection. (The mechanisms that create species have been a focal point of research in the field of organic evolution.[9]) Speciation follows most commonly from the physical division of a gene pool. The separation inhibits interbreeding between individuals in the two populations. In time, one or both populations change enough to prohibit interbreeding.

Because the fate of every species is to become extinct eventually, there must be an influx of new species for life to continue. If, presently, there are about 10 million species, and, on average, each species has a life span of 1 million years, then an average of 10 new species must originate each year. On the other hand, if the life span of a species is only 100,000 years (as many biologists ar-

gue), then 100 new species must originate each year. Scientists are often aware of a species' extinction, but the appearance of new species is difficult to observe. Because most species are insects and other small invertebrates, most new species fall into these two poorly studied groups. Thus, there are few well-documented cases of speciation.

Evolution and Resource Limitation

Certain individuals in a population survive and reproduce, while others die without leaving offspring.[10] To understand this phenomenon, scientists must first realize that there usually is not enough of some environmental resources (food, water, minerals) to go around. The availability of resources in the environment limits the rate at which individual organisms can reproduce and their ability to survive. Nucleic acid replication and cell, tissue, and organism growth and reproduction all require energy and materials, of which there are only limited amounts. Individuals that are efficient at harvesting resources have a higher probability of reproducing and surviving than do those that are inefficient. Mutations in DNA provide species with a constant input of physical and chemical variations, some of which improve the efficiency with which the species selects and harvests resources. New mutations may increase the efficiency of protein synthesis at the molecular level, or they may improve the water uptake by a plant at the organism level. Mutations may even increase the efficiency with which a mutualistic association between a plant and its specific pollinator develops at the community level.

DNA molecules that are dissimilar in the order of their nucleotides usually do not have characteristics that directly increase or decrease their own survival or reproductive capacity. Given the appropriate environment and resources, all DNA molecules reproduce at the same rate despite their nucleotide composition. DNA molecules manifest their distinctive survival rates through their effects on the living organisms in which they are embedded. These organisms, be they single cells living in pure culture or multicellular organisms living in complex communities, will inevitably differ in their ability to garner resources. Therefore, they will vary in their survival and reproductive capacities and, consequently, so will the DNA in their cells.

Natural selection is an optimizing process. It is a process that favors efficiency and produces adaptation. It is not a process that involves chance, like the appearance of mutations, but it must operate with the elements that it has. Thus, natural selection never makes a "perfect" organism. Because the environment is constantly changing, a perfect organism, if it could be produced, would soon cease to be perfect as the environment changed. Such systems, in which the optimum state is favored but cannot be attained or maintained, are called frustrated systems.

Diversity is the result of two opposite actions: the processes that produce new genotypes, new varieties, and new species and the processes that eliminate mutations, variants, and species from the system. Natural selection is primarily responsible for the reduction of biodiversity; it acts through differential reproduction and differential mortality. In other words, the probability of reproducing and the probability of dying are not the same for all organisms. Rather, reproduction and death are related to the organism's characteristics as well as to the environment in which that organism lives.

The Ecological Role of Biodiversity

How is species diversity linked to ecosystem structure and function, to the different ways in which organisms and populations interact with each other, and to those community properties that emerge from these interactions? Communities have structures and properties not possessed by the populations within them that are called emergent properties, which include trophic structure, stability, guild structure, and successional stages.

Many scientists believe that species diversity is essential for the proper functioning of communities and for the emergence of community-level properties. Just as many different DNA-encoded enzymes are needed for a complex organism to function properly, so,

scientists believe, are many kinds of species necessary to maintain community structure. But is any diversity sufficient, or are specific mixes of species necessary for the communities and ecosystems to function? This is a very old question in ecology, and two opposing views exist. One view is that a community is formed by the species that happened to arrive first—that the mix of species in a community is a matter of chance: "The vegetation of an area is merely the resultant of two factors, the fluctuating and fortuitous immigration of plants, and an equally fluctuating and variable environment."[11] According to the opposing view, "In any fairly limited area, only a fraction of the forms that could theoretically do so actually form a community at any one time. . . . The community really is an organized community in that it has a 'limited membership.' "[12]

Are these two very different views really irreconcilable? Not necessarily. If a community is defined as the sum of all the plants and animals that grow together in an area, certain patterns can

Oceans have the highest diversity of flora and fauna, and marine species, such as the tube sponge and flower coral, are among the least understood.

be observed: Areas with more than 1,000 millimeters of evenly distributed rainfall always contain a woodland; conifers prevail in areas of extremely low winter temperatures; trees in warm climates have broad leaves; and succulent plants are found in dry climates. The same is true for animals. The mix of species encountered in a community is not a random sample of all plants and animals in the world. Yet, under close scrutiny, differences that are difficult to explain can be found between similar localities. Ecologists cannot predict what species will occur in a given climate and soil type. Also, there have been cases where dominant species, such as the American chestnut, have been lost, with apparently little effect in the overall working of the biological community.

The community behaves in a "frustrated" fashion. Two types of processes are at work and neither is dominant. First, there is natural selection. Species in the community are constantly evolving to increase their ability to withstand the rigors of the environment. This process accounts for cold-hardy species near the poles, for grass-eating ruminants in savannas, and for fruit-eating bats in forests. The individuals that escape their predators and survive other environmental challenges pass on their characteristics to their offspring.

Second, there is chance. When the climate changes, characteristics that were once advantageous can become a burden. Diseases might be introduced, such as Dutch Elm disease, for which local species have no defense. Hurricanes, tornadoes, floods, and droughts are unpredictable and their frequency changes over time. Chance also plays a role in dispersal. For instance, a species might be absent from a community purely by chance. Thus, both natural selection and chance result in the steady coming and going of species through immigration, emigration, extinction, and gene mutation.

A large number of additional unanswered questions exist regarding the role of biodiversity in communities. For example, can more and more species be packed into a community, or are there upper and lower limits? A related question concerns optimal levels of di-

CENTER FOR MARINE CONSERVATION—MICHAEL WEBER

versity and the factors that control them.[13] Other issues include the role of different individuals within a population, different populations within a species, and so on.

Do the members of the community collaborate in the efficient use of energy and resources? In a variable environment, the existence of individuals with different characteristics may increase the ways in which the population can respond to change.[14] For example, if all plants of a species had similar water requirements, all of them would suffer water stress in any year that was drier than normal, and such periods would result in significantly reduced seed production. But if there is genetic variation, some individuals might perform above average each year within certain limits of environmental variation. Therefore, seed production would be satisfactory in both wet and dry years. Moreover, genetically adaptable organisms should survive in more variable environments than do genetically uniform populations.[15] Experience with crops shows that highly productive but genetically uniform varieties have more restricted environmental requirements than do less productive but more variable varieties. Also, plantations formed by uniform varieties are more susceptible to pest and disease outbreaks.

So, variability within a species seems to be important for long-term survival. Does the same hold for a community, which is formed of species that live together but do not have a common gene pool? Communities with high species diversity may also cope with long-term environmental fluctuations better than do communities with few species. However, the evidence is contradictory. Terrestrial communities in the climatically variable midlatitudes are less diverse than tropical communities in more uniform environments. Also, deep benthic communities are among the most variable communities anywhere though they exist in possibly the most equitable environment on the planet.[16]

Diversity and Niche Structure

According to modern ecological theory, every species in a community occupies a singular ecological niche, or a

Unfortunately, there are no detailed lists of the native plants and animals of many tropical countries.

role in the environment. A niche is thought of as a multidimensional space where each dimension is a characteristic of a species. Because every species has at least one physical or behavioral characteristic that separates it from other species, every species has a unique niche.

In principle, niche theory supplies the theoretical framework to explain the number and types of species in a community. Niche theory predicts that communities that vary in total resources or in the quality of those resources will contain more niches and, therefore, more species. Nevertheless, resources are sometimes partitioned differently in communities with very similar resources. In other words, communities may have large or small niches, which result in a few "generalist" species (those with broad niches) or many "specialist" species (those with small niches). Finally, communities with similar resources and niche sizes may still vary in the number of species because of different degrees of niche overlap. Niche overlap is the degree to which two species exploit the same resource.

Communities differ in niche size and overlap for a variety of reasons, such as differences in climatic stability and predictability, spatial heterogeneity, primary productivity, competition, predation, and degree of disturbance. Thus, niche theory is unable to predict the patterns of species diversity.

Using niche theory to understand why communities vary in number of species has proven very difficult in practice. Many studies have shown very elegantly how species partition various resources in space and in time,[17] but it has proven very hard to show the precise contribution of each factor. A key question is whether two species can have the same niche. In other words, is there species redundancy in a community? Present theory predicts that two species with identical niches cannot coexist, and empirical studies seem to verify this.[18] Still, the question of how similar two coexisting species can be remains to be answered.

Trophic Diversity

Additional unanswered questions concern the role of diversity in food webs. All organisms get their energy from the sun, either directly or indirectly. At the simplest level, cells of the green tissues of plants and some microorganisms capture the sun's energy. Organisms also need specific chemicals from the environment to develop. For instance, all plants obtain the materials for development from soil or water and the air. With these materials and energy from the sun, plants create energy-rich, complex chemical compounds, such as carbohydrates, cellulose, starch, proteins, fats, and wood. Animals ingest and extract the energy from these compounds and incorporate some of them into their bodies. Decomposers reduce them to simple ions that eventually are returned to the soil or water. In a cyclic fashion, then, energy and chemical substances move from the environment to plants, to animals, to decomposing microorganisms, and back to the environment. This trophic diversity, or "food web," is essential for the proper operation of ecosystems. Many studies have shown that the trophic structure of ecosystems is complex and varies from one ecosystem to another. Understanding how food webs function is essential for understanding the place of humans in the ecosystem,[19] because humans also derive their energy and materials from plants and animals. Moreover, interpretation of food webs is necessary to understand the effect of toxic compounds released into the environment and the effect of the introduction of species into new areas.

Are diverse ecosystems with complex food webs more stable than simple ecosystems? Stability, in this instance, refers to the ability of an ecosystem to withstand disturbances, including those produced by people. Scientists used to believe that simple systems are less stable than complex ones. Clearly, there is a threshold of diversity below which most ecosystems cannot function. For example, no ecosystem can function without some plants and decomposers. Without plants, solar energy cannot be captured, and without decomposers, substances are not returned to the environment. But, is there a threshold below which present complex ecosystems lose their stability? How many different sorts of plants and animals are needed? And, if there are many of each sort, does the ecosystem become more stable?

Productivity, or the total quantity of energy captured and biomass produced by an ecosystem in a year, is related to the issue of stability. Are ecosystems that are formed by many species more productive than those with few species? It would appear that simple systems, both natural (*Spartina* marshes and high-latitude marine systems) and artificial (agricultural systems), are more productive than diverse ones. However, some scientists fear that these systems are not stable and have argued that society's increasing reliance on the productivity of these simple systems could eventually put humans in jeopardy.[20]

Humans and Species Diversity

Apart from the ethical and aesthetic reasons, is there any reason to fear that human survival is at risk if biodiversity is not preserved? Can humans exist surrounded only by agricultural fields, planted forests, and the like? This question is not easy to answer. In the short term, natural ecosystems can probably lose species without any great impairment of function. For instance, many planted forests are much simpler than the natural forests they replaced. As long as the environment does not change very much, ecosystems can apparently lose many of their rare species without any visible effects. In some cases, common animal species, such as passenger pigeons, have disappeared or been drastically reduced without endangering human survival.

Yet, environments inevitably change, and sometimes drastically. It is then that the role of obscure species sometimes becomes very important. When dinosaurs became extinct, an obscure group, primitive mammals, suddenly evolved into the dominant role they still play. And now, humans seem on the verge of a drastic change, one of their own making—the result of global warming induced by increased carbon dioxide in the atmosphere.

Predicted climatic changes pose a series of specific questions relating to biodiversity for which there are presently no answers. What is the impact of elevated atmospheric concentrations of carbon dioxide on tree physiology and forest composition? How does elevated carbon dioxide affect soil organisms and soil nutrients? How is climatic change going to affect insect populations, fire frequency, and wind storms? Coastal zones, in particular, are likely to reflect phenomena occurring over a larger geographical area. Already, the North Sea coast is showing the effects of the industrial and agricultural activity of central Europe, which is hundreds of miles away. The effects of industrialization may be catastrophic for many species, which may become extinct. But a few species, and there is no way of predicting which ones, may thrive in the new environment. Therein lies the scientific importance of biodiversity: It increases the likelihood that at least some species will survive and give rise to new lineages that will replenish the Earth's biodiversity.

Possible Research on Biodiversity

A global survey of diversity is essential to understanding the distribution of life forms on the planet.[21] Biogeographical studies have shown that ecosystem diversity varies greatly from one region to another. Tropical forests, coral reefs, and deep benthic communities are very rich in species; deserts and high-latitude areas are species poor. Many studies have shown that species abundance changes naturally along latitudinal and humidity gradients.

An unprecedented transformation of the natural landscapes of the world is now occurring. Yet, different areas are being transformed at different rates. Some regions, such as the Mediterranean basin, have been altered by people for a long time. Other regions have been influenced only moderately by humans. Also, studies show that the response to disturbance varies greatly from one ecosystem to another. Therefore, it is important to determine how different biomes respond to human disturbance and to document the resulting change in biodiversity and ecosystem function.

Classical biogeography concentrated on the study of species ranges and the causes of their distribution. Given scientists' uncertainty about the number of species and the large number of undescribed species, new and more efficient approaches must be developed to complement past studies. These new approaches should take advantage of remote sensing technology and produce a geographic information system with statistical and analytical capabilities. However, because satellites cannot detect species diversity, new measures that are indicative of diversity and are more easily obtained by remote sensing must be used. For example, satellites can detect standing biomass and productivity, and research should help to establish the relationship between biodiversity and these measures.

Other problems to be studied include the effect that habitat fragmentation has on species ranges and on the probability of extinction and speciation; whether fragmented habitats are more prone to invasion; how invasions influence biodiversity; and the effect of increased geographical isolation on populations and species. Also, research should be conducted to develop a methodology for the systematic comparison of biodiversity across regions. To this end, a biogeographical information system for assessing diversity might use satellite information to identify assemblages of species at different geographical scales for long-term monitoring.

Until now, research in biodiversity has been undertaken by taxonomists at such institutions as the Smithsonian Institution, the Harvard University Herbaria and Musea, the Missouri Botanical Garden, the Royal Botanical Gardens in Kew, the Musée d'Histoire Naturelle in Paris, and the Komarov Institute in Leningrad, which are located for the most part in the north temperate zone, while biodiversity is found mostly in tropical countries. Taxonomic institutions in both the north and the south have been in relative decline over the last 50 years. Staffs have not increased or have increased only slightly, and resources allocated to taxonomy have not grown in the same proportion as those devoted to other areas of biology. Consequently, at a time when the conceptual and practical importance of biodiversity is becoming clear, there is a shortage of trained personnel and a lack of necessary funds.

Taxonomists alone cannot answer the questions that have been raised by the massive transformation of the Earth. A collaborative effort by many kinds of biologists, including geneticists, physiologists, and ecologists, is required. As new techniques of genome analysis become routine, even molecular biologists may contribute to identifying genetic variation and its biological role.

Both governmental and nongovernmental institutions should collaborate to advance scientific understanding of

To survive near the poles, modern cold-hardy species have undergone eons of genetic mutation and natural selection.

the significance of biodiversity, and many more resources will have to be made available to fund the collaborative research. The International Union of Biological Sciences, the Scientific Committee on Problems of the Environment, and the Man and the Biosphere Program of the United Nations Educational, Scientific and Cultural Organization are spearheading research in biodiversity. These organizations are in close contact with the International Geosphere-Biosphere Program and the United Nations Environmental Programme, to minimize overlap and to coordinate activities. Many national institutions also have active programs of research in biodiversity, but most of these efforts suffer from inadequate funding.

Biodiversity research and conservation figure prominently in the agenda for the 1992 United Nations Conference on Environment and Development in Rio de Janeiro, Brazil, where conference planners are seeking a sub-

stantial increase in financial support for research in biodiversity.

It is imperative that biologists and the public learn more about the importance of biodiversity and its role in ecosystem function. Only an international collaborative effort that is supported by adequate resources can accomplish this task. Increased public awareness of the serious implications of humanity's depletion of biodiversity before its importance is understood is helping to create a climate that may stimulate governments to support national and international efforts in this field.

NOTES

1. J. J. Sepkoski, Jr., "A Kinetic Model of Phanerozomic Taxonomic Diversity 1: Analysis of Marine Orders," *Paleobiology* 4 (1978):223–51; R. M. May, *Exploitation of Marine Communities* (Berlin: Springer, 1984); and A. Hoffman, *Arguments on Evolution* (Oxford, England: Oxford University Press, 1989).
2. J. F. Grassle, "Species Diversity in Deep-Sea Communities," *Trends in Ecology and Evolution* 4 (1989):12–15.

3. S. M. Stanley, *Macroevolution: Pattern and Process* (San Francisco, Calif.: Freeman Press, 1979); S. M. Stanley, "Rates of Evolution," *Paleobiology* 11 (1985): 13–26; and P. W. Signor, "The Geological History of Diversity," *Annual Review of Ecology and Systematics* 21 (1990):509–39.

4. The International Union of Biological Sciences, the Scientific Committee on Problems of the Environment (SCOPE), and the Man and the Biosphere Program of the United Nations Educational, Scientific and Cultural Organization (UNESCO) recently joined forces to develop a scientific program to study the role of biodiversity in the function of ecosystems. One of the goals of this program is to develop scientific hypotheses regarding biodiversity that will be discussed at three upcoming meetings: a workshop at Harvard University at the end of June, an international symposium sponsored by SCOPE to be held in Bayreuth, Germany, in October, and, later in October, an international symposium on biodiversity sponsored by UNESCO and the United Nations Environment Programme that will take place in Nalchik, USSR. An international program of research is expected to emerge from these meetings.

5. G. Nicolis and I. Prigogine, *Exploring Complexity* (New York: Freeman Press, 1989); and O. T. Solbrig and G. Nicolis, *Perspectives on Biological Complexity* (Paris: International Union of Biological Sciences, 1991).

6. D. L. DeAngelis and J. C. Waterhouse, "Equilibrium and Non-Equilibrium Concepts in Ecological Models," *Ecological Monographs* 57 (1987):1–21.

7. P. Schuster, "The Interface between Chemistry and Biology: Laws Determining Regularities in Early Evolution," in G. Pifat-Mrzljak, ed., *Supramolecular Structure and Function* (Berlin: Springer, 1986); P. Schuster, "Optimization Dynamics on Valuable Landscapes: Modelling Molecular Evolution," in O. T. Solbrig and G. Nicolis, pages 115–62, note 5 above; and E. Szathmary, "The Emergence, Maintenance, and Transitions of the Earliest Evolutionary Units," *Oxford Surveys in Evolutionary Biology* 6 (1989):169–205.

8. G. L. Stebbins, *Flowering Plants: Evolution above the Species Level* (Cambridge, Mass.: Harvard University Press, 1974); H. G. Andrewartha and L. C. Birch, *The Distribution and Abundance of Animals* (Chicago: University of Chicago Press, 1954); V. Grant, *Plant Speciation* (New York: Columbia University Press, 1981); E. Mayr, *Animal Species and Evolution* (Cambridge, Mass.: Harvard University Press, 1963); and S. Wright, *Evolution and the Genetic Populations*, 4 vols. (Chicago: University of Chicago Press, 1968–1978).

9. G. L. Stebbins, *Variation and Evolution in Plants* (New York: Columbia University Press, 1950); and G. G. Simpson, *The Major Features of Evolution* (New York: Columbia University Press, 1953).

10. O. T. Solbrig, "Energy, Information, and Plant Evolution," in C. R. Townsend and P. Calow, eds., *Physiological Ecology: An Evolutionary Approach to Resource Use* (Oxford, England: Blackwell Scientific Publications, 1981), 274–99; N. Eldredge, "Information, Economics, and Evolution," *Annual Review of Ecology and Systematics* 17 (1986):351–69; and J. H. Brown and B. A. Maurer, "Macroecology: The Division of Food and Space Among Species on Continents," *Science* 243 (1989):1145–50.

11. H. Gleason, "The Individualistic Concept of the Plant Association," *Bulletin of the Torrey Botanical Club* 53 (1926):120.

12. C. Elton, *The Ecology of Animals* (London: Methuen, 1933).

13. J. Roughgarden, "Evolution of Niche Width," *American Naturalist* 106 (1972):638–718; and J. Roughgarden, "The Structure and Assembly of Communities," in J. Roughgarden, R. May, and S. Levin, eds., *Perspectives in Ecological Theory* (Princeton, N.J.: Princeton University Press, 1989).

14. A. J. Cain and P. M. Sheppard, "Natural Selection in Cepaea," *Genetics* 39 (1954):89–116.

15. I. M. Lerner, *Genetic Homeostasis* (Edinburgh: Oliver & Boyd, 1954); and I. M. Lerner, "The Concept of Natural Selection: A Centennial View," *Proceedings of the American Philosophical Society* 103 (1959): 173–82.

16. J. F. Grassle, note 2 above.

17. M. L. Cody, *Competition and Structure of Bird Communities* (Princeton, N.J.: Princeton University Press, 1974).

18. R. K. Colwell and E. R. Fuentes, "Experimental Studies of the Niche," *Annual Review of Ecology and Systematics* 6 (1975):281–310.

19. J. E. Cohen, "Food Webs and Community Structure," in Roughgarden, May, and Levin, pages 181–202, note 13 above.

20. J. H. Brown, "On the Relationship Between Abundance and Distribution of Species," *American Naturalist* 124 (1984):255–79.

21. F. DiCastri and T. Younès, "Ecosystem Function of Biological Diversity," *Biology International*, special issue no. 22 (1990):1–20.

Conserving the Tropical
Cornucopia

Nigel J. H. Smith,
J. T. Williams, and
Donald L. Plucknett

NIGEL J. H. SMITH is a professor of geography at the University of Florida in Gainesville. **J. T. WILLIAMS** is a researcher at the International Fund for Agricultural Research in Arlington, Virginia. **DONALD L. PLUCKNETT** is a scientific adviser for the World Bank's Consultative Group on International Agricultural Research in Washington, D.C. This article is based on a forthcoming book, *Tropical Forests and Crop Genetic Resources*.

The plight of tropical forests has moved to the forefront of public attention around the world. Citizens of both developing and developed countries are increasingly alarmed at the widespread forest destruction that has wiped out species, accelerated soil erosion, provoked flooding, undercut the livelihood of hundreds of millions of people, and threatened to disrupt the global climate. But another serious problem arising from tropical deforestation has been largely ignored: the shrinking populations of plants in the wild, which include the near relatives of many species that yield food, beverages, oils, latex, spices, timber, and fuel wood. The loss of such resources will affect people in developed countries by driving up the price of such goods as coffee, chocolate, vanilla, and tires. People in developing countries will also suffer from reduced yields of many crops important for their subsistence and commerce.

More than 200 crop species originated in tropical forests, and most still have wild populations there.[1] Many species of perennial crops arose from disturbed, open sites in forests, such as sunlit river banks and temporary clearings created by fallen trees. Wild plant species often have genes that allow the plants to resist diseases and pests. For example, the American oil palm (*Elaeis oleifera*), is the only known palm species resistant to spear rot, a mycoplasma-like disease that is devastating plantations of African oil palm (*Elaeis guineensis*) throughout tropical South America. Breeders are using the genes from prostrate American oil palms to develop resistance to spear rot in the African species and to make oil palms easier to harvest.[2] The genes from wild plant species can also be used to raise and sustain farm yields.

In the mid 1980s, a psyllid pest (*Heteropsylla cubana*) from the Caribbean arrived in some Pacific and Southeast Asian countries and began attacking leucaena (*Leucaena leucocephala*), an important fuel and fodder crop introduced to those regions from Central America. Resistance to this spreading insect pest has been found in several of this species' near relatives, such as *Leucaena salvadorensis*, *Leucaena multicapitula*, and the Guatemalan form of *Leucaena collinsii*—all inhabitants of Central America's dwindling forests.

Bananas and plantains, now being severely attacked in many areas by the fungal disease black sigatoka, are the staff of life in numerous regions, including around the great lakes of Africa. Genetic solutions to such problems as pathogens and arthropod pests are more enduring and more environmentally benign than are pesticides.

Although most crops were domesticated by biblical times, the world may well be ready for another wave of plant domestication. One of the most logical places to turn for new crop plants is the tropical rain forest, the richest biome on the Earth's surface.

Tropical Forests in America

The American tropics are particularly rich in crop genetic resources; more than 24 perennial crop species have

From *Environment*, Vol. 33, No. 6, July/August 1991, pp. 7-9, 30-32. Reprinted with permission of the Helen Dwight Reid Educational Foundation. Published by Heldref Publications, 1319 Eighteenth St., N.W., Washington, DC 20036-1802. Copyright © 1991.

been domesticated in Amazonia alone. Mexico and the Amazon basin are key areas for crop biodiversity. Mexico has wild populations of avocado (*Persea americana*), various tropical fruits, and vanilla. Forests in Mexico and other parts of Central America also have a relative of avocado, chinene (*Persea schiedeana*), which is cultivated on a minor scale for its succulent fruit. Because chinene resists root rot, a serious problem for avocado growers worldwide, avocado branches are often grafted onto chinene stems to give the tree a resistant root system. Unfortunately, populations of chinene are rapidly disappearing as forests are felled for firewood and cleared for croplands and pastures.[3] However, in 1984, three selections of *Persea schiedeana* were released by researchers at the University of California, Riverside, for use as avocado rootstock.[4] New developments in biotechnology are likely to permit exploitation of the wider gene pool of *Persea*, but only if the avocado's near relatives survive in the wild.[5]

Vanilla fragrans, a climbing orchid, is another perennial crop, native to Central America, whose wild populations are threatened. Vanilla is undergoing a resurgence on world markets as discriminating consumers demand more natural products. Recent legislation in France and the United States requires food manufacturers to stipulate on the package whether they are using natural vanilla or its artificial imitator, vanillin. Although it is much cheaper, artificial vanillin does not encompass the nuances of natural vanilla flavor. More than 100 volatile constituents act in symphony to produce true vanilla's exquisite aroma and subtle flavor. Even if vanillin is eventually produced in tissue cultures on a commercial scale, it is not clear whether "test-tube" vanillin would rival the complexities of field-grown vanilla's flavor.

As the market for natural vanilla expands, vanilla growers in Réunion, Mauritius, the Seychelles, Madagascar, Mexico, Tahiti, and Moorea are finding fewer varieties to plant. Much

of the vanilla on the market today comes from plants descended from a single cutting introduced to Réunion in 1827 from the Jardin des Plantes in Paris.[6] More varieties are needed that resist various viral diseases as well as such fungal pathogens as *Calospora vanillae*, which causes anthracnose, and *Fusarium batatis*, which triggers root rot.

Field gene banks for vanilla contain little diversity and are of limited use to breeders. Such ex-situ germ plasm collections must be enriched with more diverse materials, including genotypes from wild plants. At the same time, wild resources of vanilla must be safeguarded in situ so that breeders will continue to have access to a wide range of genetic materials in the future.

Central America—and particularly, Mexico, Guatemala, and Honduras—is rich in tropical pines. Caribbean pine (*Pinus hondurensis*) is widely planted in the tropics for wood pulp production, and other species, such as *Pinus patula*, are planted in tropical highlands for erosion control and firewood. In collaboration with Central American countries, the Oxford Forestry Institute is surveying wild pine populations in Central America and evaluating their potential uses. More than 60 countries are involved in trials of tropical pines, and seed banks have been established for many of the species.[7] Nevertheless, many tropical pine populations in the wild are endangered by encroaching farmers and fuel wood gatherers. Gene pools must be conserved in situ so they can continue to evolve along with their pathogens and insect pests.

The Amazon basin has often been overlooked as a center of biodiversity for important plants because of the bias toward food crops, particularly cereals. Indigenous species of cereals have not been developed as food crops there. Nikolai I. Vavilov, former head of the All-Union Institute of Plant Breeding in Leningrad, did not include the region as one of eight centers of origin of crop plants,[8] and it has typically been regarded as an ethnobotanical backwater until recently. However, from the Amazon basin came rubber

WORLD BANK—RAY WITLIN

The destruction of rain forests, like this one in Papua New Guinea, for fields and plantations threatens the wild gene pools of many crop species.

(*Hevea brasiliensis*) and cacao (*Theobroma cacao*)—crops that are now very important in Southeast Asia and West Africa, respectively. Cacao is becoming an important crop in Malaysia, while Brazil's prominence as a supplier of cacao on the world market is threatened because witches-broom, a disease caused by the fungal pathogen *Crinipellis perniciosa*, has reached Bahia. Rubber and cacao are important sources of income for owners of small farms as well as for large plantation operators. The future productivity of cacao and rubber crops hinges on the survival of wild gene pools in Amazonia. Some of the rubber plant's near relatives are resistant to South American leaf blight (*Microcyclus ulei*). And more sources of resistance to witches-broom must be located among wild cacao populations.

Not only are gene pools of existing crops threatened by deforestation in Amazonia and elsewhere, but many candidate species for crops are also being lost and the genetic resources seriously eroded. The people of the Amazon collect or use sporadically a wide variety of forest plants, including piquiá (*Caryocar villosum*) and copaiba (*Copaifera* species). Piquiá is a forest giant that can grow up to 50 meters high—an ideal candidate for an overstory tree in agroforestry systems. It produces oily fruits that are avidly consumed by people throughout the Amazon basin. In addition to the tasty pulp, piquiá kernels are also edible, and the wood of the plant is hard and durable.[9]

Copaiba, which also rises above the rain forest canopy, produces resin long used for medicinal purposes. Copaiba's red resin is smeared on wounds to promote healing and is burned in lamps by rubber tappers and other rural people.[10] The resin can also serve as a substitute for diesel fuel.[11] The Amazon basin's *Copaifera langsdorfii* can produce 20 liters of resin a year, which can be burned directly in diesel engines.[12] Thirty species of *Copaifera* occur in the American tropics, and four are found in Africa. The free-flowing resin of the New World species could be used much more extensively for such industrial uses as solvents, coatings, and adhesives.[13] Copaiba could thus prove suitable for

plantations as well as for small-scale agroforestry systems.

Tropical Forests in Africa

It is sometimes remarked that Africa is not the original source of many crops. One explanation offered is that the African tropics do not have a floral population as diverse as those of the American and Asian tropics. Nevertheless, many perennials important to local people have been domesticated in Africa, such as ensete, a banana relative in eastern Africa, and various fruits in West and Central Africa. Many plants in tropical Africa are semidomesticated, and further research continues to unveil a much richer tapestry of people-plant interactions than has hitherto been suspected.

Several species of *Coffea* have been domesticated in Africa, the most famous being *Coffea arabica*. Wild arabica coffee trees are found in a few forest remnants in southeastern Ethiopia, on the Boma Plateau of the southeastern Sudan, and in the Marsabit Forest in northern Kenya. Only about 400,000 hectares of forest remain that have wild arabica coffee.[14] The survival of only small pockets of wild arabica coffee is especially worrisome considering the narrow genetic base of the world's coffee plantings. Most of the plants that produce Latin America's coffee—exports of which generate about $10 billion annually and make up more than two-thirds of the world's supply—are descended from a single coffee bush from Java.

Wild arabica coffee plants have genes that allow the plant to resist several important diseases, including coffee rust. Coffee rust wiped out commercial coffee production in Sri Lanka in the last century. Caused by a fungal pathogen, *Hemileia vastatrix*, coffee rust reached Brazil in 1970 and had spread to Panama by 1983.

The Ethiopian government is taking steps to safeguard populations of arabica coffee in the wild, including establishing in situ reserves, but population pressures are mounting in the country. In the Sudan, no efforts have been undertaken to conserve wild coffee, and

no recent surveys have been made to assess the status of forest populations. The remoteness of the southeastern Sudan does not mean that the wild arabica trees there are safe. In the 1940s, members of the Kichepo tribe cut down wild arabica trees to harvest the berries.[15] Wild coffee in Kenya's Marsabit Forest is restricted to a narrow altitudinal belt between 1,500 and 1,550 meters above sea level and is thus particularly vulnerable to clearing by farmers.[16]

In the humid parts of Africa, some 27 species of raphia palm are harvested in the wild for such useful products as building materials, fiber, wine, food, fish poison, and string for musical instruments. Important species in West Africa include *Raphia hookeri*, which is much used for wine production in swamp forests that extend from Gabon to Guinea; the bamboo or wine palm (*Raphia vinifera*), which is found from the Congo to Benin; and the thatch palm (*Raphia palma-pinusis*), which is indigenous to the forests that stretch from Ghana to the Gambia.[17]

The oily fruits of many raphia palms are eaten raw, and raphia butter is prepared by boiling the fruits. The yellow fruits are rich in vitamins A and C and can be used for various medicinal purposes, including treatment of stomach illnesses and as a liniment for muscle strains. The fruits also have some insecticidal properties, possibly because of their relatively high tannin content.[18] In addition, the kernels are edible. Because raphia palms show promise as multipurpose trees for rehabilitating degraded areas, it is important that the trees' gene pools be safeguarded in the wild so that future selections can be made from them.[19]

Tropical Forests in Asia

A bewildering assortment of fruits and nuts have been domesticated in the tropical regions of Asia. Some of them, including various citrus fruits, mangoes (*Mangifera indica*), and bananas, are well known to consumers in both North America and Europe. Others are esteemed regionally, such as rambutan (*Nephelium lappaceum*), durian (*Durio ziebethinus*), and mangosteen (*Gar-*

cinia mangostana). Breadfruit (*Artocarpus altilis*) is an important food in the Pacific countries and can still be found in the wild in New Guinea, where it was domesticated.

The forests of Southeast Asia are under increasing pressure from shifting cultivation and logging operations. The disappearance of the wild gene pools of some Asian crops would have serious repercussions for the livelihoods of many people in both developing and developed countries.

Establishing Gene Sanctuaries

Very few forest areas have been set aside specifically to safeguard the genetic resources of crop species. The plant world contains few "pandas" to galvanize the public and governments into establishing protected habitats for wild populations of cultivated plants. The payoffs must seem too remote for government leaders or the public to demand their protection. Indeed, breeding perennial crops is a protracted affair. The conservation of wild plant populations must "piggyback" on other causes to succeed, such as campaigns to set aside areas for ecotourism or to protect watersheds so that farmers and urban dwellers have sufficient water supplies.

India and Mexico have demonstrated admirable leadership in setting aside natural areas to protect the genetic resources of crop species. In 1981, the Indian government established a gene sanctuary for citrus plants in the Garo Hills of Meghalaya in the northeastern part of the country.[20] This 10,000-hectare, forested sanctuary protects diverse populations of a wild orange, *Citrus indica*.

Mexico has established a park for safeguarding a wild relative of corn in the 350,000-acre Sierra de Manantlán biosphere reserve near Guadalajara. Although not a tropical forest, this sizable park serves as a worthy example that could be emulated in the more humid regions of the tropics. The biosphere reserve protects *Zea diploperennis*, a perennial corn whose genes allow the plant to resist several diseases.

More tropical forest reserves must be established to protect the wild populations and relatives of crop species. Local people must be involved in the establishment and maintenance of germ plasm reserves. In addition, the cooperation of indigenous people who have intimate knowledge of forest resources is essential to maintaining the integrity of gene sanctuaries. If a park seems to be benefiting only coffee drinkers in North America, for example, it is unlikely to survive.

Only a small fraction of the remaining tropical forests are likely to be designated as parks or nature reserves. Even existing "protected" areas are frequently violated by loggers, gold miners, and settlers. Therefore, what happens outside biosphere reserves will largely determine the fate of the wild populations of hundreds of crop species and their near relatives that originated in tropical forests. In an attempt to better the fate of forests not included in biosphere reserves, several "extractive reserves" have been set aside in the state of Acre in the Brazilian Amazon, where people can live and derive their livelihood by exploiting the forests' resources on a sustainable basis. Extractive reserves have generated considerable interest in Brazil and in other countries and are sometimes seen as a way to conserve forests and alleviate rural poverty.

Although extractive reserves have many virtues, their ability to safeguard the forest is unproven. It seems inevitable that some forest areas will be cleared to grow food and, perhaps, cash crops. Relying on forest products alone to generate sufficient income and employment is probably impossible, at least in the short term. If new customers are found for existing forest products, such as Brazil nuts (*Bertholletia excelsa*); if markets are developed for new products, such as medicines, fruits, and oils; and if limited clearing for agriculture is permitted, extractive reserves may help save the wild gene pools of many crops.

The Global Stake

As tropical forests continue to disappear, future options to upgrade existing crops and domesticate new plants are being foreclosed. Many plants that could be developed as crops for food, drugs, oils, and biomass fuels are likely to become extinct. Genes that could be tapped by emerging biotechnologies or even conventional plant-breeding techniques are lost in the smoke that rises from many tropical forests when felled trees are torched during the dry season.

The continued erosion of genetic resources in tropical forests will affect inhabitants of the Third World as well as those in the industrialized countries. Dramatic dips in exports of tropical perennial crops will jeopardize the ability of many developing countries to repay their debts. Malaysia, for example, depends heavily on rubber and palm oil to generate foreign exchange, and many Latin American countries rely on coffee and bananas to finance much of their development and to repay loans. The livelihoods of small-scale farmers who harvest most tropical perennial crops are at stake if there is a sudden downturn in the crops' yields.

In the search for more sustainable agricultural practices in developing regions, tree crops figure prominently because they help diversify sources of income and protect the soil and water. Leguminous trees also enrich the soil with nitrogen. Spectacular increases in grain harvests will only be helpful if people have enough fuel wood to cook their meals. But reforestation, afforestation, and agroforestry schemes will only be sustainable if genetic resources are available to replace varieties that become susceptible to diseases, pests, or adverse weather.

Developing countries need assistance from the international community, including development banks, commodity groups, and foundations, to develop comprehensive conservation strategies. Such strategies should include setting up and managing reserves for crop genetic resources. Often a crop is more important commercially if it is far from its area of origin, but the region where the crop was domesticated has the lion's share of genetic resources. Thus, countries must recognize their interdependence with respect to plant genetic resources and their conservation. Substantial funds must be mobilized for managing tropical forests over the long term and for the necessary research to underpin management plans. Simply buying tracts of land or obtaining them through debt-for-nature swaps is no guarantee that wild habitats will survive. Although international funds must be funneled for such efforts, the sovereignty of nations must also be recognized, and local institutions and research facilities must be fortified.

ACKNOWLEDGMENT

The authors gratefully acknowledge the support of the Rockefeller Foundation.

NOTES

1. N. J. H. Smith, J. T. Williams, and D. L. Plucknett, *Tropical Forests and Crop Genetic Resources* (forthcoming). For the importance of biodiversity, see Otto T. Solbrig, "The Origin and Function of Biodiversity," *Environment*, June 1991, 16.

2. N. J. H. Smith and R. E. Schultes, "Deforestation and Shrinking Crop Gene-Pools in Amazonia," *Environmental Conservation* 17 (1990):227–34.

3. E. Schieber and G. A. Zentmyer, "Exploring for *Persea* on Volcano Quetzaltepeque, Guatemala," *Yearbook of the California Avocado Society* 65 (1981): 57–63; and C. A. Schroeder, "No Yas: A Threat to the Avocado," *Yearbook of the California Avocado Society* 61 (1977):37–42.

4. M. Coffey, "An Integrated Approach to the Control of Avocado Root Rot," *Yearbook of the California Avocado Society* 68 (1984):61–68; and M. Coffey, F. Guillemet, G. Schieber, and G. Zentmyer, "*Persea schiedeana* and Martin Grande," *Yearbook of the California Avocado Society* 72 (1988):107–20.

5. R. Bergh and N. Ellstrand, "Taxonomy of the Avocado," *Yearbook of the California Avocado Society* 70 (1987):135–45.

6. J. W. Purseglove, E. G. Brown, C. L. Green, and S. R. J. Robbins, *Spices*, vol. 2 (London: Longman, 1981).

7. R. D. Barnes, "Tropical Forest Genetics at the Oxford Forestry Institute," *Commonwealth Forestry Review* 67 (1988):231–41; and B. J. Zobel, G. Van Wyk, and P. Stahl, *Growing Exotic Forests* (New York: John Wiley & Sons, 1987).

8. N. I. Vavilov, *The Origin, Variation, Immunity and Breeding of Cultivated Plants* (Waltham, Mass.: Chronica Botanica, 1949); idem, "Studies on the Origin of Cultivated Plants," *Bulletin of Applied Botany, Genetics and Plant Breeding* 16 (1926):1–248; and idem, "The New Systematics of Cultivated Plants," in J. Huxley, ed., *The New Systematics* (Oxford, England: Clarendon Press, 1940), 549–66.

9. P. B. Cavalcante, *Frutas Comestíveis da Amazônia* (Belém, Brazil: Museu Paraense Emilio Goeldi, 1988); and R. E. Schultes, "The Amazonia as a Source of New Economic Plants," *Economic Botany* 33 (1979):259–66.

10. A. A. Loureiro, M. F. da Silva, and J. C. Alencar, *Essências Madeireiras da Amazônia*, vol. 1 (Manaus, Brazil: Instituto Nacional de Pesquisas da Amazônia, 1979).

11. J. H. Langenheim, "Plant Resins," *American Scientist*, January/February 1990, 16–24.

12. G. Ledec and R. Goodland, *Wildlands: Their Protection and Management in Economic Development* (Washington, D.C.: World Bank, 1988), 13.

13. J. H. Langenheim, "Leguminous Resin-Producing Trees in Africa and South America," in B. J. Meggers, E. S. Ayensu, and W. D. Duckworth, eds., *Tropical Forest Ecosystems in Africa and South America: A Comparative Review* (Washington, D.C.: Smithsonian Institution Press, 1973).

14. G. Wrigley, *Coffee* (New York: Longman/John Wiley & Sons, 1988).

15. A. S. Thomas, "The Wild *Arabica* Coffee on the Boma Plateau, Anglo-Egyptian Sudan," *Empire Journal of Experimental Agriculture* 10 (1942):207–12.

16. F. Anthony, J. Berthaud, J. Guillaumet, and M. Lourd, "Collecting Wild *Coffea* Species in Kenya and Tanzania," *Plant Genetic Resources Newsletter* 69 (1987):23–29.

17. T. A. Russel, "The *Raphia* Palms of West Africa," *Kew Bulletin* 19 (1963):173–96; and A. Rançon, *La Flore Utile du Bassin de la Gambie* (Bordeaux, France: G. Gounoilhou, 1895).

18. R. E. Heal, E. F. Rogers, R. T. Wallace, and O. Starnes, "A Survey of Plants for Insecticidal Activity," *Lloydia* 13 (1950):89–157.

19. M. O. Otedoh, "The Re-Discovery of *Raphia regalis* in Nigeria," *Journal of the Nigeria Institute for Oil Palm Research* 5 (1974):41–67; and M. O. Otedoh, "Raphia Palms: The Production of Piassava in Nigeria," *The Nigerian Field* 40 (1975):4–15.

20. B. Singh, *Establishment of First Gene Sanctuary in India for Citrus in Garo Hills* (New Delhi: Concept Publishing, 1981).

TIMBER'S LAST STAND

The timber industry is rapidly running out of natural forests to plunder. Adopting a New Forestry could help save a destructive industry from itself.

JOHN C. RYAN

John C. Ryan researches forest issues at the Worldwatch Institute.

The earth shakes as the forest giant crashes down, its crown landing hundreds of feet from the massive stump left behind. The product of countless seasons of clean air, stable climate and freedom from human disturbance, the centuries-old tree represents a windfall profit to its sawyers, one never to be realized again.

Whether it was a western hemlock or a Philippine mahogany, the tree's fall also symbolizes one of the most widespread, and visibly shocking, forms of environmental degradation: deforestation. Few images breed concern for the planet like the denuded moonscape of an Oregon clear-cut or the plot of scorched earth that was once rain forest.

Images can only partially capture the legacy of logging and the world's insatiable demand for wood products, however. Fragmented ecosystems in British Columbia, threatened native cultures in Indonesia, collapsed timber industries in western Africa, and diminished biological diversity from Germany to Java attest to the fact that deforestation is a global phenomenon.

In the tropics, the timber trade is a sup-porting player in a drama that sees the richest ecosystems on the planet grossly disrupted and then eliminated as forests shrink before cattle ranchers and desperate peasants. In Europe and North America, where the area under tree cover is fairly stable, foresters have overseen the conversion of diverse natural systems into uniform tree farms, as different from their predecessors as a corn field is from a prairie.

It is clear that the time for a new forestry—one that seeks to use and maintain the complexity of forests, rather than eliminate it—is now. Although it's hard to imagine a timber industry that treads lightly on the land, it should be possible, since wood, in theory at least, is a renewable resource. As human numbers and needs continue to rise, and as forests continue to dwindle, it is urgent that we learn to tap forests' varied riches without impoverishing their source. As the timber industry confronts the exhaustion of both woodlands and the public's tolerance of its practices, it faces the inevitable choice of living up to this challenge, or following its resource base into rapid decline.

Fortunately, in places as disparate as Peru's Palcazu Valley and Oregon's Cascade Range, small bands of researchers and activists are piecing together the beginnings of a sustainable forestry.

Mining Our Heritage

Reeling from a barrage of threats, forests in many regions of the world have a dim future. If rain-forest loss is not slowed, according to British forest watcher Norman Myers, "by early next century there will be little left of tropical forests except for a few large blocks in New Guinea, the Zaire basin, western Amazonia in Brazil and the Guyana highlands." Tropical logging degrades about 11 million acres annually and, especially in Southeast Asia and central Africa, contributes to complete deforestation—now estimated to total over 34 million acres annually (an area the size of Florida)—by making entire regions more susceptible to fire and accessible to peasants and ranchers.

In the temperate zones, which provide most of the world's timber (see Table 1), logging, farming, and human settlements have spread to such an extent that, outside of remote northern regions, little primary forest remains. The coastal rain forest of British Columbia is given 15 years before it is wiped out from logging, and the estimated 5 percent of the United States' ancient groves that still exist face rapid fragmentation and extinction.

The timber trade's attraction to primeval stands appears to be unquenchable. Much like the loggers of Siberia who set up temporary camps, cut all the usable timber from the area, and move on, the international timber industry mines one source then seeks out another. Canadian and Soviet foresters are now looking to their remote, and less lucrative, northern forests, while several U.S. companies have relocated operations from the Northwest to the pine plantations of the Southeast.

The focus of the tropical timber trade shifted in the 1960s from Africa, where overcutting has brought forests and forest industries crashing down, to Southeast Asia, where similar depletion is now occurring. According to an analysis done for the International Tropical Timber Organization (ITTO), a 69-nation trade group based in Yokohama, Japan, timber exporters Ivory Coast and Ghana will likely become net importers before the end of this decade. They follow the trail blazed by Nigeria in the 1970s.

The quality and availability of wood from

Table 1.

World's Top 15 Timber Producers, 1988[1]

Country	Volume (million cubic meters)	Share of Total (percent)
USA	417	25
USSR	305	18
Canada	173	10
China	98	6
Brazil	67	4
Sweden	48	3
Finland	46	3
Indonesia	40	2
Malaysia	36	2
France	32	2
W. Germany	31	2
Japan	28	2
India	24	2
Poland	20	1
Australia	18	1
Others	281	17
World Total	1,664	100

[1] Includes all wood products except fuelwood and charcoal.
Source: FAO, *Forest Products Yearbook 1988* (Rome: 1990)

Southeast Asia have declined, and several nations have responded with bans on log exports or on logging altogether. As one Japanese importer declared in *Nikkan Mokuzai Shimbun* (Daily Timber News) in 1988: "the depletion of tropical timber resources in Southeast Asia has become a matter of reality today, so we have to look to Brazil for a new supply . . ."

The Road to Ruin

Even if timber is managed on a "sustained yield" basis (which considers only the yield of wood, and not any of the other benefits of forests), logging can still devastate forests. Timber harvesting typically begins with a network of roads, which themselves deforest large areas and, especially in the steeper regions now being cut, can greatly increase soil erosion and sediment buildup in streams and rivers. Approximately 8 percent of logging areas in the Pacific Northwest are cleared for road building; as much as 14 percent is cleared in Southeast Asia. Myers reports that for every tree cut for timber in certain areas of Zaire, 25 are cleared making roads to get to it. In Idaho, and in northern Palawan in the Philippines, logging roads have caused erosion more than 200 times greater than on undisturbed sites.

Roads expose forests to miners, hunters,

> "*L*iquidating old-growth is not forestry, it is simply spending our inheritance. Nor is planting a monoculture . . . forestry; it is simply plantation management."

growth is not forestry," he writes, "it is simply spending our inheritance."

The Economics of Destruction

As the world's greatest storehouses of life, forests are valuable for much more than their timber. When these riches are sacrificed to wood production, logging often becomes difficult to justify on economic grounds.

Damage to fisheries and coral reefs caused by logging-induced sedimentation has been documented around the world. The harvesting of timber worth $14 million from the drainage of the South Fork of the Salmon River in central Idaho in the mid-1960s, for instance, caused an estimated $100 million in damage to the river's chinook salmon fishery. That industry has still not recovered. Fisheries in Bacuit Bay near Palawan in the Philippines were depleted after logging commenced in 1985 on surrounding hillsides. Sediment rushing into the bay killed up to half of the living coral that supported the fishery, depriving local villagers of their source of protein.

The costs of logging have usually fallen on those who depend on intact forest—forest dwellers, downstream communities, tourism-based economies, among others. But as the area of untouched forest shrinks, it is becoming clear that the timber industry is also putting itself out of business. Tropical hardwood exports, worth $8 billion in 1980, have fallen to $6 billion, and are projected to shrink to $2 billion by the end of this decade.

When diverse populations of trees are replaced with genetically uniform stands, there is a double loss to future timber harvests. Plantations can relieve pressure on natural forests by producing wood quickly, but because the natural system of checks and balances has been stripped to maximize tree growth, monocultures are prone to unravel. Widespread disease and pest outbreaks—common throughout the conifer plantations of the United States, central Europe and China, and a chronic problem in tropical plantations—can decimate entire forests, rather than localized groups of trees.

In West Germany, where forests have been logged, grazed, and raked for centuries, single-species plantations have spread to such an extent that 97 percent of forest land is covered by just three tree species. Scientists speculate (since there is virtually no natural forest left for comparison, it is impossible to prove) that the lasting damage caused by

and especially poor farmers; they also allow non-human invaders access to once-deep forest. Logging roads have accelerated the spread of destructive pests in the Northwest, including Port Orford cedar root rot, an always-fatal disease spreading rapidly into the remaining upland groves of the cedar, the region's most prized timber tree. As roads, logging and forest clearance spread, large areas of habitat are turned into islands in a sea of degraded lands. Research in both tropical and temperate forests has shown that such fragmented landscapes are unable to support the biological diversity present in continuous forest. Recent studies of isolated patches of Amazonian forest, for example, confirm that edges of forest "islands" deteriorate rapidly from exposure to damaging winds, exotic species, and dramatic changes in temperature, humidity, and light levels.

Some of the long-term degradation that follows from loss of tree cover, such as increased soil erosion and water runoff, can be minimized by reforestation after logging. But young stands cannot provide the wildlife habitat or high-quality, fine-grained wood of ancient forest. Industrial-style "reforestation"—the planting of rows of identical trees accompanied by slash burning, soil plowing, and the use of fertilizers or herbicides—provides even fewer environmental benefits than natural regrowth.

All told, these and other impacts add up to a global failure to sustain forests. A 1989 study for ITTO found that less than 0.1 percent of tropical logging was being done sustainably. Former U.S. Bureau of Land Management biologist Chris Maser argues that sustainable forestry isn't practiced outside the tropics either. "Liquidating old-

intensive forestry may have helped speed Germany's woodlands down the road to *Waldsterben*: the widespread "forest death" linked to air pollution and acid rain.

When native forests are lost, industry also loses its reservoirs of genetic variety and its scientific laboratories for uncovering the many hidden relationships that make timber growth possible. For instance, according to work done by Oregon State University entomologist Tim Schowalter and others, intact stands of natural forest are valuable as physical barriers and as sources of insect predators to stop the spread of pest outbreaks on adjacent plantations. As long as wilderness is left to tap and study, then foresters have the opportunity to learn from their mistakes. But, as industry converts native stands to plantations—putting all its eggs in the monoculture basket—its options keep narrowing.

Toward A New Forestry

How can timber be harvested without destroying forests? The answer to this question is being discovered in some very unlikely places: in a chunk of rotting wood on a forest floor, amid a buzz of insects hundreds of feet up a Douglas fir, in the fecal pellets of a flying squirrel. In these and untold other places lie the essentials of forest productivity that foresters ignore to the detriment of forests and timber production.

A small group of researchers and forest managers based at the H. J. Andrews Experimental Forest in western Oregon have been studying the lessons of natural forests and have started applying them in an attempt to reconcile the seemingly unsolvable conflict between logging and forests. The "New Forestry," as their ideas are being called, represents a fundamental change, a revolution, even, for the forestry profession, which has traditionally focused narrowly on timber production.

Perpetuation of diverse forest ecosystems has to become the focus of forestry if the forest products industry is to survive, states Jerry Franklin, an ecologist with the University of Washington and the U.S. Forest Service, and the leading proponent of New Forestry. "Already we are learning that parts of forests that we have never considered seriously are proving significant, even essential, to ecosystem functioning," he says.

One part of the forest overlooked until recently is the array of underground organisms that help keep soils fertile. Perhaps the most important of these are the "mycorrhizal" fungi that attach to the roots of 90 percent of the world's plant species and whose vast thread-like networks literally form the base of forests in the Pacific Northwest. Eaten and dispersed by the northern flying squirrel, the fungi enable trees to absorb nutrients and water from the soil and fix nitrogen. After clear-cutting, when all "host" plants and ground cover for squirrels and other mammals are removed, many of the fungus species are eliminated, robbing the land of its ability to grow more timber.

These and many other hidden linkages discovered within forest communities demonstrate the importance of maintaining intact as many of the pre-logging conditions as possible throughout the timber cutting cycle. Similarly, foresters are starting to recognize dead trees and logs on the forest floor as essential parts of a healthy forest, not a form of waste to be burned off or shredded. Besides providing important wildlife habitat, woody debris maintains soil fertility by returning organic matter and nutrients to the soil and helping to control erosion.

On a handful of sites on the Willamette and Siskiyou national forests in Oregon, U.S. Forest Service managers are beginning to actively apply these principles to their work. They leave behind live and dead trees, corridors of trees along stream banks, and small and large woody debris. The goal is to maintain the land's productivity and its diversity. They have also begun to lump timber cutting areas together to minimize fragmentation and road building.

Every Day, Every Acre

New Forestry is no substitute for protecting natural areas: no forester can create 1,000-year-old ecosystems or bring back species driven to extinction. Environmentalists are rightly suspicious of anyone trying to sell new types of logging as the solution to deforestation. Especially where the amount of wilderness left is small, preservation is still the top priority.

Nonetheless, given that logging of primary forest is not going to stop tomorrow, New Forestry promises to minimize the damage to areas that will be lumbered. The New Forestry has been researched and applied almost exclusively in the ancient forests of the Pacific Northwest, but, as David Perry, a forest ecologist at Oregon State University, notes, although particular techniques will vary greatly, "there are certain ecological principles translatable virtually anywhere in the world." The philosophical core of the New Forestry—the goal of working with the complexity of natural systems, rather than eliminating it—can apply as well to farmland and oceans as it does to forests.

Reducing the risk of future pest outbreaks and ensuring that soil is not robbed of its nutrients makes sense whether wood is harvested from a pristine rain forest, a logged-over woodland, or an intensively cropped tree farm. As evidence builds that the intensive forestry practices used today often fail to sustain timber productivity over time, timber managers may see the wisdom of restoring natural resilience and diversity to their lands.

Since most of the world's forests are already logged over or cleared, and many of the remaining areas are severely fragmented, any attempt to protect biological diversity will have to address the lands that humans use intensively. "We could never hope to protect biological diversity solely through preservation," says Franklin, "since so much diversity occurs on commodity landscapes. ... Protection of diversity must be incorporated into everything we do every day on *every* acre, whether preserve or commodity land."

Tropical Troubles

In primary rain forest, still the predominant resource for the tropical timber trade, the social, political and biological complexities of forest use raise doubts whether sustainable timbering is even possible. Removing too much wood from these forests, in which nutrients are found mostly in the plant life itself, not in the soil, leaves behind a nutritionally impoverished system that may take hundreds of years to rebound. Even selective logging is typically very destructive because of the tremendous diversity of tree species: loggers inevitably trample wide areas as they "cream" the forest—taking only a handful of desired species.

Third World governments, saddled with debt and swelling populations, typically see their rain forests as quick sources of foreign exchange or as safety valves for an expanding underclass. Unstable conditions outside the forest make long-term policy inside—such as enforcing minimum lengths of logging rotations or preventing illegal entry

in logged-over areas—difficult to enforce. The sheer number of people looking to tropical moist forests as sources of sustenance and profit may already overwhelm their carrying capacity.

Despite the numerous obstacles, a handful of projects show how sustainable tropical logging might work. The Yanesha Forestry Cooperative, the first Indian forestry cooperative in Amazonia, has been operating since 1985 in Peru's Palcazu Valley. Local people own and process the forest products; timber cutting is designed with protection of diversity in mind. By clear-cutting in narrow strips, leaving most of the forest intact, the Palcazu project seeks to mimic small-scale natural disturbances. Creating gaps in the forest canopy allows the shade-intolerant seedlings of hundreds of different species from the uncut areas to colonize the strips. Bark and branches are left in place to maintain soil fertility, rather than burned off.

Portico S.A., a Costa Rican door manufacturer, is probably the only timber company in the world researching natural forest management, according to Renee Dagseth of the Overseas Private Investment Council, a U.S. government agency helping to finance the company. Recognizing that its resource base of *caobilla,* a type of mahogany, was endangered by deforestation, Portico has been buying up forest land and trying different harvest techniques and rotations since 1988 to assure itself a steady supply of wood. Because *caobilla* cannot be grown outside its natural, swampy habitat, the company is buying from local farmers marginal farmland where the tree is found and hiring them as part-time guardians against illegal loggers.

These projects are both so new that it is not possible to call them successful yet. It would take decades of steady production to do that; unfortunately, few nations have the luxury of that much time. New approaches to logging that incorporate rather than ignore natural linkages and local people can, at the very least, prolong the useful life of logging areas, and buy time for other solutions to deforestation to be worked out.

Ending the Timber Bias

If the conflict between timber and forests is to be resolved, a two-prong strategy is needed: protection of large, viable areas of natural forest and new forestry practices on areas to be logged. Probably the greatest obstacle to achieving these goals is the commonly held view of forests primarily as timber factories. It is this timber bias, prevalent among foresters and policymakers the world over, that has already sent much of the world's natural heritage to the mill.

Most nations have laws and regulations proclaiming their commitment to sustained-

---

*S*cientists *speculate that clear-cutting and monoculture forestry may have helped speed Germany's woodlands down the road to Waldsterben—the widespread "forest death" linked to air pollution and acid rain.*

---

yield or multiple-use forestry, but almost nowhere does this translate into a balanced approach to forest use. Economist Robert Repetto of the World Resources Institute (WRI) in Washington, D.C., has documented the worldwide occurrence of government subsidies that encourage destructive logging at taxpayers' and forests' expense.

The much-heralded Tropical Forestry Action Plan, an international strategy launched in 1985 by the Food and Agriculture Organization, the United Nations Development Program, the World Bank, and WRI, is expected to accelerate deforestation because of its reliance on increased logging as a means of saving forests. The Peruvian Forestry Action Plan, for example, advocates an expansion of the road network and a four- to sixfold increase in logging, even though it recognizes that Peru's forests "are exploited in the same way as the mines of the Sierra" and describes present management as "chaos."

The timber focus also ignores the root causes of deforestation—including maldistribution of farmland, international debt, and population growth. Shifting control of forests away from exploitative users—such

as timber cutters and cattle ranchers—and toward sustainable users of forests, especially the millions of forest dwellers who have lived within the forests' limits for ages, can do much to halt forest loss. Colombia's decision in February to recognize Indian rights to half of its Amazon forests is a landmark achievement in both forest policy and social justice.

Protecting Forests, Protecting Jobs

While the crush of human demands ensures that most forests will be used in one way or another, turning natural areas of global significance into pulp—as is happening in Alaska's Tongass rain forest and Southeast Asia's last large area of coastal mangroves in Bintuni Bay, Papua New Guinea—is a travesty considering that pulpwood can be obtained with much less impact from second-growth forests and plantations. Ending subsidies for conversion of primary forests and putting in place incentives for better management on less valuable lands could help increase sustainable timber supplies.

But new supplies will take time to develop; forests will continue to be pushed beyond their limits until the world begins to curb its spiraling appetite for wood products. Because sustainable forestry will often yield less wood per acre in the near term than timber mining, and because increased recognition of the non-wood values of forest will mean fewer acres available for timber harvest, reducing the demand for wood is an inevitable part of the sustainability equation.

Many opportunities exist to reduce wasteful use of wood products—from the 50 percent of raw wood turned to chips and dust in a typical sawmill to the 25 billion disposable chopsticks consumed annually in

Table 2.

Per-Capita Paper Use, Selected Countries and Regions, 1988

Country/Region	Pounds per year	Percent recycled[1]
USA	699	29
Sweden	685	40
Canada	543	20
Japan	450	50
Norway	333	27
USSR	78	19
Latin America	55	32
China	27	21
Africa	12	17
India	5	26

[1] Amount of waste paper recycled compared to total paper consumption; 1987 figures.
Source: The Greenpeace Guide to Paper (Vancouver: Greenpeace, 1990)

Japan. Less than one-third of the paper used in the United States, the world's most gluttonous consumer of paper (see Table 2), is recycled; one-half of the total is consumed as packaging.

Wherever forests are cut down, governments can act to get the most out of each tree. In the Pacific Northwest, the timber industry blames the loss of jobs upon environmentalists, yet one out of every four trees harvested in 1989 was sent abroad as raw logs, untouched by mill workers. Timber-related employment in Oregon declined by 15 percent in the 1980s, even as timber harvests reached record levels. A ban on the export of raw logs would provide four times more wood for local mills as would be set aside under a recently announced federal plan to protect ancient forests in which the rare spotted owl lives.

Policymakers can reduce damage to lands that are logged by enforcing forestry regula-tions, which on paper are often quite sound, and by emphasizing the long-term health of forests, rather than quick profits for logging companies. As WRI's Repetto has noted, economic tools such as pricing reforms can encourage loggers not to waste the forests they are granted access to.

Ultimately, it's essential that widespread recognition of the ethical responsibility not to trade present yields for future degradation take hold. The introduction of codes of conduct among European tropical timber importers and the growth of the Association of Forest Service Employees for Environmental Ethics in the United States are two hopeful signs in this area.

Jobs and profits based on ecological destruction simply cannot last. If societies can come to grips with this fact, perhaps we can make the transition to sustainability while there are still ecosystems left worth protecting.

FORESTS UNDER SIEGE

Jeffrey L. Chapman

Mr. Chapman is a research assistant, Native Forest Council, Eugene, Ore.

''The rush to cut [America's forests] stems not from our country's lumber needs, but from the timber's cash value on the international market.''

AMERICANS clamor for an end to worldwide deforestation. Yet, while our accusing eyes turn toward the tropical regions of the globe, chainsaws scream virtually unchecked through the Earth's last vestiges of temperate rain forests in our own country. Today, less than five percent of this nation's original, native forests remain unharmed by human intrusion, and they continue to topple at the rate of 240 acres a day. While the waxing wave of environmentalism largely neglects their liquidation by a profit-crazed timber industry, the forest ecologies suffer irreconcilable damage with global ramifications.

These native forests (also known as virgin, ancient, or old-growth) are concentrated most highly in the Federally owned national forests of the Pacific Northwest. Under Congressional mandates, the U.S. Forest Service must dole out the last remnants of majestic firs, pines, spruces, hemlocks, cedars, and redwoods to private timber companies. The government's timber operations add up to an annual loss of more than $1,000,000,000, creating another taxpayer subsidy of an already enormously profitable and subsidized industry.

America's native forests exist as some of the only unaltered low-elevation ecology left in the country—untamed, natural wilderness thriving on millions of years of biological evolution. Timber advocates claim they are simply dead, decadent, and rotting—a fortune of resources just wasting away. Yet, the wealth of decomposing mass provides the very nourishment of a rich, healthy soil that supports more living vegetation per acre than any other place in the world. These self-sustaining ecosystems will continue their indigenous processes if simply left alone.

The strength of native forests lies in their biological diversity—a fundamental principle of durability and prosperity. The trees' variety of ages and species reinforce the over-all health and balance of the system. As many as 1,500 species of invertebrates may inhabit any particular grove of trees, and each individual tree may support up to 100 different species of mosses and other vegetation. The forests create their own climate that is cooler and moister than the surrounding environment. For this reason, they distinctly can be called rain forests— the last of their kind in the temperate regions of the world. Their unscrupulous removal rapidly diminishes the gene pools and threatens to destroy their unique genetic blueprints forever.

Clear-cutting—the practice of cutting down every tree on a large tract of land— used widely in the native forests has the single most devastating impact on a forest ecosystem. The canopy of trees that encloses and insulates the unique, self-contained climate and all its distinctive vegetation and wildlife is eliminated instantly and entirely. The land permanently loses rain forest characteristics, the abundance of mosses and vegetation dry up, and the temperature becomes prone to extremes. The exposed soil subsequently suffers permanent degradation. In a native forest, three percent of the rainfall is lost to surface runoff; the rest percolates into the ground. In a clear-cut, 60% of all rainfall runs off the surface, eroding away a proportionate amount of nutrients and topsoil, fouling streams with mud and silt, and spoiling local water supplies.

Man has inflicted the Earth with permanent scars by stripping it of its forests. Trees once surrounded the Mediterranean Sea. By 1000 B.C., northern Africa had lost every forest. In Greece, the soil erosion was so severe that, by 350 B.C., much of the country could no longer sustain trees. The original brown forest soil of the entire region gave way to a limestone bedrock now regarded as a typical Mediterranean soil

Reprinted from *USA Today Magazine* (Society for the Advancement of Education), March 1991, pp. 17-23. Copyright © 1991 by the Society for the Advancement of Education.

profile. Rome lost many harbors around the mouth of the Tiber River to siltation after soil from the denuded hillsides washed into the tributaries upstream. No amount of research or innovative forestry can solve erosion caused by clear-cutting. The same destruction that forever altered the Mediterranean landscape occurs every day in American forests.

Focusing on commodity resource values, the Forest Service assists in revegetating the clear-cuts with trees destined to fall again within 50 to 80 years. In place of the lush woodlands with immense age and species diversity, it plants relatively dry, even-aged, single-species crops known not as forests, but as tree farms. To facilitate the growth of only the desired species, it attempts to sterilize the land completely by removing or killing the gene pools of all other "competing vegetation." In the rush to create a monetary resource, it bypasses nature's initial stage of convalescence in which scrappy bushes and thrifty hardwoods first hold and replenish the soil for an ensuing grove of softwoods.

Environmental repercussions

This process involves burning all remaining sticks and logs (slash burning) and aerial spraying of chemical herbicides. These practices have far-reaching environmental repercussions. Ninety percent of the soil's essential nitrogen is volatilized in slash burning. In the Northwest, it accounts for 40% of all carbon emissions—a primary cause of global warming. Chemical herbicides long have been known to seep into the water table and work their way up the food chain, triggering disease and infertility in animals and humans alike.

Tree farms are physically inferior to natural forests. Aside from resembling a crop with few forest characteristics (each tree is a virtual clone of the others), they have inherent weaknesses that make them far more susceptible to disease, insects, and fire. The trees come from nurseries where every seed is germinated. In a forest, only the fittest eventually sprout. Limited to only one species, the tree farm is vulnerable to any number of diseases or insects. In the Southeast, where tree farms prevail, the budworm wreaks havoc in areas previously unharmed when the original forests existed. Fire usually damages only particular stands of trees in a native forest, while others of different age and species survive. In the dryer climate of a uniform tree farm, fire can raze the entire crop.

Never in the history of the world have humans achieved successful perpetual rotations of clear-cutting and replanting. The cumulative effects of erosion, slash burning, chemicals, and loss of natural fertilizer—the trees never are allowed to

decompose and nourish the system—deplete the soil and further diminish its ability to produce with each successive rotation. All over Europe, the forests are dying. In China, they have found the soil can not support the Chinese fir after only two or three rotations. In the southeast U.S., the dosage of chemical fertilizers continally has increased while timber output has decreased sharply.

A severe toll on wildlife closely follows the destruction of native forests. The northern spotted owl, which can survive only in this environment, has stirred a swell of controversy between conservationists and the timber industry. While the industry and pro-timber politicians maintain that the owl has no business obstructing *status quo* economics, others cite existing laws intended to guard against species extinction. Ecologists maintain that the owl population serves as an indicator for a variety of other creatures and plants. Should the owl lose its habitat, so too will the pileated woodpecker, tree vole, Roosevelt elk, Columbian black-tailed deer, and a host of other identified and unidentified species that depend on native forests for survival. Streams, kept distinctively cool in the native forests, support a variety of fish that can not survive in water above a certain critical temperature. Coupled with the suffocating effects of siltation, a rise in temperature has rendered many streams and rivers virtually deserted of fish life. Even wildlife refuges, supposedly set aside for permanent preservation, recently have suffered logging under a concept that clear-cutting enhances wildlife.

The impact outside the regions undergoing excessive logging is felt on many levels. The loss of biological diversity threatens not only the Earth's ecology and biosphere, but may eliminate potential breakthroughs in the medical field. The Pacific yew tree has shown cancer-fighting promise in recent experiments. Fifty percent of all medication comes from organic sources. In eradicating a deep reservoir of unique plant and animal species, we may be destroying the only cure for some of the world's most devastating diseases.

The temperate rain forests help counter global warming. Yet, once they are destroyed, the ensuing tree farms have a finite life span. Historical evidence suggests that harvesting the forest as a crop will exhaust the land to the point of infertility, and one more chunk of the world's carbon dioxide-consuming, oxygen-producing forests will disappear permanently.

The rush to cut this national treasure stems not from our country's lumber needs, but from the timber's cash value on the international market. As the streams of logs flowing from the forests swell, so too do the stockpiles at the ports awaiting shipment to Japan. An existing ban on Federal whole-

log exports has little effect on this lucrative overseas market since a number of loopholes render it largely ineffective and all raw-material forms such as pulp, chips, squared logs, slabs, and many others escape the ban. Less than 40% of timber cut in the Northwest is processed domestically into finished products.

Exports and increased automation deprive Washington and Oregon of an estimated 37,000 jobs in wood products processing and manufacturing. The Louisiana-Pacific timber company plans to eliminate some 1,000 potential jobs in northern California by barging rough redwood lumber to a new mill in Mexico that will employ cheap labor.

The boom/bust phenomenon of timber-dependent economies born out of the industry's cut-and-run policies threatens to destroy communities throughout the Northwest. The current unsustainable cut levels pose an inevitable shortage in timber-related employment. Historically, such economies have suffered a sudden, violent collapse, and industry shows few intentions of easing the blow. After a recent leveraged buyout, the Pacific Lumber company doubled its rate of cutting. Once hailed by workers and environmentalists alike for cutting at truly sustainable levels and selectively taking only a few trees from each stand, it now employs wide-scale clear-cutting to pay off the junk bonds used to purchase the company. The workers themselves openly oppose the new policy and hope to capture a controlling share of the company using an employee stock ownership plan.

To obscure the mounting evidence linking unemployment to poor management, automation, and exports and to propagate the image of a clean, conscientious industry, 17 major timber companies, operating as the American Forest Council, have collaborated on a $12,000,000 advertising budget. Full-page spreads in national magazines such as *Smithsonian, Life,* and *National Geographic* and abundant advertising on local television draw heavily on public sympathy for the blue-collar worker, claiming that forest preservation threatens jobs and "a way of life" in many rural communities. Yet, they fail to admit that the rate of employment continues to drop despite a steady rise in the amount of timber cut. In Oregon, the industry has eliminated 20% of the timber jobs while increasing cutting by 18%. Despite the evidence that the industry and its workers each want something entirely different—the former wanting short-term profits; the latter, long-term job security—the industry projects the illusion that it acts in the interest of regional economies and its employees.

Pro-timber policies

Feeding this insatiable appetite is a

government that continually has pursued pro-timber policies. Since 1980, native forests on Federally owned land in Oregon and Washington nearly have doubled their timber yield, reaching unprecedented levels. Pres. Bush's 1991 budget maintains the current Federal quotas.

In the summer of 1989, Congress approved Section 318 of the Interior Appropriations Bill that mandated the cutting "notwithstanding any provision of law" of more than 5,700,000,000 board feet of timber from Northwest Federal forests in 1990—more than they ever have had to relinquish in any single year. Among the laws that Section 318 countermanded were the Endangered Species, National Environmental Policy, Federal Land Policy and Management, Clean Water, and Migratory Bird Acts. It also includes a clause that effectively renders judicial review of timber sales useless, directly contradicting our constitutional concept of separation of powers. The Northwest delegation hailed it as an appropriate and innovatively conceived compromise, while conservationists, left with the prospect of losing an unprecedented acreage of native forest, feel abandoned and betrayed.

The Forest Service claims that it spends 35% of its budget on timber while recreation, fish and wildlife, and soil and water receive only two to three percent each. Conservationists' estimates of the timber budget run as high as 90%, noting that other categories such as road building and brush disposal primarily serve timber functions. In either case, both agree that the budget shortchanges non-consumptive, ecology oriented forest management.

The Forest Service sells the timber to the industry, but its annual operating deficit of over $1,100,000,000 indicates that the trees may not be fetching their true value. In Alaska's Tongass National Forest, the Forest Service grossed less than two cents in receipts for every dollar spent. It sold standing trees worth $700 each on the open market for as low as $1.48—less than the price of a Big Mac.

As one might imagine, not all Forest Service employees agree with current policy. Many rifts within the agency have surfaced and gained substantial publicity. The most notable is a Eugene, Ore.-based organization called the Association of Forest Service Employees for Environmental Ethics (AFSEEE). Formed in 1989 by a timber sale planner on the Willamette National Forest, AFSEEE has created a vehicle of expression and political leverage for employees distressed with the environmental degradation they must witness and foster in executing their jobs. Its primary focus is to promulgate this strong current of dissent and effect a shift in agency priorities from achieving Congressionally mandated, inflated harvest levels to ecologically sound stewardship of the forests. For more than 2,000 current and former Forest Service employees, AFSEEE provides a long overdue unification, and its membership continues to expand.

Possibly even more jolting to the upper echelon of the Forest Service hierarchy was the public disclosure of two November, 1989, letters to the Forest Service Chief, Dale Robertson. One came from the Region One Forest Supervisors, the other from the 63 Supervisors of Regions One, Two, Three, and Four. In them, they intimated a grave resentment over the continuing "focus on commodity resources" and expressed their exasperation with the agency's "conflicting values." In summing up their sentiments, the Region One Supervisors stated, "there is a growing concern that we have become 'an organization out of control.'"

For those who have seen the clear-cut landscapes, soiled rivers, and skylines of billowing smoke from slash fires, it is clear that something is quite amiss. Recently, a farmer from California took his family to Oregon for a vacation. "It doesn't take a scientist to realize what's going on; even my 10-year-old could see that they're cutting trees faster than they're planting them," he remarked. "I went up there to relax for a little while, and I came back madder than ever."

The lumber industry touts its ability to sustain "endless cycles of harvest and renewal." Yet, the Georgia-Pacific, Weyerhaeuser, and James River Corporation timber companies reportedly are seeking forests in the Soviet Far East, presumably because of a dwindling domestic supply.

The need not only to preserve, but to begin rehabilitating the world's forest lands is ripening in the public's mind as an issue of dire importance. We are experiencing noticeable changes in climatic trends—below-average annual snow and rainfall, rising mean temperatures, and expanding deserts—and the threat of global warming looms over the 21st century as a most fearful consequence of the rapid consumption of natural resources.

Now that the world has lost over half of its forests to development and agriculture, the time to stop their careless ruin is abruptly at hand. With nearly all of our nation's woodlands either eradicated, exploited, or converted to tree farms, the environment can not afford our jeopardizing the existence of its original, native forests. They are the last seeds of the natural, fruitful ecosystems needed to help heal and restore this neglected planet. Anything less than total and permanent preservation is environmental insolence.

Moreover, simple measures can be taken to make environmental protection economically feasible. We stringently must prohibit whole-log and raw-material exports, thereby drastically reducing demand and creating more domestic manufacturing jobs; further reduce demand by encouraging alternative, non-forest products; rechannel governmental funds currently allocated for excessive timber sales into restoration employment; and legislatively mandate proper, sustained-yield management.

Native forests remain the Earth's best hope for withstanding and resisting violent climatic change. To continue assaulting them is not only hypocritical of a country pointing fingers at Brazil, but economically illogical and environmentally disastrous.

WILDLIFE AS A CROP

Dick Pitman

Dick Pitman is an environmental consultant in Harare, Zimbabwe.

In the East African context it was almost a radical statement, a break with both the assumptions of many environmentalists and the near-universal opinion of small farmers in the region's rural districts:

"It is proposed to make wildlife an agricultural option to complement crop production and cattle raising", Zimbabwe's President Robert Mugabe announced. "Wildlife management will be rationalized to bring economic benefits to the rural communities that engage in it. Game meat will be processed in order to supplement our beef supply in the local market".

The key word in Mugabe's remarks, delivered at the opening of Zimbabwe's Parliament, was "resource". Few conservationists or planners, particularly those who approach Africa from bases in affluent Europe or North America, can bring themselves to regard wildlife as a resource. They share the exclusively aesthetic view of animals imposed on the continent by successive generations of colonists, and react to other positions with distaste.

Meanwhile, still smarting from the colonial experience, post-independence Zimbabweans from policy-makers to farmers tend to agree with a villager in the Simchembu ward of Gokwe Communal Land, who says: "Wildlife is nothing but a nuisance. Elephants destroy our crops every night. They (the government) can kill everything bigger than a hare as far as we are concerned".

The president's speech thus represented a landmark in the long evolution — often broken by radical shifts in attitudes, perceptions and legislation — of a new social and economic approach to Africa's unique animal heritage. Surprisingly, the impetus behind it stems partly from the work of several non-governmental organizations (NGOs) whose original mandates had little to do with game.

Wildlife — which in East Africa tends to mean large mammals — formed an integral part of the lives of indigenous peoples prior to colonial settlement, and the relationship wasn't always the harmonious one beloved of some romantic writers. People and big animals have always come into conflict, and crop damage is one of the less severe results. Even today, elephant, lion, buffalo, hippo and crocodile still cause many injuries and deaths in rural communities. Before colonial settlement, however, the disadvantages were at least partially offset by the "goods and services" game provided — meat, clothing and medicinal compounds.

Fundamentally aesthetic

This trade-off was disrupted by colonial governments, whose approach was fundamentally aesthetic and preservationist and admitted of few "utilitarian" inroads other than the occasional (and largely European) recreational safari. Many indigenous communities were evicted from planned wildlife areas, and themselves prohibited from any form of hunting.

The Shangaan people of the Mahenye ward, in southeastern Zimbabwe, are a classic example. They originated in South Africa's Northern Transvaal, but migrated into what is now Zimbabwe during the latter part of the 19th century. In the early 1960s their lands were selected for inclusion in Gonarezhou National Park.

The community was evicted and resettled on the edge of the new park.

From then on, the Shangaan came increasingly into conflict with wildlife authorities. They hunted elephant within the park and formed a strong bond with a notorious group of ivory poachers. After independence, believing ownership of wildlife had somehow changed hands, they went on a killing spree, using dogs to chase antelope into wires strung between trees and laying planks studded with six-inch nails on paths used by hippos. More than 100 elephants were killed by ivory poachers. The relationship between the people and park authorities deteriorated into virtual open conflict.

Eventually, faced with such examples, even preservation-oriented conservationists began to realize that classic law enforcement could not counter a situation that had its roots in a deep social injustice. The rapidly-growing rural population would continue to bring pressure on the country's spectacular fauna.

The legal key to change was a controversial piece of legislation, Zimbabwe's 1975 Parks and Wildlife Act. Under the act, wildlife remained state property, as it had been under colonial rule. But landholders were given responsibility for the wild animals on their lands and — within limits designed to prevent overuse and local ex-

Reprinted with permission from *CERES*, September/October 1990, pp. 30-35.

tinctions — permitted to use them for economic purposes.

However uncomfortable it might have been for those who believe wild game should remain physically inviolable, the practical result was plain. As a direct consequence of the economic gains made possible for landholders, the land area reserved for wildlife habitat, mostly on private farms and ranches, increased dramatically.

Most of the landholder returns accrued from sport hunting, as opposed to market hunting. Several experiments with the latter failed essentially because wild species were seen simply as replacements for cattle, providing a limited range of physical products, such as meat and hides. In competition with the highly subsidized beef industry, the logistics of hunting wild populations and transporting their products, combined with irregularity of supply and a lack of appropriate marketing institutions, made hunting for market unprofitable. Experiments with domesticated wild species, such as eland, fared little better.

Recreational values

The true economic spur was the realization that wildlife had a range of marketable values above and beyond those of meat and other products, and these values were mainly recreational. Sport hunting trophy fees can generate substantial incomes at relatively low, sustainable levels of offtake. For example, the return from an impala killed for its meat is less than US$50, while a safari operator can sell the same animal to a client for US$200 or more. Utilization based on tourism can also generate large returns — without any offtake at all.

Conventional cattle production on the Buffalo Range ranch in southern Zimbabwe witnessed a dramatic decline in productivity during drought years. A wildlife section on the ranch showed little advantage when used solely for meat cropping. But when the rancher began to market the recreational values of wildlife on the section — mostly to sport hunters — he generated returns per hectare three times as high as those of his cattle sections.

In this sector, composed mainly of private, largely European ranchers, the major institutional development has been creation of a Wildlife Producers Association (WPA), which forms part of the Commercial Farmers Union and now has the power to raise levies from its members to apply to such matters as game translocation.

However, this sector faces at least two major problems. The first has to do with land tenure and planned resettlement schemes, under which government has expressed its intention to acquire underutilized land for resettlement purposes.

Many powerful policy-makers are still unconvinced of the economic benefits that can flow from wildlife. Even if convinced, they may still prefer to see private wildlife land redistributed for a range of political and perceptual reasons. This attitude is at least partly reflected in a recent speech by Jock Kay, Deputy Minister of Lands, Agriculture and Rural Resettlement: "Game ranching is expected to expand in the 1990s, although it will be necessary for governments to ensure that this expansion is rational and does not adversely affect the livestock industry".

Kay's statement illustrates a second problem: the bias toward cattle still prevalent in Zimbabwean legislation. Game capture, translocation and the movement of products such as meat are severely hamstrung by veterinary restrictions imposed to facilitate exports of beef to the EEC under the Lomé convention. Millions of dollars have been devoted to cattle research, but little or nothing to the wildlife industry. Also, while beef is highly subsidized and marketed through the parastatal Cold Storage Commission, wildlife enjoys no subsidies and has no formalized marketing structure.

Thus, though most returns from wildlife are generated from recreation, a small but important economic potential — the sale of meat — is still denied the wildlife industry.

The situation with regard to rural communities — the people on whom the survival of wildlife ultimately depends — is somewhat different, and has resulted in a slower adoption of wildlife as a land-use option. Most rural communities occupy what are now known as "communal lands" — direct descendents of the Native Reserves established by early colonial governments. These governments often reserved the best agricultural land for commercial settler-farmers, so communal lands are generally located on poor soils, often with low or erratic rainfalls.

Many communal lands are thus unsuited to intensive agriculture, but do have significant wildlife populations. However, under the 1975 Act, rural communities do not enjoy the same rights as private landholders over their wildlife.

Even before the Act was passed, the income from sport hunting concessions

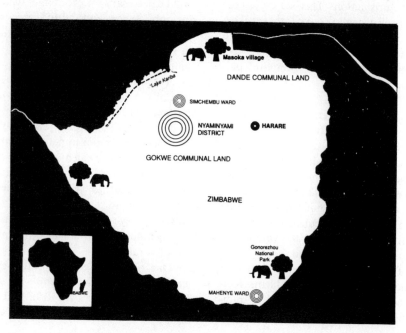

6. BIOSPHERE: Animals

in some rural areas was, at least in theory, being passed from government back to local communities. But the system suffered from several fundamental flaws. Central Treasury often retained funds for lengthy periods, or did not return them at all. Funds were often spent in areas far removed from those in which the hunts took place, thus omitting the vital link between tolerating wildlife and benefiting from its economic potentials; and communities had no involvement in planning or decision-making.

The Simchembu villagers, today so hostile toward wildlife, in fact benefited from the proceeds of an elephant culling exercise in a neighbouring National Park in 1981. But they took no part in management decisions concerning the exercise or the distribution of benefits and the project was never followed up. Vague promises of future benefits do nothing to outweigh present problems of crop damage and potential injury or death.

Campfire philosophy

There is, however, a clause in the 1975 Act that enables District Councils — a unit of local government created after independence — to become what is known as the "appropriate authority for wildlife management". Councils with this status have the same rights as private landholders and can manage, use and benefit from wildlife management in the communal lands under their jurisdiction. This status is currently awarded to councils that make a clear statement of their intention to manage wildlife by what is rapidly becoming known as the "Campfire" philosophy.

Campfire — Communal Areas Management Programme for Indigenous Resources — was evolved by the Zimbabwean Department of National Parks. Its most fundamental principle is that the rural communities that tolerate wildlife on their lands — and hence act as "wildlife producers" — should be able to take responsibility for, and benefit directly from, the wildlife resource in cash and kind.

This apparently simple statement conceals several knotty problems. Rural communities, many with a traditional bias toward cropping and cattle that is allied to their legacy of hostility toward wildlife, have to be made aware of the financial and other benefits that can flow from wildlife management.

The question of wildlife "ownership" also poses problems. Not only is it still technically state property: it is a fugitive resource, and traditionally regarded as a common resource as well. These factors create an inherent tendency toward opportunistic overuse.

For some time, the promotion of wildlife as a form of communal land-use hinged on the likely — but unproven — proposition that it is the most appropriate form of use for marginal lands in ecological and economic terms. Only now is one agency — the Worldwide Fund for Nature, in the shape of its Harare-based Multi-Species Animal Production Systems Project — examining these and other related hypotheses.

Some external aid agencies and government departments are still biased toward conventional cropping and cattle production systems on marginal communal lands — sometimes even when land-use studies recommend wildlife management as the most viable option in specific areas.

This in turn raises a couple of further points. One, not fully grasped by these agencies and departments, is that even the most enthusiastic professional proponents of wildlife schemes see them as mere adjuncts to cattle and cropping, both of which are and will still be necessary for economic (including subsistence) and cultural reasons. The key lies in sound land-use planning in close consultation with rural communities themselves.

A second crucial point lies in the nature of the agencies involved. The Department of National Parks still plays a central role in communal land wildlife projects. But a major impetus has come through the involvement of NGOs and agencies engaged in rural development — as opposed to wildlife conservation — with no real or imagined "axes to grind" beyond a real desire to facilitate genuine improvement in the quality of rural life.

One of these agencies is the University of Zimbabwe's Centre for Applied Social Sciences (CASS). In 1984 the Department of National Parks asked CASS to provide facilities for research, monitoring and evaluation of the socio-economic implications of the Campfire approach. As a result a research unit, funded by the Ford Foundation, was set up in 1985.

CASS had already identified wildlife as a potential catalyst for social, cultural and political change, and since the establishment of its specialized research unit it has focused on issues involving natural resource management and common property. After potential Campfire areas have been identified by the Department of National Parks, CASS may undertake initial socio-economic research, and may also become involved in discussions with district councils and local communities.

Nyaminyami scheme

Another agency, the Zimbabwe Trust, has become involved in institution-building at local levels. The trust had already recognized that wildlife was often the only resource that stood between many rural communities and permanent dependency on food aid. It formally established a Wildlife Community Development Programme in 1988. This programme helps rural communities to develop their institutional capacity to manage natural resources, and can also assist with project identification and appraisal, planning, monitoring and evaluation, as well as in locating initial funding.

So far, the philosophies held by such NGOs, rather than government policy, have been responsible for integrating wildlife projects into rural development. Their roles evolved during the establishment of what is regarded — not entirely accurately — as the archtypical Campfire-style scheme in the Nyaminyami district on the southern shore of Lake Kariba.

This region, one of the poorest in Zimbabwe, suffered the country's highest rate of malnutrition and protein deficiency in spite of abundant natural resources in the shape of wildlife and potential fisheries. After four years of debate, the Nyaminyami District Council became the appropriate authority for wildlife management in 1989 and generated a cash surplus of US$260 000 during its first year of operations, mainly from sport hunting.

Of this surplus, US$6 000 was placed in an operating reserve, US$26 000 was retained by the District Council, and US$194 000 was made available for participating communities.

Unfortunately, at this stage Nyaminyami ran into a bureaucratic quagmire common to several Campfire projects: having itself acquired the right to manage wildlife, the District Council became reluctant to devolve decision-making to the communities. It decided that the income should be divided equally between the 12 wards within its area — with no recognition of the principle that income should be biased toward the communities that bear the highest social and other costs of the wildlife scheme.

Unless blockages of this kind are resolved, Nyaminyami and several other Campfire schemes may run into severe problems. In the Dande communal land, several hundred kilometres east of Nyaminyami close to the Zambezi River, a similar project generated a surplus of US$101 621 for distribution to participants. There were jubilant scenes in the Masoka village when a US$200 dividend was distributed to each household — more cash than most residents often saw in a year. But in this case, other areas with an equal claim to benefits have so far received nothing at all. And the Shangaans of Mahenye, who were evicted from the newly-created Gonarezhou National Park, laboured under this burden for several years.

There is also little doubt that the Simchembu people, now hostile to wildlife, would see matters in a different light if their district council had spent the US$100 000 they recently received from wildlife management within the area — instead of erecting a beerhall in a densely-populated, semi-urban area 100 km away. But it will probably need an NGO such as the Zimbabwe Trust, to spend the time and effort needed to sit down with the community and discuss their problems and aspirations.

The effective integration of wildlife management into Zimbabwe's agricultural policy may or may not resolve problems of this kind. As yet, there are no proposals on paper to back President Mugabe's speech, except for an incipient five-year agricultural sector plan spanning the years from 1991 to 1995. The Ministry of Lands, Agriculture and Rural Resettlement has apparently suggested that a number of working groups should be formed. One of the groups will be concerned specifically with livestock and wildlife production.

This group is likely to be tasked with analysing the economic and social implications of livestock and wildlife production, and with reassessing the current animal health legislation to help promote development in the wildlife sector. If the plan deals effectively with questions such as wildlife ownership, marketing structures, land-use and — above all, in the case of communal lands — creates a legal structure that devolves decision-making down to individual communities, then the wildlife option is likely to have a bright future.

But if development simply becomes enmeshed in one more layer of governmental decision-making, this fledgling industry may find it was better off with a less formalized network of NGOs and individuals fighting and often winning battles "on the ground" instead of dealing with a centralized bureaucracy.

'I was staggered. ...'

By the early 1980s, virtual war had broken out between the Shangaan people of southeastern Zimbabwe's Mahenye ward, who had been evicted from their lands to make way for Gonarezhou National Park, and the government authorities charged with protecting the park's wildlife. Poaching was rampant, and ill-will hardening on both sides.

Then Clive Stockil, a local rancher who had grown up in the area and spoke fluent Shangaan, persuaded the Department of National Parks to provide the community with a small legal hunting quota outside the national park, on the stipulation that they halted their illegal activities.

"I was staggered when I went back to the Mahenye area a month later", says Stockil. "Not only did they comply with this stipulation: part of the community voluntarily moved off an island in the river that formed the park boundary, and declared it a wildlife management area". During the next five years the community generated a growing income from well-controlled sport hunting but, because of bureaucratic wrangling, received nothing in return except a few kilogrammes of free meat. Naturally, cynicism and suspicion mounted.

The fact that the scheme did not fall apart completely is due chiefly to Stockil's dedication and his diplomatic talents in acting as a link between the community and local and national bureaucracies.

The money was eventually dislodged and the community built a maize mill and a school, which now has 10 teachers to serve 500 local pupils. What's more, when several government agencies proposed that a 200 sq km block of land within the Mahenye area should be devoted to cattle raising, the Shangaan community — former poachers who had helped kill 100 elephants — objected strongly and insisted the land be used for wildlife management.

This shows how perceptions of wildlife can be changed by policies based on an enlightened attitude toward indigenous community needs and sensitivities.

Dick Pitman

*Private animals, public land,
and the limits of privilege.*

How the West was Eaten

George Wuerthner

GEORGE WUERTHNER, a frequent contributor to *Wilderness*, is a former park ranger and a trained biologist and botanist with wide field experience on the public lands. He is the author-photographer of several titles in the American Geographic series, including books on Alaska, Oregon, Texas, and the Southern Appalachians.

Take a moment and ponder what the following species might have in common: Grizzly bear, black-footed ferret, desert tortoise, dunes tiger beetle, Bruneau Hot Springs snail, Colorado squawfish, Mexican wolf, Uncompahgre fritillary butterfly, masked quail, Hualapai vole, Fremont cottonwood, willow-whitehorse cutthroat trout, Cusick's camas, and Bitterroot milk vetch. There may not seem to be much that a snail shares with a grizzly bear or a desert tortoise with a trout, but each of these species has suffered extirpation over most of its range to the point that it is officially listed or is a candidate for listing as an endangered species—and it is in this sorry state at least partly because of the western livestock industry.

The potential extinction of species is merely one cost of the West's love affair with the cow and the cowboy. Increasingly, however, the infatuation is going sour as the real costs of livestock grazing, particularly on the public lands, become more apparent and the industry comes un-

der fire from a broad array of people, conservative economists as well as liberal ecologists. The direct economic costs of public lands livestock grazing are relatively easy to assess. For starters, ranchers pay much less per AUM (Animal Unit Month, a measure of forage consumed by a cow in one month) to graze on public lands than on comparable private lands—in some places as much as 80 percent less. At present, stockmen pay the government $1.81 to board a cow on public lands for a month. As Rose Strickland, chair of the Sierra Club's Public Lands Committee points out, "this is far less than it would cost to feed a cat for a week."

But the public subsidy does not end with low grazing fees. Taxpayers also foot the bill in part or in full for a host of range developments and "improvements," including fencing, cattle guards, water pipelines, stock ponds, weed control, predator control, spraying, and chaining (stripping natural cover from wide areas by dragging an enormous chain across them), and seeding programs. It is not difficult to understand why many critics charge that public-lands grazing programs are little more than "cowboy welfare."

If indeed it is public welfare, who benefits? Well, most assuredly some small-time ranchers are helped. But more often, it is people like the owner of the Vail Ski Corporation, whose Roaring Springs and MC Ranches subsidiaries in Oregon lease more than 900,000 acres of federal

From *Wilderness*, Vol. 54, No. 192, Spring 1991, pp. 28-37. Copyright © 1991 by The Wilderness Society.

lands, who get the handout. Or people like J.R. Simplot, said to be the wealthiest man in Idaho and, not coincidentally, the largest beneficiary of federal grazing privileges in the state. Or corporations like Union Oil and Getty Oil, both of which run cows on federal lands.

These are not isolated instances. According to a U.S. House of Representatives Committee on Government Operations report, three percent of the livestock operators in the West utilize 38 percent of the federal grazing lands; less than ten percent of federal forage goes to permittees who are considered small-time ranchers. The reason for this disparity is simple. Access to public-lands grazing privileges is fixed to base property ownership—the more property you own, the greater the likelihood that you will have federal grazing privileges attached to it.

Arizona range-activist Lynn Jacobs, author of the forthcoming *The Waste of The West*, a book on the western livestock industry, has little patience for what he calls the "poor rancher syndrome." "Most ranchers," he says, "may be cash poor, but they hardly qualify as poverty-stricken. They live in comfortable homes surrounded by thousands of acres of land they own—land that is worth hundreds of thousands of dollars, if not millions, even after you account for their indebtedness."

While there is no doubt that permittees—rich or poor—benefit from public-lands grazing, it is less clear how the public itself—the land's real owner—profits from such programs. "For too long livestock use of public lands occurred in a vacuum," says John Zelazny, Executive Director of the Wyoming Wildlife Federation. "We made some major trade-offs between wildlife, fisheries, watersheds, soils, biodiversity, and livestock use, and most people were not aware of how extensive nor how great these trade-offs were."

The magnitude of these trade-offs is not apparent until one considers that livestock grazing is hardly a benign use of the land—and a great deal of land is involved. According to a 1989 report by the U.S. Forest Service, 44 percent of the land base in the United States, excluding Alaska, is used for grazing by livestock. The percentage of federal lands outside Alaska committed to livestock production is even higher, with grazing permitted on 89 percent of Bureau of Land Management (BLM) lands and 69 percent of Forest Service lands.

Even federal lands specifically set aside to protect wildlife values experience some level of livestock use. For instance, 103 out of 109 wildlife refuges in the Fish and Wildlife Service's Region 6—which includes Montana, Colorado, Wyoming, and several plains states—are grazed by domestic animals, as are some units in our National Park System. Even many designated Forest Service and BLM wilderness areas are open to grazing.

Altogether, more than 265 million acres of federal lands, an area larger than the combined acreage of the entire Eastern Seaboard from Maine to Florida, are leased under federal grazing programs. Despite the huge acreage involved, the Department of Agriculture estimates

that these federal lands provide the forage for less than two percent of the domestic cattle and sheep produced annually in the United States. This trifling production comes at considerable ecological cost. A 1989 analysis of BLM lands completed by the Natural Resources Defense Council and the National Wildlife Federation concluded that more than 100 million acres were considered "unsatisfactory."

An even larger amount of private land in the West is in poor condition. A 1987 National Rangelands Inventory conducted by the Soil Conservation Service estimated that 64 percent of non-federal private rangelands were in unsatisfactory condition—more than 270 million acres.

Do such figures mean that ranchers are blindly determined to ruin their own and the public's lands? No—not deliberately, in any case. The current condition of our rangelands is as much a factor of using the wrong animal in the wrong place as it is of lax management. Cows need water. Lots of it. Most cattle breeds originated in well-watered northern Europe, and they are, as a consequence of their evolutionary heritage, inefficient users of water. John Robbins, author of *Diet for a New America*, claims that each pound of beef produced requires 2,500 gallons of water, while the amount of water needed to raise an average-sized cow is sufficient to float a battleship.

Whatever the precise measure of consumption, we clearly have an animal overwhelmingly dependent on water being raised in a water-poor environment. West of an isohyetal line that roughly parallels the 98th Meridian, precipitation drops below twenty inches a year—the lower limit for many non-irrigated crops—and evaporation is high. Perhaps even more important is the fact that precipitation becomes much more erratic west of this line, and since forage production is tied almost directly to precipitation, the region's frequent droughts either require massive herd reductions or the land suffers from overgrazing.

Some of the early explorers of the West recognized these inherent limitations. John Welsey Powell, founder and second director of the U.S. Geological Survey, noted in his landmark 1879 report, *Lands of the Arid Region of the United States*, that traditional patterns of settlement and agriculture would not work in the West. Powell wrote: "Though the grasses of the pasturage lands of the west are nutritious they are not abundant, as in the humid valleys of the East. . . . These grasses are easily destroyed by improvident pasturage, and they are replaced by noxious weeds. . . . They must have protection or be ruined. . . ."

Drought or no drought, aridity limits livestock production. In moist, humid Georgia, you can raise a cow year round on an acre of ground or less. In arid Nevada, it may require more than two hundred acres to board the same cow. Not surprisingly, then, despite its size and the fact that nearly the entire state is devoted to livestock production, Nevada still only manages to raise as much meat as Vermont, a state barely larger than Yellowstone Park. It is farmers in the East, Midwest, and South who raise the vast majority of the Nation's livestock, not ranchers in the West.

Still, because of the gargantuan needs of its stock, the western livestock industry is the single greatest consumer of water. In Montana, for instance, use by industry and urban communities accounts for only 2.5 percent of the water removed from waterways, while agriculture, primarily irrigated hay fields, uses the rest—a whopping 97.5 percent.

What are the environmental impacts of this runaway consumption? One "concrete" consequence is the many dams that now plug the rivers of the West. And dams and the reservoirs behind them are more than just eyesores. According to conservation biologist Reed Noss, "dams fragment river ecosystems in much the same way that clearcuts break up forest canopies, blocking migration and genetic exchange between aquatic lifeforms." What is more, irrigation dewaters streams and rivers and concentrates pollutants, including pesticides, fertilizers, and mining wastes. As waterways shrink or become uninhabitable, the whole web is damaged. From a loss in airborne insects to feed birds and bats to fewer fish to feed mink, otter, and bald eagles, the impacts ripple all the way through the ecosystem.

In some areas, groundwater pumping for irrigation also is a problem. As water tables drop, springs dry up, with severe consequences for wildlife. Biologist Greg Mladenka, who is studying the Bruneau Hot Springs snail, a candidate species for listing under the Endangered Species Act, attributes groundwater pumping of the aquifer along Idaho's Bruneau River as the main reason behind the observed decline in snail populations.

Irrigation and groundwater pumping affect water availability, but cattle have other impacts on the watersheds themselves. Compaction of soil by countless hooves can significantly reduce water infiltration. Instead of soaking into the soil, water rapidly runs off, causing the erosion, gullying, and channel down-cutting so prevalent throughout the West. Grazing also results in greater stream sedimentation. The Environmental Protection Agency (EPA) estimates that livestock grazing on rangelands accounts for 28 percent of the annual sediment production in the West, second only to croplands in total sediment production. One Colorado study demonstrated an increase of 71 to 76 percent in sedimentation from grazed watersheds over ungrazed ones.

Nowhere is the effect of livestock grazing more apparent or damaging than in the West's fragile riparian zones, those thin, green lines of lush vegetation adjacent to streams and springs. According to a 1988 General Accounting Office (GAO) report, these biologically critical areas comprise less than one percent of the land base in the West. Perhaps because of their scarcity, their importance to wildlife and biological diversity is magnified. "In the arid West, nearly all life is dependent in one way or another on these relatively productive areas," says Steve Johnson, the former Southwest representative of Defenders of Wildlife and now a private ecosystem consultant. A 1989 study by Johnson documented the fact that as many as eighty percent of native animal species in Arizona and New Mexico are partially or fully dependent upon riparian habitat. Another study completed by the BLM in eastern Oregon found that 82 percent of 363 animal species were dependent upon such areas for their survival.

Few dispute the biological importance of riparian areas, then, but almost no one recognizes how scarce these precious corridors have become—or how mutilated. In 1988 the Arizona Fish and Game Department reported that less than three percent of the state's original endowment of riparian zones remained intact. Many factors are responsible for this decline, including highway construction and water impoundments, but according to a 1988 GAO report on riparian zones, "poorly managed livestock grazing is the major cause of degraded riparian habitat on federal rangelands." And a 1990 EPA report on riparian areas says that "extensive field observations in the late 1980s suggest riparian areas throughout much of the West are in the worst condition in history."

Often as not, such damage is trivialized as being merely isolated instances of poor management—what range managers have repeatedly told me are just "wrecks" or "bad spots." But we are not talking about a few "spots." The 1988 GAO report noted that 80 percent of the 11,867 miles of riparian habitat on BLM lands in Idaho were considered by the agency to be in "some stage of degraded condition," and officials in Arizona's Tonto National Forest estimated that 80 to 90 percent of their forest's stream riparian areas were in "unsatisfactory" condition.

Hugh Harper, who spent most of his thirty-five years of government service as a Grazing Management Specialist with the BLM, now retired in Boise, Idaho, sums up the problem. "In my observations I'd have to say that 98 percent of the livestock use occurs on one or two percent of the land—the riparian areas," he notes. "But they allocate use based upon the *whole* allotment. In most cases 50 to 80 percent is unusable because it's too steep or it's too far from water. So the riparian areas get pounded and the watershed goes to hell. With nothing to hold back the water, it runs off immediately and we get gullying, erosion, and a drop in the water table."

And because uplands are lightly grazed, if touched at all, you find range managers asserting that rangelands have improved. Such claims are often deceptive, says Barry Reiswig, manager of the Sheldon Wildlife Refuge, which spreads over portions of both Nevada and Oregon. "You'll hear a lot of range cons, [range conservationists], even people on my staff, crowing about how 99 percent of the range is in good or excellent condition, but they ignore the fact that the one percent that is trashed—the riparian zones—is really the only part important for most wildlife species."

One way cattle destroy riparian areas is by eating tree seedlings. Researchers at the University of Montana reported in early 1990 that livestock grazing was a major factor in the decline of plains cottonwood along the Wild and Scenic Missouri River in Montana. The loss has been masked to most observers because many "historic" large old trees remain—though as they die, they are not being replaced. In some areas the decline is critical to the point of extinction. The Arizona Nature Conservancy reports that Fremont cottonwood-willow forest communities, once common throughout the state, are now threatened, with only twenty occurrences known to exist—and only five of these are considered extensive.

Cottonwoods are considered a "keystone species" because the welfare of so many other plants and animals depend upon them. In many ways riparian species such as large old cottonwoods play much the same functional role that old growth species do in the temperate rainforests of the Pacific Northwest. Cottonwoods, because of their potentially large size, serve as virtual condominiums throughout much of the West too dry for extensive forest cover. Bald eagles and great blue herons roost in their large branches, woodpeckers and bluebirds nest in cavities in their boles, and even some species of bats hibernate in cavities or under pieces of loose bark on dead trees. As this use suggests, the value of large cottonwood trees does not end when they die. According to Arizona State University professor Wendell Minckley, "Once the large boles fall into streams, they provide hiding cover for fish, long term nutrient sources for aquatic ecosystems, and help to stabilize streambanks."

Cottonwoods are not the only riparian species affected by livestock use. Willow, aspen, and a host of other shrubs have all been eliminated by livestock over large portions of the West. Grazing of streamside vegetation also can change stream hydrology. Trampling and removing vegetation by livestock damages streambank stability and as a consequence banks slough off, channels widen, and streambeds become shallower. Such changes in stream morphology result in severe impacts to fish habitat. Fish populations are not only dramatically lower in grazed waterways, but their average size tends to be smaller. "This kind of damage is so widespread," says Beaverhead National Forest range technician Stephanie Wood, "that most people, including most range managers, have never seen a healthy stream channel."

In recent years there has been a general recognition of the value of riparian zones, but doing something about livestock impacts is not an easy matter. In the few areas where cattle *have* been removed, astounding changes occur. While manager of the Red Rock Lakes Wildlife Refuge in Montana, Barry Reiswig terminated five grazing permits. He says that shortly after cattle were taken off, "willow began to come back to places we hadn't seen them in fifty years, and as a result of new nesting and foraging areas we got a big in-crease in songbirds. Beaver numbers also went up. The entire biological value of the refuge increased significantly." Mere reduction in cattle numbers, however, though a step in the right direction, is often not the final solution. "It only takes a few cows to trash a riparian zone," says Barry Reiswig. "Even if you have fewer cows, they just spend more time in the riparian areas doing the same amount of damage as if you had three or four times the number."

To reduce livestock impacts on riparian areas, most government agencies propose to improve "distribution." This often means greater manipulation and development, including new fencing, construction of stock ponds, spring developments, pipelines, roads, and holding facilities—all of this to keep privately owned cattle out of public water courses and all of it at a loss to landscape integrity.

Often these range "improvements" are sold to the public as a wildlife benefit. Refuge manager Barry Reiswig claims many of the benefits are exaggerated or even suspect. "Almost any excuse is used to expand or continue livestock grazing. For example, on my refuge my range con would like to allow cows to graze wet meadows down to stubble supposedly to provide goose browse—even though we only have a dozen or so geese on the whole refuge. Never mind that an ungrazed meadow is far more valuable to most of the refuge's wildlife."

Once developments are built, they have to be maintained, and maintenance is costly. "Water developments are nothing more than mitigation for livestock," says Hugh Harper. "We are treating the symptom instead of the cause of our problems. Water developments, fencing, pipelines—I've seen a million of them—no way in the world are we going to have the staff or the money to maintain them."

Even if we wanted to spend hundreds of millions of dollars to fence the thousands of miles of riparian habitat in the West, and then transfer livestock use to upland areas by developing stock ponds and water pipelines, it would not help all that much, at least according to Harold Winegar, a retired Oregon Fish and Game biologist and now a riparian ecosystem consultant. "Watersheds are all connected," Winegar says. "If you move cattle out of the stream bottoms and into the uplands you will still be pounding to death the springs, seeps, and creeks, not to mention contributing to soil compaction over the entire uplands."

But the negative impacts of livestock do not stop with riparian areas, soil erosion, or stream dewatering. Many domestic animals carry diseases for which native species have little or no immunity. The decline and extinction of many bighorn sheep herds, for instance, is directly attributed to the introduction of a variety of diseases carried by domestic sheep.

In addition, both cattle and domestic sheep eat forage that otherwise would wind up in the bellies of native species. Too often domestic animals get the cake and icing while wildlife is left the crumbs. The Burns BLM district in

Oregon, for example, allots 252 million pounds of forage to livestock, while wildlife gets a trifling 8 million pounds. Though forage competition between livestock and big game may be obvious, dietary overlap affects other species as well. Such competition is listed as one of the major reasons why desert tortoise populations are in decline and threatened with extinction in California. In Colorado, the Uncompahgre fritillary butterfly is threatened by sheep grazing because these domestic animals eat the butterfly's primary food source, a species of alpine flower. Another more glamorous animal threatened by forage competition is the grizzly. Bears are primarily vegetarians, and a few preliminary studies indicate that domestic sheep and cattle eat many of the same plants as bears. Less food for bears may cause them to wander more widely, which of course may result in greater bear-human conflicts with the bear usually ending up the loser. Bears, butterflies, tortoises—all compete with private livestock for food on our public lands.

But cattle don't necessarily have to eat the same food to threaten the survival of other animals. Cows can trample the eggs of ground-nesting birds, and the loss of cover as a result of grazing can lead to increased predation losses to birds, small mammals, and other animals. Too often, however, this is not the predation that worries government managers. A federal agency, the Animal Damage Control (ADC) spent $29 million in 1988 to kill coyotes (76,000) as potential consumers of sheep and cattle, and prairie dogs (124,000) as competing consumers of grass. Often such "non-target" species as the golden eagle and the rare kit fox have been killed by traps and poisons aimed at "pests."

The reason such ecological mayhem is permitted is partially because of funding biases in federal agencies. Sheldon Refuge manager Barry Reiswig is quite candid about it with regard to his own bailiwick: "Nearly all of the refuge funding goes towards managing cattle owned by eight permittees. What little is spent on wildlife is mostly damage control. It's not making things better for wildlife, unless you call mitigating impacts from livestock making things better. While I have people to build and maintain fences, stock ponds, water pipelines, and other development for the permittees and a range conservationist to oversee the grazing program, I don't have one biologist on my staff—and this is supposed to be a wildlife refuge!"

Supporters of livestock grazing often justify these funding biases because, they maintain, the livestock industry produces jobs for rural economies. While in some western communities the number of direct and indirect jobs associated with the industry is indeed important, this benefit is by no means dominant—or even significant—in many other regions. Livestock operations, after all, are labor unintensive and pay so poorly that many ranchers find it necessary to import foreign workers on special three-year work visas. In some regions the job numbers are glaringly insignificant. For example, a recent review of the Greater Yellowstone Ecosystem by the Congressional Research

Service (CRS) shows that despite the fact that livestock grazing occurs on seven national forests within the ecosystem, only 144 direct jobs or 3.6 percent of the jobs attributed to Forest Service lands are the result of livestock operations. Recreation jobs, on the other hand, account for 1,200 direct jobs on the Gallatin National Forest alone.

But if the jobs involved are relatively insignificant, the power is not. The political influence of the livestock industry upon managing agencies is enormous. The 1988 GAO grazing report recognized this when it said that "the BLM is not managing the permittees, rather, permittees are managing the BLM." As an example, the report described an incident in which a BLM area manager confronted a rancher who was caught illegally cutting trees on BLM lands. The wood was confiscated. But because of the rancher's political connections, the area manager was later reprimanded by his district manager and ordered to apologize to the permittee and deliver the wood to his ranch.

Not all influence is so explicit. When I worked as a botanist for the BLM in the 1980s, one of my assignments was to inventory BLM lands for areas with outstanding biological, geological, or archeological attributes and make recommendations to protect these sites as ACECs (Areas of Critical Environmental Concern). After I submitted my first recommendation, I was called into my supervisor's office and asked to redefine my ACEC boundaries. I had included some lands that were ungrazed because of limited water supplies and as a consequence were wonderful examples of native bunchgrass rangelands. However, my supervisor said that the Grazing Advisory Board, a citizen group made up entirely of ranchers that "advises" the BLM on management decisions, would never approve the proposed ACEC as outlined. Therefore, he said, the state director would never approve it either. Thus I had to redraw my boundaries to exclude all areas that were being grazed or could potentially be grazed at some future date, or risk rejection of the entire area. As my boss said, half a loaf is better than none.

In spite of their incredible influence, the number of public lands permittees is surprisingly small. There are only 23,000 permittees spread over sixteen western states. In Nevada, for example, only 880 permittees graze livestock on federal lands. In Wyoming—the "Cowboy State"—there are only 1,607 "cowboy" permittees. There are more members of the Wyoming Wildlife Federation than there are ranchers in the entire state of Wyoming, but it is ranchers, not conservationists, who set the agenda on public lands.

As a consequence, many conservationists view with skepticism recent propaganda attempts by government agencies to paint a more positive image of the livestock industry. Lynn Jacobs, who has traveled thousands of miles in the West looking at public rangelands, doesn't believe a word of it. "Despite all the hoopla and slogans printed in nice glossy color brochures stacked in BLM and Forest Service offices talking about 'Change On The Range' and 'Grazing Successes on Public Lands,'" he writes, "except for a few expensive 'showcase' demonstration sites, when you get on the back forty all you see is the same old cow-blasted uplands, cow-nuked riparian zones, trampled, stomped, and torn streambanks."

Part of the problem is that better management in the arid West is expensive. A similar investment in Georgia or anyplace else it rains would produce far more meat for far less money. Hugh Harper explains: "Given the natural parameters of the landscape and the limitations imposed by terrain and aridity, if ranchers had to pay the full costs of operations, livestock grazing would not be economically feasible. I'm not against proper livestock grazing. I think it's possible to graze land without damaging it, but to do it right costs a great deal of money. There's just no way you can justify the supervision required—the great investment—for what you get back."

Nancy Green, Director of BLM programs for The Wilderness Society, says that The Society is not against grazing, per se, either. "The livestock industry, properly regulated, can have a place in the web of uses on the public lands of the West. But," she emphasizes, "where, when, and how much grazing should be allowed depends entirely on what the land, habitat, and wildlife resources themselves can absorb in good health. Too often and in too many places, graziers have been allowed not merely to use the land, but to abuse it. That has got to stop—and, furthermore, the public is going to have to start getting a fair return in grazing fees in those areas where livestock is allowed. As it is now, the grazing system is a bankrupt policy both financially *and* ecologically."

The question, finally, is what we want our public lands to be. Even "proper livestock management," critics of the industry point out, is not the same as maintaining natural self-renewing ecosystems. Does the public really benefit, these critics ask, by allowing its lands to function as feedlots for privately owned domestic animals? Considering the many "hidden" costs associated with livestock production in the arid West and the numerous opportunities the nation has for livestock production on lands not only in the West, but in the Midwest, South, and East, they wonder, would it not be a higher use of our public lands if they were managed for values not easily obtained or protected by the private sector—such as the preservation of biological diversity and habitat for native wildlife and plant communities?

These kinds of questions—increasingly posed by those living in the heartland of the West itself—suggest the growth of a sentiment less inclined to accept at face value the livestock industry's argument that it knows what is best for the land. That sentiment has even crept into some corners of the western press, that traditional ally of the industry. When officials in Idaho's Sawtooth National Forest announced a long-range plan to severely reduce cattle grazing in the Stanley Basin south of the Sawtooth Range, for example, *The Idaho Statesman* came down solidly on the side of the Forest Service in an editorial on October 25, 1990. The Service, the editorial said, was displaying "commendable courage," and went on to note that "grazing is coming into increasingly serious conflict with the basin's growing recreational usage, which clearly represents its future. . . . The public cannot allow grazing to continue at historic levels . . . in this most special of places."

The livestock industry, obviously, is operating in a new public environment, and if it is to survive at all in the West its proponents would do well to consider carefully the words of a politician who was once one of their own. He was Congressman Edward I. Taylor of Colorado and he was speaking in 1934 in explanation of why he had introduced and engineered passage of the Taylor Grazing Act. The legislation was the government's first real attempt to control decades of abuse and was violently opposed by the livestock industry. Taylor's fifty-seven-year-old sentiments retain a startlingly modern flavor: "I saw waste, competition, overuse, and abuse of valuable rangelands and watersheds eating into the very heart of the Western economy. . . . Erosion, yes, even human erosion, had taken root. The livestock industry . . . was headed for self-strangulation."

Environmental Information Retrieval—
On Finding Out More

There is probably more printed information on environmental issues, regulations, and concerns than on any other major topic. So much is available, from such a wide and diverse group of sources, that the first effort at finding information seems an intimidating and even impossible task. Attempting to ferret out what agencies are responsible for what concerns, what organizations to contact for specific environmental information, and who is in charge of what becomes increasingly more difficult.

To list all of the governmental agencies, private and public organizations, and journals devoted primarily to environmental issues is, of course, beyond the scope of this current volume. However, we feel that a short primer on environmental information retrieval should be included in order to serve as a springboard for further involvement; for it is through informed involvement that issues, such as those presented, will eventually be corrected.

I. Selected Offices Within Federal Agencies and Federal-State Agencies for Environmental Information Retrieval

Department of Agriculture
14th St. and Jefferson Dr., SW, Washington, DC 20250 (202) 655-4000

Appalachian Regional Commission
Public Information: 1666 Connecticut Avenue, NW, Washington, DC 20235 (202) 673-7869

Department of the Army (Corps of Engineers)
Office of the Chief of Engineers U.S. Army Corps of Engineers, Washington DC 20314 (202) 693-7000

Department of Commerce
Commerce Bldg., 14th St. between Constitution Ave. and E St., NW, Washington DC 20230 (202) 377-2000

Department of Defense
U.S. Department of Defense, The Pentagon, Washington, DC 20301 (202) 545-6700

Energy Research and Development Administration
Office of the Assistant Administrator for Environment and Safety, Energy Research and Development Administration, Washington, DC 20545 (202) 376-4000

Environmental Protection Agency
Director, Office of Federal Activities, Environmental Protection Agency, 401 M Street, SW, Washington, DC 20460 (202) 755-2673

Regional Administrator I, U.S. Environmental Protection Agency
Room 2303, John F. Kennedy Federal Bldg., Boston, MA 02203 (617) 223-7210
Connecticut, Maine, Massachusetts, New Hampshire, Rhode Island, Vermont

Regional Administrator II, U.S. Environmental Protection Agency
Room 1009, 26 Federal Plaza, New York, NY 10007 (212) 264-2525
New Jersey, New York, Puerto Rico, Virgin Islands

Regional Administrator III, U.S. Environmental Protection Agency
Curtis Bldg., 6th & Walnut St., Philadelphia, PA 19106 (215) 597-9815
Delaware, Maryland, Pennsylvania, Virginia, West Virginia, District of Columbia

Regional Administrator IV, U.S. Environmental Protection Agency
1421 Peachtree St., NE, Atlanta, GA 30309 (404) 526-5727
Alabama, Florida, Georgia, Kentucky, Mississippi, North Carolina, South Carolina, Tennessee

Regional Administrator V, U.S. Environmental Protection Agency
230 South Dearborn St., Chicago, ILL 60604 (312) 353-5250
Illinois, Indiana, Michigan, Minnesota, Ohio, Wisconsin

Regional Administrator VI, U.S. Environmental Protection Agency
1600 Patterson St., Dallas, TE 75201 (214) 749-1962
Arkansas, Louisiana, New Mexico, Texas, Oklahoma

Regional Administrator VII, U.S. Environmental Protection Agency
1735 Baltimore Ave., Kansas City, MO 64108 (816) 374-5495
Iowa, Kansas, Missouri, Nebraska

Regional Administrator VIII, U.S. Environmental Protection Agency
Suite 900 Lincoln Tower, 1860 Lincoln St., Denver, CO 80203 (303) 837-3895
Colorado, Montana, North Dakota, South Dakota, Utah, Wyoming

Regional Administrator IX, U.S. Environmental Protection Agency
100 California St., San Francisco, CA 94111 (415) 556-2320
Arizona, California, Hawaii, Nevada, American Samoa, Guam, Trust Territories of Pacific Islands, Wake Island

Regional Administrator X, U.S. Environmental Protection Agency
1200 Sixth Ave., Seattle, WA 98101 (206) 442-1220
Alaska, Idaho, Oregon, Washington

Federal Power Commission
Director of Public Information, 825 North Capitol Street, NE, Washington, DC 20426 (202) 275-4006

Great Lakes Basin Commission
Office of the Chairman, Great Lakes Basin Commission, P.O. Box 999, Ann Arbor MI 48106 (313) 769-3590

Department of Health and Welfare
Department of Health and Welfare, 200 Independence Avenue, SW, Washington, DC 20201 (202) 245-6296

Department of the Interior
Interior Bldg., C St. between 18th and 19th, NW, Washington, DC 20240 (202) 343-1100
a. Bureau of Indian Affairs, 1951 Constitution Ave., NW, Washington, DC 20245 (202) 343-5116
b. Bureau of Land Management, Washington, DC 20240 (202) 343-1100
c. Bureau of Outdoor Recreation, Washington, DC 20240 (202) 343-1005
d. United States Fish and Wildlife Service, Washington, DC 20240 (202) 343-4717
e. National Park Service, Interior Bldg., Washington, DC 20240 (202) 343-1100

Department of the Treasury (U.S. Customs Service) 1301 Constitution Ave., NW, Washington, DC 20229 (202) 566-5104

Missouri River Basin Commission
Suite 403, 10050 Regency Circle, Omaha, NE 68114 (402) 397-5714

New England River Basin Commission
Office of the Chairman, New England River Basin Commission, 53 State St., Boston, MA 02109 (617) 223-6244

Nuclear Regulatory Commission
Director, Office of Public Affairs; Washington, DC 20555 (202) 634-1645

Director, Division of Safeguards; Washington, DC 20555 (301) 427-4033

Director, Division of Site Safety and Environmental Analysis; Washington, DC 20555 (301) 492-7207

Director, Office of Inspection and Enforcement; Washington, DC 20555 (301) 492-7397

Ohio River Basin Commission
36 East 4th St., Suite 208-220, Cincinnati, OH 45202 (513) 684-3831

Pacific Northwest River Basin Commission
P.O. Box 908, Vancouver, WA 98660 (206) 694-2581

Upper Mississippi River Basin Commission
Federal Office Bldg. Room 510, Ft. Snelling, Twin Cities, MN 55111 (612) 725-4690

Susquehanna River Basin Commission
18th and C Streets, NW, Washington, DC 20240 (202) 343-4091

Department of State
Office of Environmental Affairs, Washington, DC 20520 (202) 632-9278

Office of Ocean Affairs, Washington, DC 20520 (202) 632-3262

Office of Population Affairs, Washington, DC 20520 (202) 632-2232

U.S. Water Resources Council
Suite 800, 2120 L St., NW, Washington, DC 20037 (202) 254-6303
Public Information Officer: (202) 254-0453

II. Selected State, Territorial, and Citizens Organizations for Environmental Information Retrieval

A. Government Agencies
Alabama:
Department of Conservation and Natural Resources
64 N Union St., Montgomery 36104 (205) 832-6361

Alaska:
Department of Environmental Conservation
Pouch O, Juneau 99811 (907) 465-2605

Arizona:
Game and Fish Department
2222 W. Greenway Rd., Phoenix 85023 (602) 942-3000

Land Department
1624 W. Adams, Phoenix 85007 (602) 271-4621

Arkansas:
Department of Pollution Control and Ecology
8001 National Dr., P.O. Box 9583, Little Rock 72209 (501) 371-1701

California:
Department of Education, Conservation Education Service
721 Capitol Mall, Sacramento 95814

Resources Agency, The
1416 Ninth St., Sacramento 95814 (916) 445-5656

Colorado:
Department of Natural Resources,
1313 Sherman, Rm. 718, Denver 80203 (303) 892-3311

Connecticut:
Department of Environmental Protection
State Office Bldg., 165 Capitol Ave., Hartford 06115 (203) 566-5599

Delaware:
Department of Natural Resources and Environmental Control
The Edward Tatnall Bldg., Legislative Ave. and William Penn St., Dover 19901 (302) 678-4506

District of Columbia:
Department of Environmental Services
415 12th St., NW, Washington 20004 (202) 629-3415

Florida:
Department of Natural Resources
Crown Bldg., 202 Blount St., Tallahassee 32304 (904) 488-1555

Florida Defenders of the Environment, Inc.,
622 N. Main, Gainesville 32601 (904) 372-6965

Georgia:
Department of Natural Resources
270 Washington St., SW, Atlanta 30334 (404) 656-3530

Guam:
Department of Agriculture
Agana 96910 (734-9966)

Hawaii:
Department of Land and Natural Resources
Box 621, Honolulu 96809 (548-6550)

Idaho:
Department of Health and Welfare
Statehouse, Boise 83720 (208) 384-2336

Department of Lands
State Capitol Bldg. Boise 83720 (208) 384-3280

Department of Water Resources
Statehouse, Boise 83720 (208) 384-2215

Illinois:
Department of Conservation
605 State Office Bldg., Springfield 62706 (217) 782-6302

Indiana:
Department of Natural Resources
608 State Office Bldg., Indianapolis 46204

Iowa:
Department of Environmental Quality
Henry A. Wallace Bldg., 900 East Grand, Des Moines 50319

Kansas:
State Department of Health and Environment
Forbes Field Bldg., 740 Topeka 66620

Kentucky:
Department for Natural Resources and Environmental Protection
5th Floor, Capital Plaza Tower, Frankfort 40601 (502) 564-3350

Louisiana:
State Soil and Water Conservation Committee
Louisiana State University, P.O. Drawer CS, Baton Rouge 70803 (504) 389-5017

Maine:
Department of Conservation
State Office Bldg., Augusta 04333 (207) 289-3871

Maryland:
Department of Natural Resources
Tawes State Office Bldg., Annapolis 21401 (301) 269-3683

Massachusetts:
Department of Environmental Management
100 Cambridge St., Boston 02202 (617) 727-3163

Michigan:
Department of Natural Resources
Box 30028, Lansing 48909 (517) 373-1200

Minnesota:
Department of Natural Resources
300 Centennial Bldg., 658 Cedar St., St. Paul 55155 (612) 296-2549

Mississippi:
Air and Water Pollution Control Commission
P.O. Box 827, Jackson 39205 (601) 354-2550

Game and Fish Commission
Robert E. Lee Office Bldg., 239 N. Lamar St., P.O. Box 451, Jackson 39205   (601) 354-7333

Missouri:
Department of Conservation
P.O. Box 180, Jefferson City 65101   (314) 751-4211

Montana:
Department of Natural Resources and Conservation
32 South Ewing, Helena 59601   (406) 449-3712

Nebraska:
Department of Environmental Control
State House Sta., Box 98477, Lincoln 68509   (402) 471-2186

Nevada:
Department of Conservation and Natural Resources
Capitol Complex, Nye Bldg., 201 S. Gall St., Carson City 89710   (702) 885-4360

New Hampshire:
Council of Resources and Development
State House Annex, Concord 03301   (603) 271-2155 ·
Department of Resources and Economic Development
P.O. Box 856, State House Annex, Concord 03301

New Jersey:
Department of Environmental Protection
Labor and Industry Bldg., 1390, Trenton 08625   (609) 292-2885

New Mexico:
Environmental Improvement Agency
P.O. Box 2348, Santa Fe 87503

New York:
Bureau of Environmental Protection
Dept. of Law, State of New York, Two World Trade Center, New York 10047   (212) 488-5123
Department of Environmental Conservation
50 Wolf Rd., Albany 12233   (212) 488-2755

North Carolina:
Department of Natural and Economic Resources
P.O. Box 27687, Raleigh 27611   (919) 733-4984

North Dakota:
Institute for Ecological Studies
University of North Dakota, Grand Forks 58201   (701) 777-2851

Ohio:
Department of Natural Resources
Fountain Square, Columbus 43224   (614) 466-3066

Oklahoma:
Department of Wildlife Conservation
1801 N. Lincoln, P.O. Box 53465, Oklahoma City 73105 (405) 521-3851
Oklahoma Conservation Commission
State Capitol, Oklahoma City 73105   (405) 521-2384

Oregon:
Department of Environmental Quality
1234 S.W. Morrison, Portland 97205   (503) 229-5395

Pennsylvania:
Department of Environmental Resources
Public Information, Rm. 203, Evangelical Press Bldg., Box 1467, Harrisburg 17120   (717) 787-2814

Puerto Rico:
Department of Natural Resources
P.O. Box 5887, Puerta de Tierra Station, San Juan 00906 (809) 724-8774, ext. 271

Rhode Island:
Department of Natural Resources
83 Park St., Providence 02903   (401) 277-2771

South Carolina:
Department of Health and Environmental Control
J. Marion Sims Bldg., 2600 Bull St., Columbia 29201
State Land Resources Conservation Commission
2221 Devine St., Suite 222, Columbia 29205   (803) 758-2824

South Dakota:
Board of Environmental Protection
Department of Environmental Protection, Rm. 408 Joe Foss Bldg., Pierre 57501   (605) 224-3351

Tennessee:
Department of Conservation
2611 W End Ave., Nashville 37204   (615) 741-2301
Environment Center
University of Tennessee, South Stadium Hall, Knoxville 37916   (615) 974-4251

Texas:
Parks & Wildlife Department
4200 Smith School Road, Austin 78744   (512) 475-2087
Forest Service
College Station, Texas 77843   (713) 845-2641
Texas Department of Water Resources
1700 N. Congress, Austin 78701; Mailing Address: P.O. Box 13807, Capitol Station, Austin 78711   (512) 475-3187

Utah:
State Department of Natural Resources
438 State Capitol, Salt Lake City 84114   (801) 533-5356

Vermont:
Agency of Environmental Conservation
Montpelier 05602   (802) 828-3357

Virgin Islands:
Department of Conservation and Cultural Affairs
P.O. Box 4340, St. Thomas 00801   (809) 774-3320

Virginia:
Council on the Environment
903 Ninth Street Office Bldg., Richmond 23219   (804) 786-4500

Washington:
Department of Ecology
Olympia 98504   (206) 753-2800
Department of Natural Resources
Public Lands Bldg., Olympia 98504   (206) 753-5327

West Virginia:
Department of Natural Resources
1800 Washington St., East, Charleston 25305   (304) 348-2754

Wisconsin:
Department of Natural Resources
Box 7921, Madison 53707   (608) 266-2621

Wyoming:
Environmental Quality Department
Hathaway Bldg., Cheyenne 82002   (307) 777-7391

B.   *Citizen's Organizations*

Air Pollution Control Association
4400 Fifth Ave., Pittsburg, PA 15213   (412) 621-1090

Alliance for Environmental Education, Inc.
Suite 113, 1785 Massachusetts Ave. NW, Washington, DC 20036   (202) 265-0630

American Association for the Advancement of Science
1515 Massachusetts Ave., NW, Washington, DC 20005 (202) 467-4400

American Cetacean Society
P.O. Box 4416, San Pedro, CA 90731   (213) 548-6279

American Chemical Society
1155 16th St., NW, Washington, DC 20036 (202) 872-4600

American Committee for International Conservation, Inc.,
National Zoological Park, Washington, DC 20009   (202) 381-7247

American Conservation Association, Inc.
30 Rockefeller Plaza, Rm. 5425, N.Y., NY 10020 (212) 247-8141

American Farm Bureau Federation
225 Touhy Avenue, Park Ridge, IL 60068   (312) 696-2020

American Fisheries Society
5410 Grosvenor Lane, Bethesda, MD 20014   (301) 897-8616

American Forest Institute
1619 Massachusetts Ave., NW, Washington, DC 20036
(202) 667-7807

The American Forestry Association
1319 18th St., NW, Washington, DC 20036 (202) 467-5810

American Institute of Biological Sciences, Inc.
1401 Wilson Blvd., Arlington, VA 22209 (703) 527-6776

American Littoral Society
Sandy Hook, Highlands, N.J. 07732 (201) 291-0055

American Museum of Natural History
Central Park West at 79th St., N.Y., NY 10024 (212) 873-1300

American Petroleum Institute
2101 L St., NW, Washington, DC 20037 (202) 457-7000

American Rivers Conservation Council
317 Pennsylvania Ave., SE, Washington, DC 20003 (202) 547-6900

American Society of Limnology and Oceanography, Inc.,
Great Lakes Research Division, University of Michigan, Ann Arbor, MI 48109 (313) 764-2422

Boone and Crockett Club
424 N. Washington St., Alexandria, VA 22314 (301) 843-6650

Canada—United States Environmental Council
Canada: Canadian Nature Federation, 46 Elgin Street, Ottawa, Ontario K1P 5K6; (613) 238-6154

United States: The Wilderness Society, 1901 Pennsylvania Ave., NW, Washington, DC 20006; (202) 293-2732

Center for Environmental Education
2100 M St., NW, Washington, DC 20037 (202) 466-4996

Center for International Environment Information
300 E. 42nd St., New York, NY (212) 697-3232

Citizens Committee on Natural Resources
1000 Vermont Ave., NW, Washington, DC 20005 (202) 638-3396

Ducks Unlimited, Inc.
P.O. Box 66300, Chicago, IL 60666 (313) 299-3334

Ecological Society of America
Secretary: Dr. Edward J. Kormondy, Evergreen State College, Olympia, WA 98505 (206) 866-6400

Environmental Action Foundation, Inc.
724 DuPont Circle Bldg., Washington, DC 20036 (202) 659-9682

Fish and Wildlife Reference Service
Denver Public Library Service Bldg., 2100 W. Mississippi Ave., Denver CO 80223 (303) 922-0505

Food and Agriculture Organization of the United Nations
Via delle Terme di Caracalla, Rome 00100, Italy
(Telephone: 5797)

International Council of Environmental Law
D 53 Bonn, Federal Republic of Germany, Adenaueralle 214 (02221-213452)

National Association of Conservation Districts
1025 Vermont Ave., NW, Washington, DC 20005 (202) 347-5995

National Audubon Society
950 Third Ave., N. Y., NY 10022 (212) 832-3200

National Geographic Society
1145 17th St., NW, Washington, DC 20036 (202) 857-7000

National Rifle Association of America
1600 Rhode Island Ave., NW, Washington, DC 20036 (202) 783-6505

National Wildlife Federation
1412 16th St., NW, Washington, DC 20036 (202) 797-6800

Natural Resources Council of America
Suite 914, 1025 Connecticut Ave., NW, Washington, DC 20036 (202) 293-3200

Nature Conservancy
Suite 800, 1800 N. Kent St., Arlington, VA 22209 (703) 841-5300

Population Reference Bureau
1337 Connecticut Ave., NW, Washington, DC 20036 (202) 785-4664

Sierra Club
530 Bush St., San Francisco, CA 94108 (415) 981-8634

Smithsonian Institution
1000 Jefferson Dr., SW, Washington, DC 20560 (202) 628-4400

Sport Fishsing Institute
Suite 801, 608 13th St., NW, Washington, DC 20005 (202) 737-0668

Student Conservation Association, Inc.
Box 550, Charlestown, NH 03603 (603) 826-5206

United Nations Environment Programme
P.O. Box 30552, Nairobi, Kenya

Wilderness Society
1901 Pennsylvania Ave., NW, Washington, DC 20006 (202) 293-2732

Wildlife Society
Suite 611, 7101 Wisconsin Ave., NW, Washington, DC 20014 (301) 986-8700

Zero Population Growth
1346 Connecticut Ave., NW, Washington, DC 20036 (202) 785-0100

III. **Canadian Agencies and Citizens' Organizations for Environmental Information Retrieval**
   A. *Government Agencies*

Department of the Environment
Ottawa, Ont. K1A 0H3 (613) 995-2211

Environmental Management Service
Place Vincent Massey, Hull, P.Q. (Mailing address: Ottawa, Ont. K1A 0E7 (819) 997-1459

Canadian Forestry Service
(819) 997-6555

Fisheries Research Board of Canada
580 Booth St., Ottawa, Ont. K1A 0E6 (613) 995-2171

Department of Indian and Northern Affairs
400 Laurier Ave., West, Ottawa, Ont. K1A 0H4

Environmental Protection Service
Department of Fisheries and the Environment, Ottawa, Ont. K1A 1C8 (819) 997-3131

   B. *Citizens' Groups*

Canadian Environmental Law Association
1 Spadina Crescent, Suite 303, Toronto, Ont. M5S 2J5 (416) 978-7156

Canadian Nature Federation
46 Elgin St., Ottawa, Ont. K1P 5K6 (613) 238-6154

Canadian Forestry Association
Suite 203, 185 Somerset St., West, Ottawa, Ont. K2P 0J2 (613) 232-1815

Canadian Wildlife Federation
1673 Carling Ave., Ottawa, Ont. K2A 1C4 (613) 725-2191

National and Provincial Parks Association of Canada
47 Colborne St., Suite 308, Toronto, Ont. M5E 1E3 (416) 366-3494

Nature Conservancy of Canada
Suite 611, 2200 Yonge St., Toronto M4S 2E1 (416) 486-1011

Pollution Probe/Energy Probe
43 Queen's Park Crescent East, Toronto, Ont. M5S 2C3 (416) 978-6155

IV. Selected Journals and Periodicals of Environmental Interest

American Forests
The American Forestry Association, 919 Seventeenth St., NW Washington, DC 20006

American Scientist
343 Whitney Ave., New Haven, CT 06511

Audubon
National Audubon Society, 950 Third Ave., New York, NY 10022

BioScience
American Institute of Biological Science, 1401 Wilson Blvd., Arlington, VA 22209

California Environmental Report
Business Publishers Report, P.O. Box 1067, Blair Station, Silver Spring, MD 20910

The Canadian Field-Naturalist
Box 3264, Postal Station C, Ottawa 3, Canada

Conservation Report
1412 16th St. NW, Washington, DC 20036

Conservation Directory
1412 16th St. NW, Washington, DC 20036

Ecology USA
Business Publishers, Inc., P.O. Box 1067, Silver Spring, MD 20910

Environment
4000 Albemarle St., NW, Washington, DC 20016

Environment Action Bulletin
Rodale Press, Inc., 33 E. Minor St., Emmaus, PA 18049

Environment Reporter
Bureau of National Affairs, Inc., 1231 25th St., NW, Washington, DC 20037

Environmental Information Handbook
Simon and Schuster, Inc., 630 Fifth Avenue, New York, NY 10020

Environmental Science and Technology
American Chemical Society Publications, 1155 16th St. NW, Washington, DC 20036

Annual Report on the Council of Environmental Quality
Superintendent of Documents, U.S. Government Printing Office, Washington, DC 20402

The Futurist
P.O. Box 30369, Bethesda Branch, Washington, DC 20014

Journal of Soil and Water Conservation
7515 Northeast Ankeny Road, Ankeny, IA 50021

Journal of Wildlife Management
The Wildlife Society, 3900 Wisconsin Ave., NW, Washington, DC 20016

The Living Wilderness
729 15th St., NW, Washington, DC 20005

Mother Earth News
P.O. Box 70, Hendersonville, NC 28739

Nature
MacMillan (Journals) Ltd., Brunel Road, Basingstoke Hampshire, England

Natural Resources Journal
University of New Mexico School of Law, 1117 Stanford NE, Albuquerque, NM 87131

National Wildlife
1412 16th St., NW, Washington, DC 20036

Nature Canada
46 Elgin St., Ottawa, Canada K1P 5K6

Nature Conservancy News
1800 N. Kent St., Arlington, VA 22209

Oceans
7075 A, Mission Gorge Road, San Diego, CA 92120

Omni
Omni Publications International, Ltd., 909 3rd Ave., New York, NY 10022

Pollution Abstracts
Data Courier, Inc., 620 S. Fifth St., Louisville, KY 40202

Quest
Comac Communications, Ltd., 2300 Yonge St., Toronto, Ontario Canada, M4P 1E4

Science
1515 Massachusetts Ave., NW, Washington, DC 20005

Sierra Club Bulletin
530 Bush St., San Francisco, CA 94108

Smithsonian
P.O. Box 5300, Greenwich, CT 06830

Technology Review
Room 10-140, M.I.T., Cambridge, MA 02139

U.S.News and World Report
2300 N St. NW, Washington, DC 20037

This glossary of 168 environment terms is included to provide you with a convenient and ready reference as you encounter general terms in your study of environment which are unfamiliar or require a review. It is not intended to be comprehensive but taken together with the many definitions included in the articles themselves it should prove to be quite useful.

Abiotic Without life; any system characterized by a lack of living organisms.

Abortion Expulsion of a fetus from the uterine cavity prior to birth.

Acid Any compound capable of reacting with a base to form a salt; a substance containing a high hydrogen ion concentration (low pH).

Acid Rain Precipitation containing a high concentration of acid.

Acre Foot Unit used to measure the volume of water equal to the quantity of water required to cover one acre to a depth of one foot; equal to 325,851 gallons.

Adaptation Any characteristic that aids an organism to survive and reproduce in its environment.

Additive A substance added to another in order to impart or improve desirable properties or suppress undesirable properties.

Aerobic Environmental conditions where oxygen is present; aerobic organisms require oxygen in order to survive.

Aesthetic Pertaining to a sense or feeling of the beautiful.

Age Distribution The proportion of a population in each age class.

Agriculture Production of crops, livestock, or poultry.

Air Quality Standard A prescribed level of a pollutant in the air that should not be exceeded.

Alkali Soil A soil of such a high degree of alkalinity (high pH) that growth of most crop plants is reduced.

Alpha Particle A positively charged particle given off from the nucleus of some radioactive substances: it is identical to a helium atom that has lost its electrons.

Anaerobic Without oxygen; environmental conditions where oxygen is absent.

Ammonia A colorless gas composed of one atom of nitrogen and three atoms of hydrogen; liquified ammonia is used as a fertilizer.

Anthropocentric Considering man to be the central or most important part of the universe.

Atom The smallest particle of an element, composed of electrons moving around an inner core (nucleus) of protons and neutrons. Atoms of elements combine to form molecules and chemical compounds.

Atomic Energy Energy released by changes in the nucleus of an atom, either by the splitting of the nucleus or by the joining of nuclei.

Atomic Reactor A structure fueled by radioactive materials which generates energy usually in the form of electricity; reactors are also utilized for medical and biological research. Plutonium, a radioactive substance, is also produced by reactors and has been used for the production of atomic devices.

Autotrophic Organisms capable of using chemical elements in the synthesis of larger compounds; green plants are autotrophic.

Background Radiation The normal radioactivity present; coming principally from outer-space and naturally occurring radioactive substances in the earth.

Bacteria One-celled microscopic organisms found in the air, water, and soil. Bacteria cause many diseases of plants and animals; they also are beneficial in agriculture, decay of dead matter, in food industries and in chemical industries.

Biochemical Oxygen Demand (BOD) The oxygen utilized in meeting the metabolic needs of aquatic organisms.

Biodegradable Capable of being reduced to simple compounds through the action of biological processes.

Biogeochemical Cycle The cyclical series of transformations of an element through the organisms in a community and their physical environment.

Biological Control The suppression of reproduction of a pest organism utilizing other organisms rather than chemical means.

Biomass The weight of all living tissue in a sample.

Biome A major climax community type covering a specific area on earth.

Biota The flora and fauna of any region.

Biotic Biological; relating to living systems.

Biotic Potential Maximum possible growth rate of living systems under ideal conditions.

Birth Rate Number of live births in one year per 1,000 midyear population.

Breeder Reactor A nuclear reactor in which the production of fissionable material occurs.

Carbon One of the most common elements; compounds of carbon are the chief constituents of living systems.

Carbon Cycle Process by which carbon is incorporated into living systems, released to the atmosphere and returned to living organisms.

Carbon Dioxide A gas, CO_2, making up about 0.03 of the earth's atmosphere. It is an end product of burning (oxidation) of organic matter or carbon-containing substances.

Carbon Monoxide A gas, poisonous to most living systems; formed when burning occurs in the absence of much oxygen.

Carcinogen Any substance capable of producing cancer.

Chlorinated Hydrocarbon Insecticide Synthetic organic poisons containing hydrogen, carbon, and chlorine. Because they are fat soluble they tend to be recycled through food chains eventually affecting non-target systems. Damage is normally done to the organism's nervous system. Examples include DDT, Aldrin, Deildrin and Chlordane.

Clear-Cutting The practice of removing all trees in a specific area.

Climax Community Terminal state of ecological succession in an area; the redwoods are a climax community.

Coal Gasification Process of converting coal to gas; the resultant gas, if used for fuel, sharply reduces sulfur oxide emissions and particulates that result from coal burning.

Commensalism The relationship between two different species in which one benefits while the other is neither harmed nor benefited.

Community All organisms existing in a specific region.

Competition The struggle between individuals of the same or different species for food, space, mates or other limited resource.

Competitive Exclusion Resulting from competition; one species forced out of part of an available habitat by a more efficient species.

Conservation The planned management of a natural resource to prevent over-exploitation, destruction, or neglect.

Contraception Process of preventing conception.

Crankcase Smog Devices (PCV System) A system, used principally in automobiles, designed to prevent discharge of combustion emissions to the external environment.

Death Rate Number of deaths in one year per 1,000 midyear population.

Decomposer Any organism which causes the decay of organic matter; bacteria and fungi are two examples.

Demography The statistical study of populations; related principally to human populations.

Desert An arid biome characterized by little rainfall, high daily temperatures, and low diversity of animal and plant life.

Detergent A synthetic soap-like material that emulsifies fats and oils and holds dirt in suspension; some detergents have caused pollution problems because of certain chemicals used in their formulation.

DNA (Deoxyribonucleic Acid) One of two principal nucleic acids, the other being RNA (Ribonucleic Acid). DNA contains information used for the control of a living cell. Specific segments of DNA are now recognized as genes, those agents controlling evolutionary and hereditary processes.

Dominant Species Any species of plant or animal that is particularly abundant or controls a major portion of the energy flow in a community.

Dust Bowl The name usually associated with the south-central region of the United States; applied during the 1930s during periods of droughts and dust storms that occurred in Colorado, Kansas, New Mexico, Texas, and Oklahoma.

Ecological Density The number of a singular species in a geographical area; including the highest concentration points within the defined boundaries.

Ecology Study of the interrelationships between organisms and their environments.

Ecosystem The organisms of a specific area, together with their functionally related environments; considered as a definitive unit.

Effluent A liquid discharged as waste.

Electron Small, negatively charged particle; normally found in orbit around the nucleus of an atom.

Eminent Domain Superior dominion exerted by a governmental state over all property within its boundaries that authorizes it to appropriate all or any part thereof to a necessary public use, reasonable compensation being made.

Environment The physical and biological aspects of a specific area.

Environmental Protection Agency (EPA) Federal agency responsible for control of air and water pollution, radiation and pesticide problems, ecological research and solid waste disposal.

Erosion Progressive destruction or impairment of a geographical area; wind and water are the principal agents involved.

Estuary Area formed where streams, or rivers, enter oceanic zones.

Eutrophic Well nourished; refers to aquatic areas rich in dissolved nutrients.

Evolution A change in the gene frequency within a population; sometimes involving a visible change in the population's characteristics.

Fallow Cropland that is plowed but not replanted; left idle in order to restore productivity mainly through water accumulation, weed control, and buildup of soil nutrients.

Fauna The animal life of a specified area.

Feral Animals or plants that have reverted to a noncultivated or wild state.

Fertilizer Any natural or artificial substance added to soil to promote growth.

Fission The splitting of an atom into smaller systems.

Flora The plant life of an area.

Food Additive Substance added to food, usually added to improve color, flavor, or shelf life.

Food Chain The sequence of organisms in a community, each of which uses the lower source as its energy supply. Green plants are the ultimate basis for the entire sequence.

Fossil Fuel Coal, oil, natural gas, and/or lignite; those fuels derived from former living systems; usually called nonrenewable fuels.

Fuel Cell Manufactured chemical systems capable of producing electrical energy; usually derive their capabilities via complex reactions involving the sun as the driving energy source.

Fusion The formation of a heavier atomic complex brought about by the addition of atomic nuclei; during the process there is an attendant release of energy.

Gamma Ray A ray given off by the nucleus of some radioactive elements. A form of energy similar to X-rays.

Greenhouse Effect The effect noticed in greenhouses when shortwave solar radiation penetrates glass, is converted to longer wavelengths, and is blocked from escaping by the windows. It results in a temperature increase. The earth's atmosphere acts in a similar manner.

Ground Water All water located below the earth's surface.

Habitat   The natural environment of a plant or animal.

Heterotrophic   Obtaining nourishment from organic matter.

Herbicide   Any substance used to kill plants.

Hydrocarbon   Organic compounds containing hydrogen, oxygen and carbon. Commonly found in petroleum, natural gas, and coal.

Hydrogen   Lightest known gas; major element found in all living systems.

Hydrogen Sulfide   Compound of hydrogen and sulfur; toxic air contaminant, smells like rotten eggs.

Immigration   Movement into a new area of residence.

Ion   An atom or group of atoms, possessing a charge; brought about by the loss or gain of electrons.

Ionizing Radiation   Energy in the form of rays or particles which have the capacity to dislodge electrons and/or other atomic particles from matter which is irradiated.

Irradiation   Exposure to any form of radiation.

Isotope   Two or more forms of an element having the same number of protons in the nucleus of each atom but different numbers of neutrons.

Kilowatt   Unit of power equal to 1,000 watts.

Leaching   Dissolving out of soluble materials by water percolating through soil.

Limnologist   Individual who studies the physical, chemical, and biological conditions of aquatic systems.

Malthusian Theory   The theory that populations tend to increase by geometric progression (1, 2, 4, 8, 16, etc.) while food supplies increase by arithmetic means (1, 2, 3, 4, 5, etc.).

Metabolism   The chemical processes in living tissue through which energy is provided for continuation of the system.

Methane   Often called marsh gas (CH_4); an odorless, flammable gas that is the major constituent of natural gas. In nature it develops from decomposing organic matter.

Migration   Periodic departure and return of organisms to and from a population area.

Monoculture   Cultivation of a single crop, such as wheat or corn, to the exclusion of other land uses.

Mutagen   Any agent capable of causing a mutation.

Mutation   Change in genetic material (gene) that determines species characteristics; can be caused by a number of agents including radiation and chemicals.

Natural Selection   The agent of evolutionary change by which organisms possessing advantageous adaptations leave more offspring than those lacking such adaptations.

Niche   The unique occupation or way of life of a plant or animal species; where it lives and what it does in the community.

Nitrate   A salt of nitric acid. Nitrates are the major source of nitrogen for higher plants. Sodium nitrate and potassium nitrate are used as fertilizers.

Nitrite   Highly toxic compound; salt of nitrous acid.

Nitrogen Oxides   Common air pollutants. Formed by combination of nitrogen and oxygen; often the products of petroleum combustion in automobiles.

Oil Shale   Rock impregnated with petroleum. Regarded as a potential source of future petroleum products.

Oligotrophic   Most often refers to those lakes with a low concentration of organic matter. Usually containing considerable oxygen; Lakes Tahoe and Baikal are examples.

Organic   Derived from living systems.

Organophosphates   A large group of nonpersistent synthetic poisons used in the pesticide industry; include parathion and malathion.

Ozone   Molecule of oxygen containing three oxygen atoms; shields much of the earth from ultraviolet radiation.

Particulate   Existing in the form of small separate particles; various atmospheric pollutants are industrial produced particulates.

Peroxyacyl Nitrate (PAN)   Compound making up part of photochemical smog and the major plant toxicant of smog type injury; levels as low as 0.01 ppm can injure sensitive plants. Also causes eye irritation in man.

Pesticide   Any material used to kill rats, mice, bacteria, fungi, or other pests of man.

Petrochemical   Chemicals derived from petroleum bases.

pH   Scale used to designate the degree of acidity or alkalinity; ranges from 1-14; a neutral solution has a pH of 7; low pHs are acid in nature while pHs above 7 are alkaline.

Phosphate   A phosphorous compound; used in medicine and as fertilizers.

Photochemical   Pertaining to the chemical effects of light.

Photochemical Smog   Type of air pollution; results from sunlight acting with hydrocarbons and oxides of nitrogen in the atmosphere.

Photosynthesis   Formation of carbohydrates from carbon dioxide and hydrogen in plants exposed to sunlight; involves a release of oxygen through the decomposition of water.

Physical Half-Life   Time required for half of the atoms of a radioactive substance present at some beginning to become disintegrated and transformed.

Plutonium   A heavy, radioactive, manmade, metallic element. Used in weapons and as a reactor fuel. Highly toxic to life forms and possesses an extremely long physical half-life.

Pollution   The process of contaminating air, water, or soil with materials that reduce the quality of the medium.

Polychlorinated Biphenyls (PCBs)   Poisonous compounds similar in chemical structure to DDT. PCBs are found in a wide variety of products ranging from lubricants, waxes, asphalt, and transformers, to inks and insecticides. Known to cause liver, spleen, kidney, and heart damage.

Population   All members of a particular species occupying a specific area.

Predator Any organism that consumes all, or part, of another system; usually responsible for death of the prey.

Primary Production The energy accumulated and stored by plants through photosynthesis.

Rad (Radiation Absorbed Dose) Measurement unit relative to the amount of radiation absorbed by a particular target, biotic or abiotic.

Radioactive Waste Any radioactive by-product of nuclear reactors or nuclear processes.

Radioactivity The emission of electrons, protons (atomic nuclei), and/or rays from elements capable of emitting radiation.

Recycle To reuse; usually involves manufactured items, such as aluminum cans, being restructured after use and utilized again.

Redwood *Sequoia sempervirens;* world's tallest tree. Used extensively for lumber; grows in coastal areas of California and Oregon.

Riparian Water Right Legal right of an owner of land bordering a natural lake or stream to remove water from that aquatic system.

RNA Ribonucleic acid; nucleic acid most often located in the cytoplasm of cells; used principally in the manufacture of proteins.

Scrubber Anti-pollution system which uses liquid sprays in removing particulate pollutants from an air stream.

Sediment Soil particles moved from land into aquatic systems as a result of man-caused activities.

Seepage Movement of water through soil.

Selection The process, either natural or artificial, of removing or selecting the best or less desirable members of a population.

Selective Breeding Process of selecting and breeding organisms containing traits considered most desirable.

Selective Harvesting Process of taking specific individuals from a population; removal of trees in a specific age class would be an example.

Sewage Any waste material coming from domestic and industrial origins.

Smog A mixture of smoke and air; now applies to any type of air pollution.

Solid Waste Unwanted solid materials usually resulting from industrial processes.

Species Populations, or a population, capable of interbreeding and producing viable offspring.

Species Diversity A ratio between the number of species in a community and the number of individuals in each species. Generally, the greater the species diversity comprising a community, the more stable is the community.

Strip Mining Mining in which the earth's surface is removed in order to obtain sub-surface materials.

Strontium-90 Radioactive isotope of strontium; results from nuclear explosions and is dangerous, especially for vertebrates, because it is taken up in the construction of bone.

Succession Change in the structure and function of an ecosystem; replacement of one system with another through time.

Sulfur Dioxide (SO₂) Gas produced by burning coal and as a byproduct of smelting and other industrial processes. Very toxic to plants.

Sulfuric Acid (H₂SO₄) Very corrosive acid; produced from sulfur dioxide; found as a component of acid rain.

Sulfur Oxides (SOₓ) Oxides of sulfur produced by the burning of oils and coal which contain small amounts of sulfur. Common air pollutants.

Technology Applied science; application of knowledge for practical application.

Tetraethyl Lead Major source of lead found in living tissue; produced to reduce engine knock in automobiles.

Thermal Pollution Unwanted heat; the result of ejection of heat from various sources into the environment.

Thermocline The layer of water in a body of water that separates an upper warm layer from a deeper colder zone.

Threshold Effect The situation in which no effect is noticed, physiologically or psychologically, until a certain level or concentration is reached.

Tolerance Limit The point at which resistance to a poison or drug breaks down.

Toxic Poisonous; capable of producing harm to a living system.

Toxicant Any agent capable of producing toxic reactions.

Trophic Relating to nutrition; often expressed in trophic pyramids in which organisms feeding on other systems are said to be at a higher trophic level; an example would be carnivores feeding on herbivores which, in turn, feed on vegetation.

Turbidity Usually refers to the amount of sediment suspended in an aquatic system.

Uranium 235 An isotope of uranium that when bombarded with neutrons undergoes fission resulting in radiation and energy. Used in atomic reactors for electrical generation.

Zero Population Growth The condition of a population in which birthrates equal death rates; results in no growth of the population.

abortion, U.S. policy against in world aid, 64–67; in Romania, 64
acid rain, 23, 26, 35, 42, 69, 118, 148, 156, 157, 181
Africa, and food production, 70–76; population growth in, 48; and rain forests, 209; and wildlife, 222–225
Agency for International Development (AID), 64
Agent Orange, 27, 29
agriculture: clearing land for, 9, 165; development of, 6, 8; effect of decline on society, 8, 9; effect of global warming on, 15; practices in U.S., 171; subsidies, 170–179; swidden, 9, 10; in tropical areas, 9–10; in USSR, 148
agroforestry, 33
agrometeorology, 78
AIDS, 32
air pollution, 12, 119, 120, 123, 141, 143, 144, 148–151, 156
Alar, 14
American Petroleum Institute (API), 91
animal, casualties induced by war, 25, 87
animals, domestic, effect on land, 9, 10, 30, 50, 163–169, 170; see also, livestock grazing
Antarctica, 21, 22, 30, 41, 77, 174, 193
Aral Sea, pollution of, 148, 149, 182
Arctic, the, 30, 151
Arctic National Wildlife Refuge (ANWR), 84, 91, 93, 94, 159
Asia, growth of population of, 48; rain forests of, 209
automobile, 55, 92–94; emissions, 84, 156

Baby Boom, and population growth, 56
balance of nature, 199
Bangladesh, 174, 175
Benign Demographic Transition Theory, 57
Bhopal, India, 131, 135
biodiversity, 14, 15, 32, 33, 34, 80, 155, 158, 198–206, 210, 212–218, 219–221
Biofuels Feedstock Development Program, 110
biogeography, 204
biomass fuel, 85, 86, 93, 104, 109–115
biosphere, 23, 31; Biosphere-2, 13, 15
birth control pill, 53
birth rate, 50, 57
Black Sea, 142, 144, 148
Bulgaria, 143, 144
Bush, George, 28; administration policy and UN and abortion, 64–67, 76; energy policy, 84–89, 90–94; environmental policy, 154–162, 190

campfire philosophy of wildlife management, 224
carbon dioxide (CO_2), 18–23, 35, 39, 43, 77, 80, 110, 123, 184–189, 190–194, 219, 220; carbon tax, 85–107
carcinogen, 120, 137, 138
carrying capacity, 14, 23, 54, 141
Carter, Jimmy, energy policy of, 84, 89, 90
cedar trees, exploitation of, 8
cellulose production, in biomass fuel, 112–114
Chernobyl, 13, 15, 144, 148, 154
chlorofluorocarbons (CFCs): 12, 15, 27, 34, 36, 77, 80, 81, 122, 128, 190
Clean Air/Water Acts, 123, 156, 157
clean technology, 126–129

clear cutting, of forests, 148, 219, 220
climate change, 15, 16, 34, 42, 70–76, 77–81, 183, 184–189; climate control of Earth, 19
coal, 35, 81, 93, 107, 109, 181; brown coal, 144, 145
Coalition for Environmentally Responsible Economies (CERES), 16, 17
coastal lands, sensitivity of, 172–179, 204
coffee, environmental impact of demand, 209
colonialism, as root of environmental problems, 164
Commission of European Communities, environmental regulations, 126–129
Commoner, Barry, 123, 194
commons, creation of, 14; "tragedy of the commons," 166
competition, for food supplies, 9, 10, 11
conservation, and preservation, 14, 15; of energy, 84, 86, 92, 93
contraception, 51–53, 64–67
Corporate Automobile Fuel Efficiency (CAFE) standard, 85–89, 91–94
corporate responsibility, 13, 16
Costa Rica, use of hazardous chemicals, 135–140
cottonwood tree, importance of in West, 229
cropland, 30, 62; loss of, 68–76
cultivation, contribution to environmental degradation, 10, 11, 30
Czechoslovakia, 141, 144, 181

DDT, 119, 120
defoliation, 24, 25, 27
DNA (deoxyribonucleic acid), 200–203
deforestation, 6, 8, 9, 10, 29, 30, 33, 36, 50, 60–62, 70–76, 155, 163, 191, 195, 207–211, 212–218, 219, 221
"Demographic Transition," 57
Denmark, environmental restrictions, 122, 126, 128
Department of Energy (DOE), 84, 110
Depo-Provera, 53
desert, 13, 28, 41
desertification, 33, 35, 61, 70, 80, 163
developing countries: environmental issues in, 59–67, 133, 210, 216; food production in, 68–76, 135; population growth in, 48–53
dibromochloropropane (DBCP), 136–139
dimethylsulfide (DMS), theory of climate change, 21–22
drainage, and agricultural practices, 10, 11
drought, 154, 165, 194
dynamite fishing, 28

Earth Day, 54, 118, 154
Eastern Europe, environmental issues in, 141–147, 149, 181
ecological interdependencies, 13
ecological stability, of hunter/gatherer societies, 6; lack of in settled societies, 6–11
econvironergy policy, 91–94
education, importance of, 38; in family planning, 51, 52, 76
Ehrlich, Paul, 54, 55
El Niño current, 42, 43
electricity, 85–89, 90–94, 95–100, 101–108, 149, 151; in California, 101; and photovoltaic technology, 102

electrolysis, in formation of hydrogen, 105, 107
emissions limit, 125; automobile, 144, 156; industrial, 149, 156
Endangered Species Act, 28, 155
Environmental Defense Fund, 119
environmental degradation: cost of, 147–151; Eastern Europe, 140–147; Mediterranean, 6, 8; of Mesopotamia, 6, 7; Near East, 8; Soviet Union, 148–151
environmental liability, 16
Environmental Protection Agency (EPA), 16, 41, 119, 123, 131, 132, 143, 172, 177, 194
environmental regulations, 122–129, 130–139, 140–147
environmentalism, 15
erosion, 174, 175, 214–218, 219–221; see also, soil erosion
ethanol, 93, 109, 112
Ethiopia, 59
eutrophication, 180, 181
extinction, of species, 25, 39
Exxon Valdez oil spill, 13, 26, 86, 151, 154, 159

family planning, 48–53, 60, 64–67, 75, 76
Farm Bill of 1985, 171
"feebate," 92, 93
feedback mechanisms, of Gaia hypothesis, 19–23
fertility rate, 50, 63–67
fertilizer, use of, 69–76, 130, 131, 148, 160, 180
fleet vehicles, 93
flooding, 172–179, 191, 194, 207
food: contamination, 136, 161; production of, 49, 59, 62, 135; supply, 8–11, 62, 68–76
food chain, toxic chemicals in, 123, 158
Food Security Act, 62
food web, see trophic diversity
fossil fuels, 11, 77, 84–89, 90–94, 99, 100, 101, 102–109, 144–146, 184
fuel, 35, 51, 84–89, 90–94, 101–105; alternate fuels, 84–89, 93; fuel efficiency, 86, 88, 91, 159;
fungi, in sustenance of forest species, 215

Gaia hypothesis, 18–23, 31, 33, 190
gene mutation, 200–203
gene pool, 201, 208, 209, 220
gene sanctuary, 210
Geneva Convention Protocols, environmental component, 29
geophysiology, 19–23
geothermal energy, 89, 93, 104
glaciation, 173, 187
global change ecology, 39–45
global change movement, 22, 23
global circulation model (GCM) of greenhouse warming, 190–195
global climate model, of greenhouse warming, 185–189
global economy, 11, 36, 37, 100, 105
global temperature fluctuation, 18, 77–82, 172, 184–189, 190–194
global warming, 15, 16, 18, 42–45, 62, 70–76, 77–81, 88, 109, 118, 172–179, 184–189, 190–195, 204, 207, 220, 221; beneficial effects of, 188, 189; models of, 185
grain production, 68–76

Green Initiative (Big Green) of California, 161
green machines, 106
Green politics, in Holland and Germany, 15
Green Revolution, 61, 72
greenhouse effect, 18–23, 42, 81, 184–185, 190–195
greenhouse gases, 15, 19, 34–36, 39, 42–44, 88, 108, 172–179, 184–189, 190–195; see also, carbon dioxide; CFCs; methane; sulfur
Greenpeace, 84, 87, 130, 131, 140
gross national product (GNP), 85, 91, 92, 195
Gulf War, see Persian Gulf War

habitat depletion, 25, 155, 158, 181, 204, 210
Hardin, Garrett ("Tragedy of the Commons"), 14, 17, 54–59, 166
health effects of pollution, 135–140, 141, 142, 156
Hiroshima, atomic bombing of, 26, 27, 28, 29
human settlement, effect on environment, 8–11
Hungary, 143, 145
hunger, and population growth, 59–67, 68–76
Huxley, T. H., 19
hydroelectric power, 85–89, 107, 151
hydrogen, as power source, 93, 94, 101, 105–107, 109

ice age, 22, 186, 187
ice-cover research, and global warming, 42–44
Impact Equation, 55, 57, 58
industrialized countries, food production in, 62–63, 68–76
interglacial warming, 187–189
International Geosphere Biosphere Programme, and LTER, 45
International Panel on Climate Change (IPCC), 78, 184–189
intrauterine device (IUD), 53
invisible present, in terms of ecological research, 42–44
Iraq, 13, 14, 26, 61, 90
irrigation, detrimental effect of, 7, 8, 11, 30, 61, 68–76, 81, 180, 183, 228

Japan, use of power, 94

kilowatt-hour, 95–100, 102
Kircher, James, challenge to Gaia hypothesis, 20–23
Kuwait, 25–29, 87, 90

land, ownership/use, 163–169, 178, 222–225
land subsidence, and sea level rise, 175, 189
landfills, 161
Lappé, Francis Moore, 54
Latin America, population growth, 48
lead, 119, 131
lighting technologies, 95–100, 195
Little Ice Age, 42, 186, 187
livestock grazing, environmental impact, 226–231
logging, 148, 212–218
Long-Term Ecological Research (LTER) program, 39–45

Love Canal, 118, 120, 121
Lovelock, James E., Gaia hypothesis of, 18–23, 31
Luz International, solar energy showcase, 89, 103

malnutrition, and population growth, 73–76
"managing planet Earth," challenge of, 13–17
manure, use of in agriculture, 8
Margulis, Lynn, Gaia hypothesis of, 18–23
Márquez, Gabriel García, 135, 140
mass transportation, 84–89, 92, 106
Mayan society, agriculture in, 9–11; development/decline of, 9–11
media hype, by environmental advocates, 14, 186, 190
Mediterranean, environmental degradation of, 8, 9, 11, 39, 219, 220
mercury, 118, 123, 128, 131, 181
Mesopotamia, as first settled society to modify natural environment, 7, 8, 11
methane, 35, 51, 78, 81, 190, 192
Mexico City, air quality of, 12, 60
Mobil Oil Co., 91
Montreal protocol, 15, 80
Muir, John, 14

Nader, Ralph, 91
National Aeronautics and Space Administration (NASA), 18, 186, 189, 195
National Energy Strategy (NES), of George Bush, 84–89, 90–94
National Wildlife Federation, 154–162
natural gas: industry, 85, 87, 90–94; as transition fuel, 89, 91, 93, 105; use of, 84, 85, 144, 145, 148–151; as vehicle fuel, 106
Natural Resources Defense Council (NRDC), 84, 131, 174
natural selection, 200–203
negawatts (saved watts), 95–100
nematodes, control of, 136–138
Netherlands, The, environmental regulations in, 128, 129; water supply of, 176–179
New Forestry, 212–218
niche theory of biodiversity, 203
Norplant, 53
nuclear energy, 84, 85, 90, 93, 94, 107, 144, 149, 154, 194; and the military, 26
nuclear winter scenario, 26
nutritional deficiency, in ancient civilizations, 10

Office of Technology Assessment, 14, 126, 128
offshore drilling, 93, 94, 159
oil, exploration, 93, 94; dependence on, 84, 85, 92; industry, 85, 87, 90–94; in USSR, 148–151
optimization, of life, 22–23
Outer Continental Shelf (OCS), 84, 94
ozone layer, depletion of, 14, 15, 18, 23, 34, 50, 68–76, 77, 122

Pacific Gas & Electric Company, 92, 102
Pacific yew, 220; see also, taxol
paper industry, and biomass energy, 111, 114; and timbering, 150, 158
payback period, in switch to energy-efficient fuel/technology, 98

peacetime, vs. wartime, impact on environment, 24–29
Persian Gulf War, environmental impact of, 24–29; and U.S. energy policy, 84–89, 90–94, 195
pesticides, 120, 121, 135–140, 148, 149, 181
pharmaceutical industry, 15
photosynthetic efficiency, limits on crops, 72
photovoltaic cells, 102, 103
phytoplankton, role in Gaia hypothesis, 19, 193
Planned Parenthood Federation of America, 58; International, 64, 65, 76
Poland, 141, 143, 144, 181
pollution, abatement, 16
pollution, voluntary prevention by industry, 16
politics, environmental, 15, 35
polychlorinated biphenyls (PCBs), 34, 119, 120, 121, 131
polystyrene, 12
polyvinyl chloride (PVC), 122, 123
population, global, 30, 32, 35, 48, 59, 64
population, growth of, 7, 8, 9, 26, 32, 35, 37, 48–51, 54–58, 59–67, 68–76, 163, 172–179, 180; rate of increase, 48, 49, 54
Population Institute, 59
poverty, correlation with population growth, 35, 36, 48, 49, 57, 60, 62–67
precipitation, increase with global warming, 184–189
Prince William Sound, 13
public lands grazing program, 226–231

radioactive waste, 131
radon, 118, 120
rain forest, 6, 13, 15, 24, 27, 28, 29, 33, 39, 41, 207–211, 213, 216, 217
rainfall, 32, 33, 34, 55, 56, 68–76, 78
Reagan administration, energy policy, 85–89, 157, 192; population politics, 63–67
recycling, 161
Reilly, William, EPA head, 157
remote sensing, use in global change ecology, 43–44, 78, 80
renewable fuels, 90–94, 101–108
research and development (R&D), need for in search for energy alternatives, 93, 94, 106, 108
Rhine river, pollution of, 13, 181
riparian areas, ecological importance in U.S., 229–231
risk assessment, 124
Rocky Mountain Institute, 97, 99, 182
Romania, 142, 145
Roosevelt, Theodore, 14
runoff, 68–76

Sadik, Nafir, 48, 60
Sagan, Carl, 19
salinization, effect on agriculture, 7, 8, 11, 149
satellites, use of in remote sensing, 44; in biogeography, 204
Seabrook nuclear power plant, 88
sea-level rise, 34, 71–76, 78, 80, 172–179, 188, 189, 191, 194
SEDUE, Mexican environmental protection agency, 12

silt, effect of buildup, 8, 9, 11, 149, 165, 214, 219, 220
slash and burn, method of land clearing, 28, 165, 219, 220
Smith, Adam, 54, 92
social effect of rise in cost of food, 73, 74, 75
society, effect of on ecosystem, 7–11
soil, impact of development on, 30, 62, 68–76, 160, 163–168, 210, 214, 228
soil erosion, 8, 9, 10, 11, 33, 61, 68–76, 79, 111, 142–148, 151, 160, 163–169, 170, 171, 214–218, 219–221
solar aquatics, 183
solar power, 88–89, 93, 101–108
sorghum, 111
Soviet Union, environmental conditions in, 148–152
speciation, 201–203
species: diversity of, 198–206, 219, 221; loss of, 195, 198–206, 207
sport hunting, of wildlife, 222–225
spotted owl, 158, 220
S-shaped curve, and farming practice, 71–73
sterility, 137, 138, 139
sterilization, 53
Strategic Petroleum Reserve (SPRA), 90, 159
sulfur dioxide (SO₂), 26, 35, 42, 119, 141, 144
sulfur emissions, 35, 99, 105, 181
Sumeria, and recorded history of societal changes, 7, 8; Early Dynasty period of, 7, 8
Superfund, 118, 120
Support for Eastern European Democracy (SEED), 143, 144

Sweden, energy use in, 115; environmental restrictions in, 122, 126, 128
swidden agriculture, 10
Switzerland, environmental restrictions, 122, 126

taxol, and cancer, 158
temperature of Earth, see global temperature fluctuation
terracing, practice of agriculture, 10
thermal expansion, of sea water, 173, 192, 194
Third World, environmental issues in, 36, 37, 50–53, 65–67, 68–76, 133, 135–140, 176, 207–210, 216–218
Tigris and Euphrates rivers, effect on agriculture in Mesopotamia, 7, 180
timber, in USSR, 148; in U.S., 150, 158, 212–218, 219–221
tourism, environmental impact, 151; and wildlife industry, 223
toxic chemicals, 135, 140, 148–152, 157, 160
toxic waste, 14, 120, 122, 130–134; dumping of, 130–134, 135; trade in, 130, 132; toxic terrorism, 131
transhumance, practice of, 8
tree cutting, in Mediterranean, 8
tree farming, 111, 220
trophic diversity, 203
tropical land, agriculture in, 9, 10
tropical rain forest, see rain forest

UNICEF, 73
United Nations Environment Programme, 31, 36, 133, 134, 176, 181, 186, 205
United Nations Population Fund (UNFPA), 48, 64, 65, 70, 76

urban growth, in developing countries, 49, 50, 60
urban heat island, 186, 188, 193
U.S. Department of Agriculture (USDA), 150, 170, 188
U.S. Forest Service, 219–221, 226–231
utilities, regulations on, 85–89; and energy efficiency, 98, 102, 107

"Valdez Principles, The," 17
vanilla, environmental impact of demand, 208
Vietnam War, environmental impact of, 25–29

war, effect on environment, 24–29, 87
water: quality, 119, 123, 141, 143, 144, 151; supply, 55, 61, 68–76, 157, 172–179, 180–183; trading, 182
Watkins, Admiral James, 84–89, 159
wildlife management, in Zimbabwe, 222–225
wildlife refuges, in U.S., 227
wind, as power source, 89, 103, 164
World Bank, 73, 134, 146, 164
World Commission on Environment and Development, 31
World Conference on Environment and Development (1992), 36, 37
World Conservation Union, The (IUCN), 30
World Health Organization (WHO), 53, 142
World War II, environmental impact of, 26–29
World Wildlife Fund, 31, 141

Yugoslavia, 145

Zero Population Growth, 56–58

Credits/ Acknowledgments

Cover design by Charles Vitelli

1. Global Environment
Facing overview—Courtesy of NASA.

2. World's Population
Facing overview—United Nations photo by John Isaac.

3. Energy
Facing overview—Courtesy of NASA.

4. Pollution
Facing overview—EPA-Documerica.

5. Resources
Facing overview—United Nations photo by U.S. Department of Agriculture.

6. Biosphere
Facing overview—CIRIC for World Bank photo by Alain Prott.

ANNUAL EDITIONS ARTICLE REVIEW FORM

■ NAME: _____ DATE: _____

■ TITLE AND NUMBER OF ARTICLE: _____

■ BRIEFLY STATE THE MAIN IDEA OF THIS ARTICLE: _____

_____

_____

_____

_____

■ LIST THREE IMPORTANT FACTS THAT THE AUTHOR USES TO SUPPORT THE MAIN IDEA:

_____

_____

_____

_____

_____

■ WHAT INFORMATION OR IDEAS DISCUSSED IN THIS ARTICLE ARE ALSO DISCUSSED IN YOUR
TEXTBOOK OR OTHER READING YOU HAVE DONE? LIST THE TEXTBOOK CHAPTERS AND PAGE
NUMBERS:

_____

_____

_____

_____

_____

■ LIST ANY EXAMPLES OF BIAS OR FAULTY REASONING THAT YOU FOUND IN THE ARTICLE:

_____

_____

_____

_____

■ LIST ANY NEW TERMS/CONCEPTS THAT WERE DISCUSSED IN THE ARTICLE AND WRITE A
SHORT DEFINITION:

_____

_____

_____

_____

*Your instructor may require you to use this Annual Editions Article Review Form in any number of ways:
for articles that are assigned, for extra credit, as a tool to assist in developing assigned papers, or simply
for your own reference. Even if it is not required, we encourage you to photocopy and use this page;
you'll find that reflecting on the articles will greatly enhance the information from your text.

We Want Your Advice

ANNUAL EDITIONS: ENVIRONMENT 92/93
Article Rating Form

Here is an opportunity for you to have direct input into the next revision of this volume. We would like you to rate each of the 35 articles listed below, using the following scale:

1. **Excellent: should definitely be retained**
2. **Above average: should probably be retained**
3. **Below average: should probably be deleted**
4. **Poor: should definitely be deleted**

Your ratings will play a vital part in the next revision. So please mail this prepaid form to us just as soon as you complete it.
Thanks for your help!

Annual Editions revisions depend on two major opinion sources: one is our Advisory Board, listed in the front of this volume, which works with us in scanning the thousands of articles published in the public press each year; the other is you—the person actually using the book. Please help us and the users of the next edition by completing the prepaid article rating form on this page and returning it to us. Thank you.

Rating	Article	Rating	Article
	1. Historical Perspectives on Sustainable Development		20. Will the Circle Be Unbroken?
	2. Managing as If the Earth Mattered		21. Eastern Europe: Restoring a Damaged Environment
	3. Debating Gaia		22. Environmental Devastation in the Soviet Union
	4. War & the Environment		
	5. The Environment of Tomorrow		23. 23rd Environmental Quality Index: The Year of the Deal
	6. Global Change Ecology		
	7. World Population Continues to Rise		24. A New Lay of the Land
	8. Sheer Numbers: Can Environmentalists Grasp the Nettle of Population?		25. U.S. Farmers Cut Soil Erosion by One-Third
	9. Population Politics		26. Holding Back the Sea
	10. Feeding Six Billion		27. Water, Water, Everywhere, How Many Drops to Drink?
	11. Forecast: Famine?		
	12. Energy for the Next Century		28. Global Climate Changes: Fact and Fiction
	13. Balance Sought: Energy, Environment, Economy		
			29. Gazing Into Our Greenhouse Future
	14. The Negawatt Revolution		30. The Origin and Function of Biodiversity
	15. Here Comes the Sun		31. Conserving the Tropical Cornucopia
	16. Energy Crops for Biofuels		32. Timber's Last Stand
	17. It's Enough to Make You Sick		33. Forests Under Siege
	18. The Greening of Industry		34. Wildlife as a Crop
	19. From Ash to Cash: The International Trade in Toxic Waste		35. How the West Was Eaten

(Continued on next page)

ABOUT YOU

Name_____ Date_____

Are you a teacher? ☐ Or student? ☐

Your School Name _____

Department _____

Address _____

City _____ State _____ Zip _____

School Telephone # _____

YOUR COMMENTS ARE IMPORTANT TO US!

Please fill in the following information:

For which course did you use this book? _____

Did you use a text with this Annual Edition? ☐ yes ☐ no

The title of the text? _____

What are your general reactions to the Annual Editions concept?

Have you read any particular articles recently that you think should be included in the next edition?

Are there any articles you feel should be replaced in the next edition? Why?

Are there other areas that you feel would utilize an Annual Edition?

May we contact you for editorial input?

May we quote you from above?

ANNUAL EDITIONS: ENVIRONMENT 92/93

BUSINESS REPLY MAIL

First Class   Permit No. 84   Guilford, CT

Postage will be paid by addressee

The Dushkin Publishing Group, Inc.
Sluice Dock
DPG **Guilford, Connecticut 06437**